DEEP LIFE

DEEP
LIFE

The Hunt for the Hidden Biology
of Earth, Mars, and Beyond

TULLIS C. ONSTOTT

PRINCETON UNIVERSITY PRESS
Princeton and Oxford

press.princeton.edu
Jacket image: In a side tunnel of Beatrix Gold Mine, the
southernmost mine in the Witwatersrand Basin of South
Africa, a borehole one mile beneath the surface intersects with
an ancient fault zone. High-pressure water containing viruses,
bacteria and *H. mephisto*, the "Worm from Hell," shoots out
from the fault. Photo courtesy of Dr. Gaetan Borgonie,
Extreme Life Isyensa, Belgium, and Olukayode Kuloyo and Dr.
Borja Linage of the University of the Free State, South Africa.

ISBN 978-0-691-09644-5

Library of Congress Cataloging-in-Publication Data

Names: Onstott, T. C. (Tullis Cullen)
Title: Deep life : the hunt for the hidden biology of Earth,
Mars, and beyond / Tullis C. Onstott.
Description: Princeton : Princeton University Press, [2017] |
Includes bibliographical references and index.
Identifiers: LCCN 2016027791 | ISBN 9780691096445
(hardcover)
Subjects: LCSH: Adaptation (Biology) | Extreme environments.
| Space environment.
Classification: LCC QP82 .O57 2017 | DDC 612/.0144—dc23
LC record available at https://lccn.loc.gov/2016027791

British Library Cataloging-in-Publication Data is available

This book has been composed in Sabon Next Pro & Montserrat

Printed on acid-free paper. ∞

Printed in the United States of America

1 3 5 7 9 10 8 6 4 2

To **ELEANORA ANDERSON**, my Princess, my one true
love, always and forever, in this universe and beyond.

To my colleagues and friends—**TOM KIEFT,
TOMMY PHELPS, SUSAN PFIFFNER, ESTA
VAN HEERDEN,** and **GAETAN BORGONIE**—
with whom I've had the privilege of sharing many
good times as we explored this universe together.

To **DR. FRANK J. WOBBER**, whose forethought, tenacity,
managerial skills, and geological insight changed my life
and those of many others and who kept all of us in the
Subsurface Science Program focused on the prize.

To the memories of **DAVID BOONE** and **D. C. WHITE**,
both of whom provided guidance and inspiration
for this geologist as to what could be achieved.

To the memory of **PIET BOTHA**, our jolly
jester of the Crocodilian Estate, whose bright
light was tragically extinguished.

To the memories of **ROB** and **SYBIL HARGRAVES**,
who encouraged me to pursue the answers to questions I
wanted to ask, no matter how unconventional they were.

Finally, to the memory of **GENE SHOEMAKER**
and to **CAROLYN SHOEMAKER**, who
were so nurturing to this young techie.

CONTENTS

FOREWORD

In September 1996 I was stumbling through a pitch-black tunnel in Western Deep Levels, one of the AngloAmerican company's deepest gold mines, west of Johannesburg in South Africa. At two miles beneath the surface of the Earth, the heat and humidity were almost all I could think about. Earlier that morning my guide and I had learned that the ventilation system was off where the train was supposed to have taken us to collect samples. So instead, he picked an alternate route, and I followed him at a walk/run through a maze of intertwining tunnels. After an hour of hiking, we finally reached the stope, a narrow tilted tunnel into the rock where the miners had been removing the freshly blasted gold and carbon-rich rock. I had not, however, anticipated that the entrance to the stope would be the little rabbit hole in the side of the tunnel that I now faced. We squeezed into the hole and shimmied down the 30° slope of the three-foot-high stope. There my guide barked orders to the miners in a language unintelligible to my ears, and they unplugged the detonators to the explosives in the rock. After they had finished, we gingerly levered out a couple of large, thirty-pound chunks of the carbon-rich layer, which I wrapped in sterilized bags and shoehorned into my backpack. I tried not to think of the weight of the rock above my head every time my helmet banged against the ceiling of the stope. With my samples bagged, my guide began crab crawling his way further down the slick, steeply dipping scree slope. "We're going deeper?" I shouted down at the helmet lamp of my guide below me. With my backpack in one gloved hand, I started sliding further down the stopes on my butt after him. "Hurry, they have to blast soon!" my guide shouted up from below me. The air smelt of burnt rock, literally a fire-and-brimstone odor. It pervaded the stope. There had been no signs of water anywhere, except for the tiny, rusty stains on a rock that I had collected half an hour earlier in one of the tunnels. As I caught up to my guide, he quickly grabbed my arm and pulled me aside as I almost fell into a dark pit. "That's a 100-meter drop, my friend," he said in his stiff Afrikaans accent. As we continued downward, we suddenly popped out of another rabbit hole and into a deeper and even hotter tunnel, where we trudged for an hour until we came to what I had hoped to find. Water! It was dribbling out of a drill hole in the ceiling of a tunnel intersection, and my nostrils started stinging from the acrid smell of ammonia. I climbed up the rusting rock bolts that held wire mesh against the rocky wall and stretched my gloved hand up toward the dripping borehole. I was able to collect enough water to make some pH and temperature measurements in the little cup that I held. "Fantastic! It's 50°C and a pH of 9," I shouted down to my guide. After readjusting my small knapsack to my chest, I climbed higher, and balancing precariously on the bolts, I filled every sterilized serum vial I had with me. As soon as I dropped to the tunnel floor we were off again. "Hurry, we've got to catch the cage," he shouted back to me, and I hurried behind him the best I could in my ill-fitting gumboots. I continually pulled up the loose belt that held my headlamp battery and struggled to keep up with him while lugging seventy pounds of rock and water samples.

We finally reached the station for the cage. We had minutes to spare, so I took some quick pictures. The miners, some of whom had just walked in from their shift, were reclining in various positions on the concrete floor away from the cage entrance while the shift bosses sat on benches right next to it. I was clicking away when my guide pulled down his coverall and "mooned" me, and the shift bosses started laughing. The miners looked on, stoically bemused. I nervously laughed as well, but as we waited for the cage to arrive, I silently asked myself, "What in hell am I doing here miles beneath the surface of the Earth collecting microbial samples?" The results from the samples in my backpack would tell me if it was worth the trip. I might find bacteria that had been isolated from the surface of the Earth for two billion years or I might even discover the secrets to the origin of life itself. But first I had to isolate the rocks from air and then get them back to the USA.

Finally the cage arrived, and I snapped back to the present as I crammed into the cage with thirty other hot and sweaty miners. My arms were pinned to my sides, my backpack was wedged between my feet. With a bang the cage door was slammed shut and locked from the outside. It was dark and strangely quiet except for the sounds of mechanical chirping birds. Then the cage jerked and began accelerating toward the cool, sunny surface so far above. I had walked, run, climbed, and crawled 11,000 feet beneath the surface of the Earth in search of a hidden universe of life. In ten minutes I would once again be safe on top.

ACKNOWLEDGMENTS

This book is populated by the scientists, miners, drillers, and various professionals intimately involved in the events portrayed. The source material is extracted from personal remembrances, faxes, emails, field notes, and copious photographs from all those involved, for which I am eternally grateful. Nonetheless, this is a small fraction of people with whom I have worked these past twenty-five years and who played key roles in and made so many important contributions to this story. Because time and space did not permit me to include all of my coworkers in the plot line, I wish to apologize in advance and to offer them my sincerest thanks.

I also want to sincerely thank all the mine managers, mine geologists, miners, and drillers in the mines of South Africa, United States, and Canada. Words cannot express how much I appreciate the time and support you have generously donated to me and my colleagues over the past twenty years. Without that support this story would have been impossible. I also appreciate the care you always took to ensure our safety.

ABBREVIATIONS

AEC   Atomic Energy Commission
AMS   accelerator mass spectrometer
ATP   adenosine 5′-triphosphate
BHA   bottom-hole array
CBD   Convention on Biological Diversity
DIC   dissolved inorganic carbon
DLO   *Desulfotomaculum*-like organism
DNA   deoxyribonucleic acid
DOC   dissolved organic carbon
DOE   Department of Energy
DTS   distributed temperature sensor
DUSEL   Deep Underground Science and Engineering
          Laboratory
EPA   Environmental Protection Agency
ESRC   Environmental Science and Research Center
EXLOG   Exploration Logging Company
FISH   fluorescent in situ hybridization
FSU   Florida State University
GC-MS   gas chromatography–mass spectrometry
GeMHEx   Geological Microbial Hydrological Experiment
GPR   ground-penetrating radar
HPLC   high-performance liquid chromatography
IMT   inherent microbiological tracer
INEL   Idaho National Engineering Laboratory
IPC   intermediate pumping chamber
IPR   intellectual property rights
ISSM   International Symposium on Subsurface
          Microbiology
JOIDES   Joint Oceanographic Institutions for Deep Earth
          Sampling

LANL	Los Alamos National Laboratory
LBNL	Lawrence Berkeley National Laboratory
LExEn	Life in Extreme Environments (NSF Program)
MMEL	Mobile Microbial Ecology Laboratory
MLS	multilevel samplers
MOU	memorandum of understanding
MSR	Mars Sample Return
MWX	Multi-Well Experiment
NAI	NASA Astrobiology Institute
NASA	National Aeronautics and Space Administration
NCBI	National Center for Biotechnology Information
NMT	New Mexico Tech (New Mexico Institute of Technology)
NOSR	Naval Oil Shale Reserves
NRF	National Research Foundation (South Africa)
NSF	National Science Foundation
NTS	Nevada Test Site
ODP	Ocean Drilling Program
OHER	Office of Health and Environmental Research
ORNL	Oak Ridge National Laboratory
PCR	polymerase chain reaction
PFC	perfluorocarbon
PFT	perfluorocarbon tracer
PI	principal investigator
PLFA	phospholipid fatty acid
PNAS	*Proceedings of the National Academy of Sciences*
PNL	Pacific Northwest Laboratory
PNNL	Pacific Northwest National Laboratory
PP	planetary protection
ppm	parts per million
PVC	polyvinyl chloride
QA/QC	quality assurance/quality control
REU	Research Experiences for Undergraduates
RNA	ribonucleic acid
rRNA	ribosomal RNA
RWMC	Radioactive Waste Management Complex

SCWRC   South Carolina Water Resources Commission
SEM   scanning electron microscopy
SIP   science implementation plan
SLiMEs   subsurface lithospheric microbial ecosystems
SMCC   Subsurface Microbial Culture Collection
SPIE   International Society for Optics and Photonics
SRB   sulfate-reducing bacterium
SRP   Savannah River Plant
SSP   Subsurface Science Program
TCE   trichloroethane
TDEF   time domain electromagnetic field
TEM   transmission electron microscopy
TEP   triethylphosphate
T_{max} maximum temperature
URL   underground research laboratory
USGS   United States Geological Survey
UV   ultraviolet·
WIPP   Waste Isolation Pilot Plant

DEEP LIFE

INTRODUCTION

WHAT ARE THE LIMITS OF LIFE ON EARTH?

Currently we do not have a complete picture of how life originated on this planet. We do have pieces of the puzzle, but the entire picture has not yet been assembled. Only within the last couple of decades have we been able to place constraints on the limits of life on the Earth, but much still needs explanation as to why those limits exist. These gaps in our understanding of the origin and extent of life on Earth make it especially challenging to explore other planetary bodies in our solar system for life. This book is about the exploration of the limits of life in the rocky crust deep beneath the surface of our planet and how what we have discovered over the past twenty-five years informs our exploration for life in the solar system, and in particular, the planet Mars.

We do not normally think of rock as harboring life. We quarry granite for building stone, volcanic rock for road gravel, and marble for table tops and great works of art. I would wager that the last thought on your mind as you gaze upon Michelangelo's *David* towering above you in the Accademia Gallery in Florence is the fact that within microscopic pores buried inches beneath the smooth surface of Carrara marble are living bacteria. These bacteria may have been trapped inside the marble for a million years and are slowly reproducing, releasing carbon dioxide (CO_2) and making nanocrystalline calcite, the same mineral phase that forms the marble. I can personally guarantee that as miners tunnel into the Earth to remove the coal from beneath the Appalachian Mountains or extract metal-rich veins from miles beneath the surface of South Africa, they do not typically think of rocks as crawling with life. I have never encountered any roughnecks on a drill rig who have taken the time to stop and ponder whether the oil or gas reservoir into which they were drilling a mile or more beneath their feet happens to be something's domicile.

I am a geologist by training, and like most geologists, I too have viewed rocks as inanimate entities. We examine the microscopic details of rocks primarily to infer the most probable explanation of how and when they came to be what they are. We try to infer the underlying physical processes that formed them and how those processes relate to geological history at the global scale. Paleontologists focus on those rocks that contain fossilized life forms that existed on the surface of the Earth at the time the materials forming the rock were being deposited. Some paleontologists who study rocks from the Precambrian, a time prior to the rise of multicellular organisms, look for traces of microorganisms that once existed on the surface. They do this by analyzing the composition of relic organic molecules, their stable isotopic composition, such as the ratio of carbon-13 to carbon-12 ($^{13}C/^{12}C$), or some physical structures in the rock that may represent a fossilized surface biofilm. Their goal is to understand the evolution of surface life and how that evolution relates to the evolution of Earth's oceans and atmosphere over time. Even from their perspective, those ancient microorganisms supposedly quickly died as the rock was formed from the sand or mud that contained them. Geologists used to view rock as being dead.

Of course, scientifically speaking, microbiologists are the antithesis of geologists, even though microbiologists travel to the same regions of the world as geologists in order to carry out fieldwork. Microbiologists gather samples, take them to the lab, isolate microorganisms from them, and then perform experiments on the microbes to deduce what they are doing in their environment and how they are doing it. During the past one hundred years, a subset of these scientists has been steadfastly isolating microorganisms from environments that at first glance would not be what we would consider likely places for life to exist. They have isolated bacteria from the Arctic that can grow at temperatures below freezing (psychrophiles), and in the hot springs of Yellowstone National Park, they have discovered bacteria that can thrive at temperatures close to the boiling point of water (thermophiles and hyperthermophiles). Also in Yellowstone, microbiologists have found microbes that can live in acid (acidophiles). In the hypersaline, soda volcanic lakes of East Africa, they have isolated bacteria that can flourish at pH 10 (alkaphiles). One of the pioneers of microbiology, Antonie van Leeuwenhoek, performed experiments in the seventeenth century that led to the discovery of microbes that can live in the absence of oxygen (anaerobes). Microbiologists have even found bacteria, known as gram-positive bacteria, that show a remarkable tenacity to survive by remaining dormant as spores. Deprived of any sustenance for decades, the spores germinate as soon as they are provided with some nutrients in water. In fact, microbiologists have been hard-pressed to find any corner on the surface of the Earth, no matter how harsh the environment, that has not been colonized by some form of microbe that could be grown in the lab.

Geologists first joined forces with microbiologists in the 1920s to explore the one realm into which microbiologists had not yet ventured: the subsurface. Edson Sunderland Bastin, an economic geologist from the University of Chicago, teamed up with a bacteriologist, Frank E. Greer, from the Department of Hygiene and Bacteriology at the University of Chicago to test a straightforward hypothesis. They proposed that sour oil (oil containing hydrogen sulfide, or H_2S, the gas that exudes from rotting eggs) was created by anaerobic sulfate-reducing bacteria (SRBs). When Greer grew SRBs at high temperatures from the oil-field water collected from geological formations dating from the Pennsylvanian Epoch provided by Bastin, they were able to demonstrate that their proposed answer was definitively correct! But it raised a second, more interesting question. Had those SRBs been living in the same formation beneath the surface for the 300 million years since they were first deposited in the oceans? This was a question the bacteriologist could not answer by deduction. Nor could the answer be inferred by the geologist.

Bastin and Greer were not the only ones to speculate about the existence of subsurface ancient life. Before World War I, German microbiologists had begun exploring their deepest coal mines. They were looking for bacteria inhabiting the coal, hypothesizing that they could be living fossils of the bacteria that had inhabited the 300-

million-year-old peat bogs from which the coal had formed. Why did microbiologists at that time believe in the longevity of bacteria? One of those microbiologists, Charles Bernard Lipman, of the University of California Berkeley, explained in a 1931 article in the *Journal of Bacteriology*: "I have been asked frequently since the inception of the studies under consideration to state what led me to make such an investigation when *a priori* one would not expect anything but negative results. My answer to this question is that for twenty years prior to the initiation of these experiments I had been accumulating more and more evidence on the persistence of the life of bacteria and bacterial spores for periods of forty years, as a maximum in the latter case, as authentic facts. The unabated virulence of pathogenic organisms grown from very old spores and the remarkable viability of bacteria from very old and very dry soils preserved in unopened bottles for about forty years have furnished me with much food for thought for well nigh a quarter of a century."[1]

But the many bacteria grown from coal and oil reservoir samples from the 1920s to the 1930s, including Lipman's remarkable spore-forming bacteria, were largely dismissed as modern bacterial contaminants from the surface. After all, a drill rig or a mining operation cannot be sterilized. Undeterred, microbiologists reported finding bacteria that produced alkanes and began speculating on the bacterial genesis of petroleum deposits. Geologists, however, were not receptive to this idea, primarily because they had a geological model based upon thermal alteration of dead organic matter that explained the origin of oil quite nicely. This geological model was solidly built upon observations of when and where in the geological record that oil formed and migrated, and upon inductive reasoning about the common steps required to produce all petroleum and natural gas deposits around the world and over a wide span of geological ages. Subsurface bacteria were simply not required to explain oil or gas. Sporadically over the subsequent decades, microbiologists would report the discovery of bacteria from deep within our planet, only to have the discovery summarily repudiated and dismissed as surface contamination. During the 1950s Soviet microbiologists remained the last holdouts for deep-dwelling life and the central role of bacteria in forming many economic ore deposits. They introduced the term "geological microbiology" in the early 1960s, the practitioners of which were geomicrobiologists.[2] However, there was not an enormous waiting list of scientists signing up to become geomicrobiologists. By the 1970s even soil microbiologists were certain that "depth is another secondary ecological variable that affects the bacteria. In temperate zones, these organisms are almost all in the top meter, largely in the upper few centimeters."[3] The perception was widespread in both the geological and microbial communities that even a lowly bacterium would have to struggle to live within the tiny cracks and pores of rocks forever separated from sunlight and survive on the occasional dribs and drabs of organic matter that might descend downward past the soil layer.

At this same time, the first communities of organisms were discovered at black-smoker vents at the bottom of the Pacific Ocean not far from the Galápagos Islands, the birthplace of Darwin's theory of evolution by natural selection. Before *Riftia pachyptila* tube worms were discovered at black smokers, no one had imagined that such complex organisms or ecosystems could survive on chemical gradients in the utter darkness of the seafloor. But miles beneath the surface of the ocean, there was at least plenty of room to grow and evolve complex ecosystems. It seemed highly improbable that complex ecosystems could exist in the tiny pores a thousand feet beneath the rocky surfaces of Yosemite, the forest floor of the Amazon jungle, the desert dunes of the Sahara, the ice sheets of Antarctica, or the wheat fields of Kansas. By the 1970s, it was accepted that, with the exception of some shallow aquifers, the deep subsurface of our planet was generally a sterile environment. There was simply not enough energy or room down there to support life for long.

In many respects this belief was reassuring. After all, the Atomic Energy Commission (AEC), the predecessor to the U.S. Department of Energy (DOE), had contaminated the subsurface beneath the laboratory complexes where they had been concentrating enormous amounts of radioactive material for nuclear warheads. These contaminants were a witch's brew of carcinogenic organics, toxic metals, and radionuclides. Modeling their geochemical behavior and hydrological transport was inherently simpler in the age of punch-card, mainframe computers if biology was not part of the equations. The AEC had also been exploding these warheads underground in Nevada since the early 1950s and then solely underground after John F. Kennedy, Nikita Khrushchev, and Harold Macmillan agreed to the Limited Nuclear Test Ban Treaty in 1963. The AEC had become so adept at subsurface atomic detonations that they began to detonate A-bombs beneath the surface of New Mexico and Colorado to help release natural gas from tight formations as part of the Plowshare Program. In today's parlance, we might call this practice fissionogenic-fracking and be grateful that it was abandoned. The U.S. DOE was also busily working on burying high-level radioactive waste in the "sterile" subsurface salt, granite, and volcanic formations, as were other countries. These formations had to retain the radioactive waste for a hundred thousand years. The possibility that subsurface life forms might interact with the radioactive waste and, hence, complicate their calculations seemed a very remote possibility.

Then, in the mid-80s an obscure program within the U.S. DOE, called the Subsurface Science Program (SSP) and led by Frank J. Wobber, started claiming that the subsurface was not sterile after all but instead was inhabited by an abundant and diverse community of bacteria and eukaryotes (e.g., protists and fungi). Unlike the earlier subsurface studies, the scientists working in the SSP had developed physical and chemical tracers to screen out the surface microbial contaminants. They were confident that they were finding indigenous bacteria everywhere, even in Triassic rocks from two miles beneath the surface of a farmer's field in Virginia. They claimed the abundance of this subsurface life rivaled that of surface life on our planet, and they soon began talking about a subsurface biosphere. The reports made big news with the general public, who found them as intriguing and fantastical as Jules Verne's journey to the center of the Earth. The broader scientific community, however, was extremely skeptical of the claims made by these DOE scientists and their program manager and also of the significance of the discovery. But the SSP microbiologists had isolated thousands of species of subterranean bacteria, which they planned to use to remediate the toxic underground legacy of the Cold War, a legacy that had started to migrate away from the DOE laboratories and toward the drinking-water supply of the surrounding communities. How ironic it is that a DOE program would not only discover a subsurface biosphere but also enlist its aid when its predecessor, the AEC, had unknowingly killed quintillions of subsurface bacteria in their homes with underground nuclear blasts. The SSP scientists also discovered that entire microbial communities in the subsurface were being powered from hydrogen (H_2) gas generated by the weathering of iron (Fe) minerals in water. Very soon thereafter, scientists started finding ecosystems in every subsurface corner of the planet. Some were three miles down and powered by radiation. Others were carving enormous underground cavities beneath mountains. Some were found trapped in ice thousands of feet beneath the surface of Antarctica. Even predatory, deep-dwelling worms were being discovered a mile beneath the surface. By then geomicrobiology had its own dedicated publication, and geomicrobiologists had founded a society for subsurface microbiology.

LIFE BENEATH THE SURFACE OF OTHER PLANETS

Now, as a geomicrobiologist, I look at all rocks as microcosms of tiny microorganisms, some of which may have been living in the rock since its formation hundreds of millions of years ago. When you look at the surface of Mars, you get the impression that it was once habitable for life. But currently, its surface is not habitable by life as we know it. Still you can't help but wonder if Mars, having had a surface biosphere at one time, still maintains a subsurface biosphere like that of the Earth's.

Could there be life beneath the surface of Mars? If you had openly asked a question like that in the sixteenth century you would have been burned at the stake, like the Italian Dominican friar and philosopher, Giordano Bruno. More than a hundred years ago, Percival

Lowell believed that he could see evidence of an extraterrestrial civilization living on the surface of Mars, but that optical illusion, along with his scientific credibility, was dispelled within fifteen years of his discovery. In 1901 H. G. Wells, a contemporary of Lowell, published *The First Men in the Moon*, in which he constructed a fantastically sophisticated society of insect-like creatures, Selenites, living beneath the lunar surface. Since then, the concept of subsurface life has figured frequently in science fiction novels and movies, even episodes of *Star Trek* and *Star Trek: The Next Generation*, but is there any scientific basis supporting it? It is certainly true that if you burrow miles into the Moon or Mars, you will reach a temperature at which fresh liquid water, the essential ingredient for life as we know it, can exist. But can life survive miles beneath a planet's surface even when that surface is inimical to it? With the discovery by SSP in 1996 that the H_2 generated from rocks and water could support a subsurface ecosystem, the answer seemed to be yes. Life beneath the surface of Mars was now no longer restricted to the realm of science fiction and could be treated as a real, tangible target for space probes. The National Aeronautics and Space Administration (NASA) quickly became engaged with designing drilling systems that could be sent to Mars.

All this narrative may sound to you like the script of a B-grade sci-fi film, and it could easily be one. But it is historic fact that in the space of twenty years scientific consensus concerning deep subsurface life on Earth and all its extraterrestrial implications switched from incredulity to active scientific endeavor. This book is in part about that remarkable scientific and philosophical transition as seen through the eyes of its protagonists, including myself.

But mostly in this book, you and I will actively explore the limits of life beneath the surface of our planet and search for anything that would prohibit life from existing beneath the surface of other planets, especially Mars. During our quest I will be taking you to places that few surface dwellers have seen. While we are hunting on our subterranean safaris, I want you to keep the following questions in mind, because my colleagues and I asked ourselves these same questions, not knowing what the answers would be.

Is it possible that subsurface life exists in such abundance that it exceeds the entire surface biosphere of our planet?

Just how deep down into the Earth do living organisms occur?

How long can a single subsurface microbe live? Can it live for a hundred years, a hundred thousand years, a million years?

Can microorganisms live on radiation?

How long can ecosystems survive beneath a planet's surface? Billions of years?

If it is for billions of years, can life exist beneath the surface of Mars?

Where did the microbes come from and how did they get down there? Can they travel for hundreds of miles beneath the surface?

Could life have originated beneath the surface of our planet, or any planet for that matter?

Is it possible that complex life forms and complex living ecosystems can exist beneath the surface of any planet?

To help us with the answers to these questions, I will be bringing plenty of scientific background with us.

In chapter 1 we begin with the discovery by SSP scientists of bacteria inhabiting rocks 9,000 feet beneath the surface of the Earth. We then flash back to the earlier years, when the SSP first started drilling and developing the tracer techniques that were pivotal in proving that subsurface bacteria were not contaminants.

In chapter 2 we follow the SSP to northern New Mexico, where we drill into an ancient volcano, and to western Colorado, where we drill more than a mile down into a tight gas formation near an underground nuclear test site. We will also begin our first Jules Vernian subsurface expedition down into the Waste Isolation Pilot Plant near Carlsbad, New Mexico. There we will explore the Permian salt formations in search of fluid inclusions containing 250-million-year-old seawater and, hopefully, living bacteria. We will also learn how scientists began to conceive of radiation-supported subsurface life.

In chapter 3 we will see how researchers used geological history to determine the origins of the microbes collected from the various SSP drilling campaigns. They deduced how long the microbes had resided down there, how far they had traveled to get there, and how quickly they moved. We will explore the boundary between life and death and what that means for a microbe through the use of the "Death-o-Meter." Finally we will learn of the discovery of the subsurface lithoautotrophic microbial ecosystems (SLiMEs), supported by H_2 gas derived from basaltic volcanic rocks.

The termination of the SSP in chapter 4 forces us to journey to the deep mines of South Africa in search of a way to obtain deep and potentially very ancient microbial ecosystems supported by radiation and to search for the origins of life itself. With the discovery of SLiMEs we will participate in the NASA meetings where the plans were made for drilling on Mars with the goal of detecting life beneath the Martian surface. We will see how, coincidently, that search effort was bolstered by the reports of fossil subsurface Martian life in a meteorite called ALH84001.

In chapters 5 through 8, we will explore the maze of tunnels in the ultradeep gold mines of South Africa and the challenges that greeted the scientists there. As we journey downward more than two miles in depth, we will learn what life must be like for a thermophilic subsurface bacterium. Finally, we reach our goal with the discovery of *Candidatus* "Desulforudis audaxviator", a novel new bacterium that travels beneath the surface of South Africa, living indefinitely on a chemical battery supported by the radioactive decay of uranium embedded in its rocky environments.

In chapter 9 we fly north of the Arctic Circle in Canada in order to answer the question of whether life could exist deep beneath the frozen surface of Mars. There we travel through the ice caves of the Lupin gold mine down to salty water a half mile below the frozen tundra to find chemolithoautotrophic bacteria, microorganisms that are able to obtain the carbon and nitrogen for biosynthesis by metabolizing inorganic elements, that cycle sulfur compounds. We also test the feasibility of drilling through permafrost to sample a deep biosphere on Mars.

Given the difficulties the researchers encountered drilling through permafrost, in chapter 10 we explore the possibility of using caves to gain access to subterranean life. We will go into the deep caverns of New Mexico, Romania, and Mexico in search of chemolithoautotrophic life and complex ecosystems. We see firsthand how microbial life has carved out its own habitat and what this could mean for Mars. Finally, we return to South Africa to discover multicellular life in the form of a predatory, hermaphroditic nematode named *Halicephalobus mephisto*, the worm from hell, living one mile beneath the surface of our planet.

In the epilogue, we will review the questions that scientists have answered and address the questions that remain to be answered and how they are going to do that. With each chapter, I also provide historical accounts of the earlier discoveries of subsurface life stretching back to 1793, which had been discounted but now, placed in the modern context of our current understanding, make sense. In the appendixes I also provide a chronology of subsurface life investigations and summaries of the seminal U.S. DOE meetings. The locations of subsurface microbiological field campaigns along with photographs from those expeditions can be found as a Google Earth document at http://press.princeton.edu/titles/10805.html.

CHAPTER 1

TRIASSIC PARK

It would seem probable that any waters entombed in the rocks as long ago as Pennsylvanian times must have undergone repeated changes in concentration and at least minor changes in composition by contact with other waters and with the rocks. Whether the bacteria found in these waters today are lineal descendants of forms living on the sea-bottom at the time the sediments were laid down or have been introduced later by ground waters descending from the surface to the oil-bearing horizons is an interesting question that may never be possible to answer.

—Bastin et al. 1926

MARCH 12, 1992, THORN HILL FARM, KING GEORGE COUNTY, VIRGINIA

The cerulean blue Schlumberger vans were lined up beneath a Texaco Murco 54 drilling rig on the fields of Thorn Hill Farm, Virginia (figure 1.1). After having punched through the gas-rich target zone at 9,100 to 9,200 feet, Texaco's drillers had been the first to reach a total depth of 10,213 feet and the bottom of the hidden Taylorsville Triassic basin. They continued to circulate the drilling mud to clean out and smooth the borehole wall in preparation for the insertion of Schlumberger's logging and sidewall coring tools. The roughnecks then began the tedious task of hauling out the drilling rods, or drill string, a hundred feet at a time. Hours later, they finally yanked the drill bit out of the one-foot-wide borehole. The roughnecks then raised the string of long and shiny, cylindrical Schlumberger logging

FIGURE 1.1. Texaco drill rig at the Thorn Hill Farm in Virginia, spring 1992 (courtesy of F. J. Wobber).

tools on a steel cable, the wireline; centered them over the hot, roiling, mud-filled borehole; and slowly lowered them to begin imaging the layers of rock in the gas-rich zone deep below.

Back in their vans filled with the smell of burnt coffee and stale cigarette butts the Schlumberger field engineers began probing the rock around the borehole with low-energy neutrons, electrons, seismic waves, and gamma rays emitted by their logging tools.[1] As the two-dimensional images of the rock at 9,000 feet came into focus, the engineers conferred with the Texaco geologists and a representative of the DOE's SSP, Tim Griffin, all of whom were peering down at the paper logs strung out on the table. Based on the Schlumberger logs, what the Texaco geologists had seen in the drill cuttings, and the gas plays detected by the gas chromatographs hidden in the Exploration Logging Company (EXLOG) trailer,[2] the choice targets for the microbial sampling appeared to be alternating sandstone and shale of a paleo-lake sequence at a depth of 8,900 to 9,200 feet.[3]

Tim could only hope they were right, because at 9,000 feet beneath the surface, they were exploring for life at a depth ten times greater than anyone else had done before them and no one, including Schlumberger, had a logging tool for detecting deeply dwelling bacteria.[4] The DOE team had arrived at the site resolved to collect pristine cores from different sedimentary facies, sedimentary rock that is visibly distinguishable and represents specific depositional environments. In the case of these Triassic lake beds, they wanted some cores from the dark, organic-rich shale, some cores from the reddish silts and sands, and some cores from the interfaces between these two facies. These sediments had been quickly buried in a fault-bound basin as North America struggled to tectonically disengage itself from Africa and the rest of Gondwanaland.[5] Then as Pangaea split asunder and the mid-Atlantic ridge started pouring basalt into the new seafloor, the basin became buried by an Early Cretaceous, newly formed Atlantic Ocean. The lake sediments and the bacteria that were trapped within them had been completely hidden from the surface for 230 million years, until today. If the bacteria or their descendants were still alive, they could be living fossils from the dawn of the age of the dinosaurs.

The caliper logs indicated that there had been few washouts in these zones,[6] that is, the hole gauge was uniform and ideal for coring into the side of the borehole wall. This meant that the sidewall core tool could be rammed tightly against the borehole wall with few gaps. The engineer reassured Tim that recovery of the core would probably be pretty good. That was an encouraging thought given all the challenges they had faced to get this far. The gas logs revealed minor gas plays, which disappointed the Texaco geologists, who were looking for natural gas, not deep-dwelling bacteria. The data analyses had suggested that they had a "dry" hole, that is, one that was not a commercially viable source of natural gas. The Texaco geologists were already receiving instructions from Houston to wrap it up and demobilize the rig as soon as they had finished their deal with the U.S. DOE.

It was twilight on Friday, March 13, 1992, when Tim stepped outside to start pumping the perfluorocarbon tracer (PFT) into the drilling mud and help prepare the sidewall coring tool as the roughnecks finished hauling the Schlumberger logging tools up and out of the two-mile-deep borehole.[7] An hour later the rotary sidewall corer,[8] along with its gamma and caliper logging tools, started sliding down the steaming hole to the target depth. It wasn't long before the Schlumberger team stopped the corer's descent. They then started the coring into the side of the borehole at the deepest shale/sand transition. At 11 p.m. the coring tool returned to the surface with a disappointing eight of the twenty-five attempted core samples. At 3 a.m., as the tool started back down to collect the DOE microbial cores, Tim called Tommy Phelps at the Hampton Inn in nearby Fredericksburg, who then woke Rick Colwell and Todd Stevens.[9] They quickly drove out to the drill site. In the dark, the site was easily visible from miles away because the Murco 54 was lit up like the shuttle launch pad at Kennedy Space Center. They had prepped the argon-filled glove bag before dinner that evening in anticipation of getting cores during the night. When they pulled onto the drill site, Tim was out with the Schlumberger engineers on the rig deck breaking down the corer beneath the floodlights of the rig. They went to work with sterile gloves, scooping out the samples and the drilling mud from the collection baskets and placing them in sterile Whirl-Pak bags. The cores were still warm to the touch, which meant that they were hot in their center, holding on to their primal heat against the frosty night air. The microbiologists dashed back to their trailer with their cherished loot, where they quickly transferred the cores into the glove bag. Rick and Todd photographed and logged them. They had only three short cores out of twenty-five core attempts and a lot of worthless mud-cake samples.

Meanwhile the post mortem on the drill deck by the Schlumberger engineers indicated that shale fragments had lodged in the rotating drilling mechanism, making it difficult for the tiny bit to rotate fully to the proper drilling angle. The samples that had been collected were also highly fractured, increasing the likelihood of contamination from the drilling mud. The third coring run returned to the surface at 9:30 a.m. with a few more cores than the second run, but again the core quality was very poor, with a high risk of contamination from the drilling fluids. The fourth coring run returned empty, this time due to an electrical short. Bad luck for Texaco. The fifth and final run began at 11:30 a.m., and Tim was informed by the Texaco site manager that they would be allowed only six of the twenty-five, with Texaco insisting that they keep to a fifty-fifty split deal even though they had only nineteen successful attempts so far at coring. Tim wondered how a fifty-cores-apiece deal turned into 50% of what cores you could get. At 12:40 p.m. the coring tool was pulled back out of the hole because the gamma logger had failed and needed to be repaired. Tommy, Rick, and Todd continued working on mud samples, and by 2 p.m. the coring tool was sliding back down the borehole for the final attempt. At 3 p.m. the rotary sidewall corer was pulled out of the borehole, and upon examination had collected only one core, at 9,161 feet deep.

The drill manager told the DOE research team that that was their last run. Tim expressed their dismay at getting only twenty attempts and recovering only ten cores, half of which were of marginal quality, leaving only five intact samples after the blood, sweat, and money they had shed over the past six weeks. Tim made a request for one last core run, but the request fell on what he thought were deaf ears.

About an hour later, the Texaco geologist returned to their trailer after talking to the drill manager, who had contacted Texaco in Houston. He offered an opportunity to attempt a coring run with the percussion sidewall corer, the tool that literally shoots the core barrels into the rock. If the researchers were interested, they could keep all the samples.

The sun was setting outside, the Schlumberger crew members were beyond fatigue, and the rest were strung out on caffeine. Although the chances of successfully collecting microbial samples by the percussion method were slim, they all felt that having come this far, they must exhaust all of the options. Texaco was gracious enough to offer the percussion tool, and they took them up on their offer without hesitation.[10] While Schlumberger prepped the percussion tool for the final coring attempt, they picked their five target depths. By 7:40 p.m. the core gun was sliding down the Thorn Hill no. 1 borehole to 9,188 feet.

A HIDDEN BASIN? OCTOBER 1991, LEWES, DELAWARE

The idea of looking for bacteria 9,000 feet beneath the surface in Virginia had begun six months earlier, when Dr. Frank J. Wobber sat down to work on a crossword puzzle at his shore home near Lewes, Delaware, decompressing from the intense schedule of the past four months. Although he was pleased with the progress made in the four field programs that he had been supporting—at the Pacific Northwest Laboratory (PNL), Idaho National Engineering Laboratory (INEL), the Savannah River Plant (SRP), and the Nevada Test Site (NTS)—he was frustrated with not being able to constrain the ages of the various novel microorganisms being grown from the rock cores at these locations. But as he opened up his *Washington Post*, he noticed near the bottom of the front page a small footnote of an article, no more than six or seven lines—one of those column fillers, local fluff—saying that Texaco was drilling natural-gas exploration wells into a completely buried Triassic-age basin, the Taylorsville

Basin, starting in Virginia, then moving to Maryland, and then into Delaware.

Frank was a geologist, a sedimentologist/stratigrapher by training. Supported by a Fulbright Scholarship in the early 1960s, he had mapped the rugged, windswept, wave-cut beach outcrops of the Bristol Channel. He had also spent a summer working for Texaco Oil Company after graduating from the University of Illinois. From this geological training and his past ten years of experience as a program manager at DOE working on groundwater contamination, he knew of only one type of "basin" that had features that might isolate deep, indigenous, subsurface bacteria for millions upon millions of years. That basin would have to be completely buried so that recent groundwater never penetrated its formations. Its porous sedimentary compartments would need to be sealed by low-permeability rock formations, perfect for trapping the natural gas being sought by Texaco.

Most important, he needed a basin that had never been drilled before so that there was no risk of drilling contamination from oil exploration. As an undergrad, Frank had heard of the geologist at the University of Chicago, Edson Sunderland Bastin, who back in the '20s had studied the source of H_2S and bicarbonate in water from oil fields.[11] His colleague, Frank Greer, was aware that certain microorganisms, SRBs, could utilize sulfate for respiration in place of O_2 in these anoxic environments.[12] Bastin reasoned that the H_2S and bicarbonate were being produced by SRBs as they degraded organic components in oil. In an experiment reported in *Science* in 1926,[13] Bastin and Greer cultured SRBs from groundwater samples collected from oil wells in the Waterloo oil field of Illinois that had been drilled to depths of 500 to 1,700 feet. One question asked by Bastin and Greer was whether those SRBs were descendants of those buried more than 340 million years ago, when the sediments were deposited, or whether they had been introduced during drilling. The latter explanation was favored by many scientists who viewed the existence of microorganisms deep beneath the surface with guarded skepticism, if not outright disbelief. After all, drilling an oil well was a filthy, dirty process that made it difficult to obtain samples that were uncontaminated by microorganisms from the surface, and Bastin had known this. Their

Ralph Boyle–Graphic House
MARTIN D. KAMEN

Ralph Boyle–Graphic House
SOL SPIEGELMAN

With radioactive tracers, controlled competition.

Sun Laze

in the *winter mildness of*

SAN DIEGO

Come Now There are plenty of things to do and see in balmy San Diego, while old winter blows itself out in other climes. The most equable seacoast climate in Southern California offers play days out of doors ... sightseeing in panoramas of unsurpassed beauty! Live in a lively city, historically rich in the lore of Spanish Missions. Visit exotic Tijuana in Old Mexico, 18 miles away. Begin *your* winter vacation where *California* began ... in San Diego .... and come now!

ATTACH THIS CHECK LIST TO YOUR LETTER

[NOTE: Do not come to San Diego seeking employment]
(Please print name and address clearly on your letter)

1. I desire information covering:
   Vacation _____ Permanent home _____
   Business _____ Agriculture _____
2. Will start now _____ in 6 mos. _____
3. Send Hotel list _____ Auto Courts _____

Write

SAN DIEGO CALIFORNIA CLUB
Room 6, 499 W. Broadway, San Diego 1, California

Hotel Accommodations now - No time limit
CONSULT YOUR TRAVEL AGENT

50

favor certain plasmagenes above the others, allowing the favored ones to multiply abnormally, as rabbits once did in Australia. A change in the balance of plasmagenes affects the cell's chemical behavior, as rabbits affected Australia by eating much of its rangeland bare of grass. The cell's genes take no part in the transformation.

New Control. What good was the discovery? The biological revolutionists were reluctant to say. But they admitted (with a gleam in their eyes) that it gave a new, promising method of controlling cell life and growth. They had already controlled yeast cells by regulating competition among plasmagenes. Future biologists might do the same with bacteria cells or man cells.

One vital problem might be cleared up. Cancer is caused when a cell of the human body loses its normal function and starts multiplying lawlessly. This perversion cannot be blamed on true "mutation," which involves the genes themselves. It happens too often for that.

But plasmagenes are much more numerous than genes. If something goes wrong with the division of one of them, a lawless plasmagene may be produced. It multiplies wildly at the expense of others. The corrupted cell turns into a deadly cancer.

Drs. Spiegelman and Kamen emphasized that they had no cure for cancer. But they may have charted a course for future cancer physiologists.

Ferrets in the Oilfields

Bacteria found in deep sea mud might soon make oil wells as buggy as vinegar works. Last week Dr. Claude E. ZoBell, of the Scripps Institution of Oceanography, La Jolla, Calif., announced that he was well along on a process to infect exhausted oilsands with these bacteria. Snuffling underground like fierce microscopic ferrets, they would chase residual oil toward waiting wells.

Dr. ZoBell has busied himself for years with the microflora of oil strata, including sea-bottom muds where oil is thought to be formed. His original idea was to study how bacteria modify crude oil (TIME, Dec. 17, 1945). But in 1943, he found in sea mud a comma-shaped bacterium which he named (he was only 38 at the time, and feeling in the pink) *desulphovibrio halohydrocarbonoclasticus.* He put it in a test tube filled with material to simulate a limestone oilsand. Four days later, oil bubbled out of the test tube's mouth. A little later, all the oil in the sand was gone.

By sheer multiplication, the bacteria push oil particles off the grains of oilsand. They dissolve limestone, making the formation more porous. They generate carbon dioxide, which pushes oil particles ahead of it by gas pressure. The bacteria also produce a "detergent" (soaplike substance) which makes clinging oil films gather into free globules.

To the Last Drop. At first Dr. ZoBell did not think much of his find, but oilmen heard about it and jumped at it eagerly. Their interest was based on a sad and simple fact: no known extraction process gets all the oil out of an oilfield. Much of it stays below ground, sticking to rock particles. But if the bacterial process works, this oil can be got at, and exhausted oil pools will yield a valuable "second crop."

For two years, the Pennsylvania Grade Crude Oil Association has been experimenting (more or less in secret) with the practical application of Dr. ZoBell's discovery. At present many Pennsylvania oilmen glean their underground fields by forcing water through worked-out strata. They plan to introduce Dr. ZoBell's bacteria along with the water. The bacteria should hunt out and bring to the surface the last drops of oil.

FIGURE 1.2. Claude E. ZoBell's "Ferrets in the Oilfields" from *Time* magazine, January 1927.

results were ignored for two decades, until the late '40s and early '50s. A microbiologist by the name of Claude ZoBell began renewing interest in subsurface bacteria, this time in marine sediments, by claiming that they not only ate oil (figure 1.2), but also produced petroleum.[14] But ZoBell failed to prove that the bacteria in oil wells were nothing more than contaminants from drilling or reservoir flooding, so the scientific community at large had dismissed the possibility of deep subsurface life again.

By 1991 the oil and gas wells in the United States numbered over two million.[15] But the oil and gas companies had somehow missed Taylorsville Basin, which made it the one-of-its-kind basin that Frank had been seeking. Texaco would be drilling into sediments far older, 150 million years older, than the ones Frank's team of scientists had cored at the SRP four years earlier. The Taylorsville Basin sediments were Triassic in age, and the Texaco drilling campaign would offer a window into a previously hidden Triassic basin uncontaminated by the oil and gas industry.

Frank quickly contacted a geologist friend who worked for the Delaware Geological Survey, but his friend hadn't heard about any drilling by Texaco and had difficulty pinning down any details. Frank then phoned Tim Griffin and Brent Russell, Tim's boss, at Golder Associates in Richland, Washington, who quickly volunteered to make inquiries off the clock, so to speak. Tim contacted an old friend working at the New Orleans office of Texaco, who in turn directed Tim to a colleague in Texaco's exploration office in Houston. The Houston office told him that the Delaware drilling was a "no go." "We recently took a few shallow slim cores from the Triassic Taylorsville Basin on the Virginia side, but we gave those to Paul Olsen at Lamont-Doherty Geological Observatory. Would those be helpful? You should check with him," said the manager over the phone. Tim thought to himself, "Nope, those would be contaminated." The manager went on to say, "And we ARE drilling deeper in the Taylorsville Basin. Down near Fredericksburg, Virginia, at a place called Wilkins. Would you be interested in samples from that site?" Tim's ears perked up and he barely constrained his enthusiasm, "Sure we would be interested." The Houston exploration manager carried on, "In May we will be completing our final and deepest hole near Thorn Hill Farm. It is about an hour's drive south of D.C. We could talk to our management. Maybe we could get you folks a few core samples. We don't know about you coming on site though. We CANNOT guarantee you anything, of course." After he hung up the phone, Tim thought

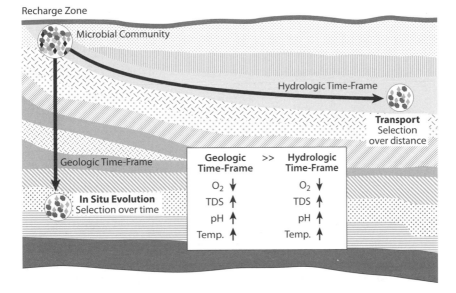

FIGURE 1.3. End-member origins hypotheses presented at Las Vegas DOE SSP meeting in January 1992 (courtesy of E. Murphy).

to himself, "We need to be on site to collect those samples, and we have a lot to do in a short time-frame."

Between January 14 and 16, 1992, Frank held a Deep Microbiology Investigators stocktaking meeting at the only non-gaming motel in Las Vegas,[16] during which all his investigators were to report on the results of the analyses and experiments performed on the samples they had just collected months before from INEL, PNL, and NTS. They were also asked to present hypotheses concerning the origin of deep subsurface bacteria in different types of sedimentary formations (figure 1.3). How could they distinguish those bacteria that had been deposited with the sediment from those that might have migrated into the sediment with groundwater flow?

It was at the end of the meeting that Frank announced to the group that an opportunity had presented itself to gain access to sedimentary rock samples from a buried Triassic basin with the help of Texaco, which was drilling a gas exploration well in Virginia starting next month. He was looking for volunteers to help with the field campaign. In fact, Tim and Brent had been working feverishly behind the scenes and managed over the course of several weeks to track down the right managers at Texaco to obtain permission to be on site to collect core samples. Frank and his team now had only six weeks left to mobilize a team of scientists and equipment to Thorn Hill Farm, where the drilling would begin. Recognizing that the window of opportunity to search for bacteria from the deepest and oldest sediments yet would be quickly closing, Frank and his team moved resources to the site at a breakneck speed.[17] Faxes were flying in and out of DOE headquarters in Germantown, Maryland. Frank's agreement with Brent and Tim was that if at any point they hit a wall, they would stop. But amazingly, the two of them always managed to find a way through or around or over one wall after another. They mobilized a research trailer with sampling gear at the biology department of Mary Washington University, in Fredericksburg, which served as the staging area. Frank had just finished sending the $40,000 award to INEL to subcontract Golder Associates for the work they were already performing out of pocket. The effort was moving so fast that it was hard to get the subcontracts through the bureaucracy of DOE quickly enough. They worked with Rick Colwell to have two large plywood shipping boxes constructed to handle the anaerobic glove bag and core splitter that they had used at the INEL WO1 borehole the previous year.[18] So heavy was the box carrying the core splitter that some driver along the shipping route had scrawled on the outside, "The box from Hell."[19] The deal that Brent and Tim had made with Texaco was that for $50,000 they would get fifty sidewall cores from five different zones each with ten core points, and that they would split them fifty-fifty. Texaco and DOE would collect their respective cores from alternating core runs, beginning with Texaco's.[20]

The research trailer was already on site at Thorn Hill on Monday, March 9, when Tommy returned to his home in Knoxville at midday following an extended trip to Germany.[21] While Tommy had been gone, Susan Pfiffner, his microbiologist girlfriend, had absolutely stuffed his Saab with sampling gear from their labs at Oak Ridge

National Laboratory (ORNL) and the University of Tennessee. She had even managed to wedge in an anaerobic glove bag. There was just enough room for Tommy and his suitcase. He slid behind the wheel and closed the door as Susan reached in to give him a kiss and a six-pack of Coca-Cola and sent him off to work. He drove like a demon possessed toward Fredericksburg, Virginia. Tommy relished every opportunity to sort out methods to obtain subsurface samples for microbiology with the least amount of contamination, but the Taylorsville project would be his most challenging yet.

THE MEETING THAT BEGAN IT ALL: FEBRUARY 25, 1986, SAVANNAH RIVER PLANT, SOUTH CAROLINA

As Tommy drove up Interstate 81, he reflected on the first meeting six years earlier in the lunchroom of the SRP,[22] a DOE facility near Aiken, South Carolina. Frank, who had just started the SSP, and Jack Corey, an environmental manager at SRP, had organized that meeting.[23] About a dozen scientists from SSP and SRP were sitting around a table discussing how to collect microbial samples from the 600-foot-deep aquifer that Jack was planning to drill at various locations around the SRP.[24] Although Carl Fliermans, an environmental microbiologist at SRP, had been leading the discussion, it was Tommy who dominated conversation. Back then, Tommy was a brash, freshly minted Ph.D. microbiologist. He had been working as a postdoc for D. C. White of Florida State University (FSU) and had been retrieving subsurface core samples for phospholipid analyses from several monitoring-well installations at SRP.[25] By the time of the meeting, Tommy had had months of drilling experience. He proposed readily available modifications to local technologies using existing practices, equipment, and personnel for this new type of drilling campaign. In his usual vibrant manner, he outlined how to obtain uncompromised microbial samples: how to minimize the contamination from drilling, how to trace the contamination from drilling, how to handle the core samples, how to minimize their exposure to O_2, how to distribute the samples without contamination or O_2 exposure, and finally how much it would cost and how long it would take. Every-one sitting around the table had started jumping into the discussion, offering to contribute one kind of analysis after another. They had assembled a laundry list of analyses that included radiolabeled aero-bic (O_2-using) and anaerobic activity measurements, cell counts, phospholipid fatty acid analyses, geochemistry, and carbon utiliza-tion assays. Tommy recalled, as if it were yesterday, that when Jack Corey had heard enough, he stood up and walked over to whisper in the ear of Carl Fliermans. Tommy had nudged D. C., who was sitting next to him and told him, "It's a done deal." At that moment, Jack Corey committed $250,000 to the project, $200,000 of which would support different subcontracts to analyze the microbial content of cores. That two-hour meeting in the lunchroom of the Savannah River Plant on February 25, 1986, had reignited the search for life deep beneath the Earth's surface, which had lain largely dormant since ZoBell's work three decades earlier.

Three months later, three drilling sites for the new DOE project, which had been named Deep Probe by Carl, were located near the P monitoring wells, which were scattered across the SRP, and carefully avoided the melon fields.[26] The three P-well sites were spread along eight miles of the SRP and had been selected because they were as far removed from the potentially contaminated sites on SRP as was pos-sible. Small two-inch-diameter wireline cores had already been col-lected at two sites down to a depth of 900 feet. The cores had revealed terrestrial and marine sediments belonging to the Tertiary and Cre-taceous periods, dating from 40 to 70 million years ago. From these cores, the microbiologists had determined which formations, litholo-gies, and depths to target. The Deep Probe team, led by Tommy, had recognized early on that the microorganisms they hoped to isolate from great depth might be indigenous, that is, either deposited with the sediments or transported to their current residence by groundwa-ter flow (figure 1.3).

Coring at these depths for microorganisms, however, had never been done before.[27] Because coring at any significant depth required rotary coring with a core barrel and a hollow cutting drill bit on the end, they had to use drilling mud to carry away the dirt cuttings carved out by the metal drill bit as it rotated its way down.[28] For the P-series drilling campaign, Tommy had had to defeat three sources of contamination: (1) the drilling and sampling hardware, (2) the bentonite (clay) drilling mud, and (3) the sample processing.

The drillers drilled without taking cores, down to the depth where Tommy and his team wanted to collect a microbial core. Then they pulled up the drill pipe, thirty feet at a time, disconnecting each pipe section until they reached the drill bit at the bottom. Then they attached the coring tool and bit and lowered them down as they screwed on each thirty-foot pipe section, one at a time. It was truly filthy, dirty work. To reduce the contamination, Tommy had steam-cleaned each drill pipe before it followed the coring bit down the hole.

Then there had been the drilling mud, all one thousand gallons of it. The drilling mud would undoubtedly contain tens of thousands of times more bacteria per gram than the sediments that they were coring, if not more. That mud would contaminate the coring barrel as it was being lowered and retrieved from the target zone. How do you get the sediments into the core barrel without contaminating them with the drilling mud? Tommy tried a variety of coring tools to limit the penetration of the drilling mud at the coring bit (figure 1.4),[29] but he had no means of preventing the drilling mud from contaminating the top and bottom of the core as it was being pulled up. So they had to remove ends of the cores and their outside surfaces in a sterile, anaerobic environment.

To accomplish this, Carl Fliermans had converted a Blue Bird mobile home into what he called the Mobile Microbial Ecology Laboratory (MMEL). It contained a six-cubic-foot, double-layered polyethylene glove bag with a manual core extruder, a propane-operated refrigerator and freezer for sample storage, propane torches and alcohol for sterilizing all the tools, and high-pressure gas cylinders for flushing the glove bag with nitrogen (N_2). To process the core samples, they parked the MMEL near the drill rig, and Tommy had been able to get some help with the core processing: Susan Pfiffner and a crew of FSU students had joined him.[30] As the microbial cores were being raised to the surface by pulling up the rods, Tommy and Su-

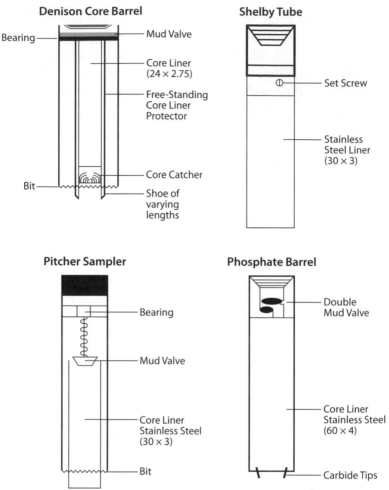

FIGURE 1.4. Different types of coring tools used during the C10 microbial coring project (from Phelps et al. 1989). The phosphate barrel and Shelby tubes did not have separate core liners. Note that the Denison core barrel had an extended shoe that penetrated the soft sediment ahead of the coring bit to reduce drilling contamination.

san's team disinfected the glove bag, alcohol-flame-sterilized all the tools, autoclaved the mason jars and lids, and flushed the glove bag so that it was as anaerobic as possible before the core arrived at the surface. As soon as the core barrel was on the surface, they removed the core barrel, capped the ends, and ran it into the MMEL. There it was placed in the core extruder and pushed into the N_2 glove bag while its outside surface was scraped away. Using sterile knives, chisels, spatulas, and spoons, the team removed any evidence of drilling-fluid-impacted sediment, then subdivided the apparently pristine core into samples,[31] scooped the samples into sterile Whirl-Pak bags, and double bagged them. They then weighed, labeled, and placed the bags into the N_2-flushed mason jars. The mason jars were packed into Styrofoam coolers with Blue Ice and shipped out overnight. The sample jars had arrived at their various receiving laboratories across the country by the next morning.[32]

Once the field team had finished processing that sample, they had started the whole process all over again. They had to completely remove the pared-off, contaminated sediment left behind in the glove bag. The glove bag and tools had to be resterilized, and the bag had to be flushed again with N_2 before the next core was shuttled in from the rig. All of this need for speed had been to minimize O_2 contamination, which would be lethal to the anaerobic bacteria. The speed had also cut down on the growth of microbial "weeds" that may have crept into the samples during the sample processing. Well, that was how it had worked most of the time. Occasionally the core extruder intermittently *had not* extruded the core, or the glove bag leaked or the core barrel arrived at the surface without any core, or a host of other inconvenient surprises arose that required "McGyvering" by the chief driller and Tommy and Susan on a daily basis.[33]

The chief driller had become quite interested in all their scientific perspiration and, being adept at modifying coring tools, started making new ones, such as the phosphate barrel, designing new drilling shoes for the Denison core barrel, and working continuously with Tommy. It had been widely joked that when sober, the chief driller was the second best in the state, and on those other occasions, was the best in the state. Occasionally members of Tommy's team had been called upon to assist the second-string roughnecks. When the drill crew had left the site on errands and progress was impeded, the scientists had stepped up to help. Usually this had entailed the less dangerous activities, such as refueling the Blue Bird or other trucks at the site, pumping water, and checking viscosity of the drilling fluids. It also included preparing sampling tools for deployment and wrenching sampling tools on and off the drilling rod. Occasionally their help had been sought to assist in recovering core samples, pulling drill rods, or even racking drill rods. In the last, Tommy strapped himself into a harness and then climbed up to the crow's nest at the top of the wobbly derrick. There he would sling a hammer to disengage the quick-release subassembly used to pull the rods from the core hole. Then he had to roll each 30-foot drill rod to the side to rack them vertically. The unacceptable alternative was to have left the cored sample at the bottom of the hole, bathed in drilling fluids, until the roughnecks had returned. By July 6 and over a period of four weeks, Tommy, Susan, and their team had distributed fifteen microbial subcores from each of the three drill sites to seven different labs at a remarkably frugal cost of $20 per sample (shipping and handling not included).

The results from these cores had been positively stunning as each investigator reported their findings at Frank's stocktaking meeting three months later at DOE headquarters.[34] By all measures, whether it had been the substrates upon which the microbes grew, their morphologies, or the composition of their lipid membranes, the microflora yielded by the sediment cores were more diverse and more numerous than had ever been encountered in the very shallow aquifers and vadose zones of previous studies.[35] Algae and protozoa were also present in the deep sandy aquifers. The clay-rich samples yielded lower microbial counts and diversity than did the sandy samples. These conclusions were independently confirmed by seven different microbiology laboratories. The properties of the microbial isolates, which numbered in the thousands, had varied with depth and had been quite distinct from those of the soil samples collected at the drill sites. It was as if the scientists had penetrated a surface veil to reveal a hidden universe of life beneath. The microbiology textbooks

FIGURE 1.5. The C10 drill site viewed from U.S. Forest Service fire tower during summer of 1988 (courtesy of T. J. Phelps).

would now have to be rewritten once the results from the P wells were published.[36] Not bad for four weeks of fieldwork.

THE FIRST MICROBIOLOGY DRILLING PROJECT: AUGUST 1988, FOUR MILES NORTHWEST OF ALLENDALE, SOUTH CAROLINA

Frank immediately pushed Tommy and the other principal investigators (PIs) into developing a drilling project on subsurface microbiology that would go even deeper than the P wells had.[37] By the summer of 1988, a forty-foot-tall, tan-colored Gardner Denver derrick pinned the tail end of a ten-wheel diesel truck at a drill site they had called C10. C10 was located fourteen miles southeast of the SRP in a sandy meadow in the midst of a lush green Loblolly Pine forest. One lone U.S. Forest Service ranger, looking down from his fire watchtower, had observed them as they walked back and forth from the Blue Bird MMEL and the derrick (figure 1.5).

The only place to stay that was close to the C10 drill site was the sleepy South Carolina town of Allendale, with its really not-so-good motels filled with migrant workers who picked fruit during the harvest season. So Tommy had to rent a small two-bedroom house in town. Tommy and Susan had shared the office bedroom, and the guys stayed in the laboratory bedroom, complete with a laminar flow hood. The living room served for shipping and receiving, storage, and experiment incubation space. Most of the women stayed next door in a bedroom rented from a female prison guard. Upon arrival of cool September evenings, they realized that the landlord had removed the heating stove from the living room to provide more desperately needed space. The extra refrigerator she provided had probably been used by a taxidermist, but once cleaned, it was adequate for storing sealed containers.

The original coring plan had been scheduled for four weeks, during which they would drill down more than 1,650 feet, deeper than they had ever attempted before, and collect fifteen microbial cores. This depth would have taken them through the Cretaceous sedimentary sequence and possibly even into the Triassic sedimentary rocks of Dunbarton Basin, which lay beneath.

They adapted a new coring system to speed up core retrieval, but it had to be modified to reduce the drilling-mud contamination as they penetrated the different lithologies.[38] Brent Russell had been brought on board to help Tommy implement new quality assurance/quality control (QA/QC) protocols. At C10 all tool decontamination and sample processing had to take place upwind from the drill site to avoid airborne contamination from the exhaust of the drill's motor. The mud tanks had been completely covered for the same reason, and the drill pad and pipes were regularly steam cleaned with chlorinated drinking water. A wireline core of the Tertiary/Cretaceous sequence had been collected 450 feet down gradient from the microbial borehole so as not to risk contamination between the boreholes.[39]

The drill pad had been roped off and was accessed only by the Graves drillers and two members of the microbial team (clad in helmets, safety glasses, and steel-toed boots).[40] The sterilized microbial core barrels had an inner barrel filled with chlorinated water that was held in place by a plunger. The plunger prevented the inner barrel from being contaminated by the drilling mud on its way down the inside of the drill pipes. The inner barrel also contained a Whirl-Pak bag filled with magnetic micron-size spheres that would break open when penetrated by the core as it entered the tube, releasing the spheres to mix and settle throughout the length of the core.[41] After the coring barrel was prepared by the microbial team, they then carried it to the drillers, who connected it to the wireline messenger and sent it down the inside of the drill pipe to lock in place in the core barrel at the bottom of the drill string.

Government regulations prohibited the use of microbial tracers, but the drilling mud had quickly grown a healthy population of nonfecal coliform bacteria that acted as a natural microbial tracer.[42] The SSP began referring to the drilling mud microbial community when used as a tracer as an inherent microbiological tracer (IMT).[43] Because it had been impossible to sterilize or keep sterile 2,000 gallons of mud,[44] Tommy added rhodamine dye until its concentration reached 30 parts per million (ppm).[45] The concentration was determined empirically as the maximal color of pink stains on exposed skin that was tolerated by the drilling crew and their families (to everyone's delight, dilute bleach neutralized the dye). When the rhodamine was added to the grayish-brown mud, it turned into a pool of bubble-gum-pink, bubbling goo. For the first time, the SSP drilling team also used PFC tracers, which were provided by Brookhaven National Laboratory. That laboratory had successfully used PFCs as atmospheric tracers, but this was the first time they were tried in water as tracers. They simply had to hook up a vintage high performance liquid chromatography (HPLC) pump to the mud line in order to inject the PFCs. It worked like a charm.

Carl Fliermans had had a custom glove bag built by Dick Coy with a twelve-inch-diameter, three-foot-long, aluminum-tube air lock at one end and a four-inch-diameter sleeve into which a mechanical core extruder was fitted at the other end.[46] The sampling team mounted the core with its liner into the tube, cranked the handle, and slowly pushed the inner core into the N_2-filled glove bag. It was like squeezing a tube of brown toothpaste. The research team delicately pared away the outer layers of the core samples and then used a fluorimeter to immediately test for the presence of rhodamine-infused drilling mud in the parings and inner samples. This approach provided four orders of magnitude of contamination detection on site. In other words, if the drilling mud had 10^5 cells per gram of bacteria, then the absence of detectable rhodamine in the interior of the cores meant that fewer than 10 cells of drilling-mud bacteria were present in a gram of that sample. For the Congaree Formation, a sandy aquifer consisting of large sand grains,[47] every sample, even the inner sub-samples, had been contaminated with rhodamine dye. Consequently, they had not shipped any samples of that formation to the off-site scientists.

Tommy and the sampling team not only had to implement the new QA/QC protocols, but they also had to cater to the demands of twenty different microbiologists, chemists, and geochemists. Almost all of them had never been near a drill rig. Each had wanted their samples processed their own particular way and had spoken a language not easily understood by the drillers. Brent and Tommy had their hands more than full. To make matters worse, Tommy had run the sampling operation for more than sixty days, twice as long as planned, writing personal checks that had wiped out $10,000 of his personal savings, while waiting for additional funds from Frank. They needed more money and more help.

The latter soon arrived in the form of Tim Griffin, a geologist who had worked in Exxon's Offshore Production Division in the Gulf of Mexico and now worked for the South Carolina Water Resources Commission (SCWRC), a partner with DOE in the C10 drilling program. The SCWRC had footed the bill for the installation of monitoring wells after the coring was done. Tim was a sedimentologist and stratigrapher by training and knew nothing about "bugs in rocks" but was excited about working with the DOE research team. Tim had a deceptively easygoing and polite Southern manner, but before Tommy knew it, Tim was busy not just identifying the target zones for the microbiologists but also communicating to the drillers about the stringent decontamination protocols, holding regular morning briefings, keeping wandering microbiologists from hurting themselves and each other, and communicating with the management, aka Frank, on their progress. Tim had essentially brought a professional touch to this scientific drilling project, telling "y'all" what they had to do to keep civil order from spontaneously disintegrating into a cataclysmic disaster. Still, even Tim had not foreseen that the drill pipes would clog, causing overpressurization of the borehole, and the eruption of a hundred-foot-tall pink geyser spewing from the subsurface, painting the drill rig and drillers pink. Neither had they anticipated losing the 2,000 gallons of drilling mud that had flowed into one of the more permeable formations just as Frank arrived for a site visit. But after only eighty-nine days, the drillers had cored through the Cape Fear Formation and collected twenty-three microbial cores, all delivered to their receiving labs in less than seventy-two hours. They had intersected granite at the bottom of the hole, not the Triassic sedimentary rock of the Dunbarton Basin that they had expected. Tim, fulfilling his role as the site's spokesperson, in a National Public Radio interview stated in his best southern drawl that "we must have entered the Triassic."[48] Though mistaken, it was in fact a prescient comment, because Tim, Tommy, and others they had not yet met would be coring the Triassic sediment at Thorn Hill no. 1 four years later (figure 1.6).[49]

STERILIZE A DRILL RIG? MONDAY, MARCH 9, 1992, KING GEORGE COUNTY, VIRGINIA

Tommy in his overstuffed Saab finally arrived in the parking lot of the Hampton Inn in Fredericksburg after an eight-hour commute. He was greeted in the lobby by Rick, who proceeded to brief him on the status of the Thorn Hill no. 1 drilling. After an hour and a half of intense cross-examination, Tommy was satisfied that they had thought of just about everything and then called Frank, who was at the Los Alamos National Laboratory (LANL) at the time. "Are you still in Germany?" Frank asked. "No, Frank, I flew in this morning and drove up today. I'm in Fredericksburg with the rest of the team,

FIGURE 1.6. Frank Wobber pointing to the bottom of core barrel C-10. It was really only three feet of granite, most of which was left in the hole and only six inches was recovered, but that six-inch core would be given to Frank by Tommy Phelps as a set of bookends twenty-five years later (courtesy of T. J. Phelps).

and I am calling you to tell you we are a go for the Taylorsville campaign." Frank guiltily acknowledged that his feet had been propped up in his motel room, as he relaxed after only two hours of time change and a six-hour flight. Frank was clearly impressed and thankful that the team had so rapidly and efficiently deployed.

The next morning Tommy, Tim, Rick, and Todd drove out to the drill site, taking Virginia Route 218 east out of Fredericksburg for about twenty miles. They then turned south before reaching Dahlgren and followed the county roads across the tidewater countryside. From the bridge crossing on Windsor Drive, they spotted the top of the Murco 54 drill rig several miles to their right, poking above the tree line. After driving by several small gentleman farms and rural homes that dotted the sides of the road, they came to a big white "Texaco Inc. No Admittance" sign. They pulled into an empty two-acre farmer's field completely covered by 8-foot-long, rough-cut, two-by-eight oak boards laid two to three layers thick. An earthen berm separated the road from the wooden platform, on top of which were six single-wide 50-foot-long trailers; a bank of three red-orange, truck-size, 1,700 horsepower mud pumps; a football-field size platform holding hundreds of rusty 30-foot-long drill pipes; and a diesel generator "farm" large enough to power a small city. All of this surrounded the base of the Murco 54. Just as they pulled up, a man-sized winch was hauling one of those pipes up a three-story ramp until it was just above the platform, where the roughnecks grabbed it and then used immense hydraulic tongs to screw it onto the top of the drill string sticking up from the deck. Beneath the platform sat the orange hydraulic jacks that made up the blowout preventer. If they had the misfortune of hitting a pocket of high-pressure methane (CH_4), these jacks would crush the drill pipe and seal the borehole. Off to one side sat the EXLOG trailer that was monitoring the gas levels of the drilling mud, looking for volatile hydrocarbon plays. An enormous yellow traveling block, which by itself was as big as the entire drill rig they had used at the C10 site, held the drill string as the bushing began to rotate.[50] Immediately they heard the deep-throated moan, the pitch of which increased until it was a whine as the pipe revolved faster and faster, and they could feel the ground vibrating beneath their feet. A strong chill wind had crept down from the north and filled the red Texaco windsock.

They then decided to retreat to the warmth of the field lab, a lonely 50-foot-long trailer at the end of the trailer park. They drove onto the wooden platform, parked near the trailer, and stepped into it and closed the door to hear themselves think. Inside was Tim's pride and joy—a bright, shiny double glove bag with a green torpedo tube sitting in the middle of the trailer. There was not a speck of dirt on the outside or inside. The bag had been overinflated so that the five pairs of arms stuck out straight into the room. The white tile floor had been mopped and bleached. A desk with a fax machine, telephone, and power strips for computers sat at one end. A table with shipping supplies, balances, plumbing parts, syringes, and other sundry microbial instruments sat at the other. Tim, Todd, and Rick looked nervously at one another as Tommy walked around inside the lab, inspecting the bag, power connections, and gas connections to the high-pressure cylinders. He went through his mental checklist, and then he nodded his head, smiling. He reached over to pick up some mysterious cores wrapped in aluminum foil that were sitting on top of the desk. Rick quickly explained that he had baked those at 550°C overnight and would process them through the anaerobic glove bag to provide some negative controls on the contamination levels. Tim then explained to Tommy that Texaco had set up the site in just one month using sixty-seven tractor-trailers to haul in all the gear. By the time the scientists had arrived on site, Texaco had already drilled through 2,300 feet of the overlying coastal plain aquifers, which were stratigraphically similar to what they had sampled at C10. They had then cased that section off before they started drilling into the underlying Triassic. They were deepening the hole at a rate of 10 to 30 feet per hour, hoping to eventually reach around 10,000 to 11,000 feet. The lab inspection complete, they all stepped back outside so Tommy could get a better look at the drill rig and pad.

It was one of the biggest land rigs that existed and was certainly taller than any that they had worked on before, with the exception of Tim, who had worked on the offshore drill rigs in the Gulf of Mexico. It even had an elevator to take you up to the drill deck. It was taller than a twenty-story building, and even when they walked to the edge of the farm, they still could not get it into one frame of their camera. With a smirk, Todd peered out from beneath his black Oregon State cap and shouted to Tommy above the whine, "How do you propose we sterilize that?"

Not far away in Germantown, Maryland, Frank had been doing everything he could to keep the SSP moving forward. He had made numerous presentations to the upper management of DOE, emphasizing the potential impact on remediation, and had provided weekly progress reports on his projects. It had been the brief report on the Texaco collaboration in the Triassic Taylorsville Basin that had caught the eye of the director of the Office of Science, Will Happer.[51] This report was definitely not your normal DOE-sponsored research. What had intrigued Happer was the word "Triassic," and he had wondered if those sediments were the same as those near his home in Princeton that were deposited when North America and Africa separated to form the Atlantic Ocean. It had been 6 p.m., but on a whim he had called Frank's office and had been pleasantly surprised to catch Frank still at his desk. After a brief discussion, Frank was quite impressed with how well Happer, a physicist by training, was so attuned to New Jersey geology. Frank asked Happer if he or some staff member would like to take a day to go visit the drill site and see the operations. Harper enthusiastically said yes. The next week they met downtown, and Frank drove Happer to the Thorn Hill Farm drill site.

It was a typical gray, overcast, early March day when Frank arrived on site with Happer in tow. Texaco's drilling manager was waiting for DOE's upper management and gave them a guided tour of operation, speaking above the steady droning of the Murco 54. The drilling crew explained that Texaco and Exxon were looking for some gas plays in the organic-rich Triassic lake sediments that were located 8,000 to 9,000 feet beneath them, hence the huge drill rig. Happer was familiar with Frank's SSP reports on bacteria from 600 feet beneath the surface, but how could they hope to find bacteria 9,000 feet underground with this monstrous drill rig? After Texaco's drilling engineer explained to Happer how the rig operated (figure 1.7), Tim showed him the PFC tracer system that they had set up near the inlet of one the blue, high-pressure pump feeds for the drilling-mud delivery system. He explained how the sheer volume of the drilling-mud system, some 100,000 liters (about 26,000 gallons), precluded the use of rhodamine dye, bromide, and even deuterated water,[52] but the highly sensitive detection limit for PFC meant that only 200 milliliters of PFC was required. The high-pressure pumps pushed 1,500 liters of drilling mud per minute down into the borehole and back out again. They had to pump only 1.7 milliliters of PFC per minute into the pump line. Next to the massive mud-delivery line was the tiny lunch-box-sized HPLC pump in a small polycarbonate box with a one-sixteenth-inch transfer line for delivering the PFC to the drilling mud just prior to its entering the high-pressure pumps.

Tim had then shown Frank and Happer the two different sampling tools that Schlumberger had offered to them. Both were thirty-foot-long, shiny metal sidewall corers. One contained a single one-

FIGURE 1.7. William Happer talking to Texaco employees with Frank in the background at the Texaco drill rig at Thorn Hill Farm, Virginia, spring 1992 (courtesy of T. Griffin).

inch-diameter diamond bit at the top of the core tool. Once at the targeted sample depth, the bit rotated 90° toward the borehole wall, drilled a two-inch-long core sample, and then rotated back into the core barrel, where a rod pushed the sample into a collection chamber. The other core sampler, a percussion core gun, contained a long line of explosive core tubes, one-inch-diameter hollow bullets, connected to the core gun by cables. Once the core tube was fired into the borehole wall like a bullet, it was pulled free from the borehole wall with a cable and left "dangling." Sidewall corers weren't the best of core sampling options because of their small sample sizes, but they were the only options.[53]

The DOE team had also decided to adopt the IMTs they had used during coring at SRP. They would characterize the microbiology of the drilling mud as Texaco drilled progressively deeper and hope that the microbial characteristics were distinctive enough from those of any indigenous bacteria in the sidewall cores to provide a useful tracer. So the team had been collecting drilling-mud samples daily, processing and preserving them for microbial enrichments and assays. They had also collected samples of soil from the farmer's field and water from a nearby well that was used to make the drilling mud. The well drew its water from the Patapsco Formation, the same formation that Francis Chapelle, another pioneer in subsurface microbiology, had sampled not more than some twelve miles to the north, across the Potomac River in Maryland five years earlier.[54] Tim had collected and labeled bags of drill cuttings, using some for IMT and archiving other cuttings just in case someone in their group would need them for something other than microbiology.

Saving the best for last, Tim had guided Frank and Happer to the trailer where the core samples would be processed. They knocked on the door with the sign "Subsurface Science Microbiology Research Program—U.S. DOE—Headquarters Program Manager Dr. Frank Wobber." "C'mon in, Will!" Rick shouted as they stepped through the door. Rick was inside working in the glove bag on some rather dubious mud and water samples and pulled his arms out of the gloves to greet Happer. Rick explained to Happer and Frank how the glove bag worked and the details of the sample processing and shipping. Their inspection of the microbe lab trailer completed, Frank and Happer joined the research team at a local restaurant and then returned to Washington, D. C. "Come on in, Will?" Tim had said in a mock air of incredulity later that evening and shook his head laughing. "Gee, Rick, I didn't know you were on a first-name basis with the director of the Office of Science."

HUNTING FOR SUBSURFACE BACTERIA:
MARCH 14, 1992, THORN HILL FARM, VIRGINIA

The high from Happer's visit forty-eight hours ago had been replaced by uncertainty gnawing at their self-confidence and resolve as Texaco fired the percussion cores at the five selected target depths. At 10 p.m.

the roughnecks pulled the tool from the dark borehole with steaming drill mud dripping from it. Tim and Tommy grabbed the core "bullets," brought them to the trailer, and quickly placed them in the argon-filled glove bag. Haggard-looking from not having slept for more than twenty-four hours, they all stood around the glove bag gazing upon the meager fruits of their labors. After all the effort and expense, they were facing only five tiny rotary cores, each one no more than two inches long and no more than 50 grams apiece, and they were lucky to have those. The percussion core had harvested twenty-one cores, but eleven of those were mostly mud cake from the borehole wall and had little intact sediment. Tim merely removed those from the glove bag and archived them. As bone-tired as they were, now was not the time to sleep.

With grim determination they stuck their hands into the gloves of the glove bag and set about making thin slices of the rotary cores. They shaved off the obvious contaminated ends and then pared off the outsides, leaving behind a tiny internal cube that was least likely to have been exposed to the drilling fluids and other sources of contamination. By the time they were finished, they had a dozen tiny, precious rock nuggets and dozens of core ends and parings. They all wondered how they could make do with so little. They then processed one of Rick's control cores in the same manner and labeled the subsamples as if they had been cored from a depth of 10,000 feet. They would use this to characterize potential contamination from the glove-bag processing, from the shipping, and from processing in the receiving lab. Finally, they carefully parsed out the inner cores so that everyone would get at least something and shipped them in argon-filled mason jars on Blue Ice to eight different research laboratories across North America. Within seventy-two hours after their removal from the subsurface, the inner cores were placed in a variety of growth media in labs across the United States, all at temperatures, salinities, and pHs designed to reproduce the environment at depth. Tim had estimated those parameters from the analyses of the Schlumberger geophysical logs and from the drilling-mud measurements. Three days after they had obtained their cores, Texaco had plugged the well and demobilized their Thorn Hill village.

Frank got the full report from an apologetic Tim, who explained the unfortunate series of breakdowns. It wasn't anyone's fault. No one could have pulled off what they had done. Frank was just finishing a letter to James Kinnear, the president of Texaco, thanking him profusely for all the support at Thorn Hill and making sure that everyone who helped was named. If there were any significant results coming out of those samples, Frank was going to need to publish them quickly and in high-impact journals. Being able to do this while not upsetting Texaco's proprietary concerns was going to be a delicate balancing act. A humble letter of gratitude from the director of the Office of Science of U.S. DOE to the CEO and president of Texaco would certainly help remediate that problem if and when it arose. The final outcome from the Thorn Hill campaign now rested in the hands of the microbiologists and what they could find in those few tiny, precious nuggets.

THE BACTERIUM NOT FROM HELL: MARCH 16, 1992, PORTLAND, OREGON

David Boone loved microbes that produced methane—methanogens.[55] He was a microbiologist from Portland State University who had a penchant for anaerobes, but he really, truly loved methanogens. He was particularly enraptured if they were from exotic and extreme settings. An academic microbiologist, he was as academic as they come, serving on the committee that defines microbial species.[56] He had a remarkable, yet perhaps unappreciated, insight into microbial processes. He had heard about the SSP and had been invited to one of the early meetings in Germantown. David Balkwill,[57] a microbiologist from FSU, had both his hands and feet full with the Subsurface Microbiology Culture Collection (SMCC) for aerobic bacteria and had become a nursemaid to thousands of isolates.[58] With the program moving into deeper environments, they needed a similar facility for anaerobes, and Boone was a logical curator for that collection. David was excited about the Triassic samples that Frank had mentioned at the Las Vegas meeting. Tommy had steered $25,000 of his money to Boone to demonstrate to Frank, David's skills at culturing and archiving anaerobes. Nonetheless, David was a bit surprised upon opening the cooler delivered from Mary Washington University to find only a few tiny kernels of dark, flinty rock.

Over the weeks following the Thorn Hill coring campaign, the other microbiologists had reported vast quantities of aerobic bacterial colonies rapidly growing from the soil and drilling mud at mesophilic temperatures, that is, bacteria that love to grow at temperatures from 10°C to 45°C. However, little growth appeared in the enrichments from the rock nuggets, and none from the sandstone from 10,000 feet. Much to Boone's surprise, methanogens were growing out of the drilling muds. As the months passed, several strains of thermophilic (heat-loving), saline-tolerant, anaerobic bacteria slowly began to emerge in David's, as well as Tommy's and Todd's, lab cultures. Most of these anaerobic bacteria used oxidized forms of manganese, iron, and sulfur to oxidize organic matter for energy. They were quite distinct from the bacteria present in the drilling mud. Only 1 to 10 of these bacteria existed in one gram of the Taylorsville sediment compared with 10^9 bacteria per gram typical for soils and 10^7 cultivatable cells in one gram of the drilling mud.[59] Each one needed to be treated with tender loving care, so David used the data provided by Tim to fine-tune the temperature, salinity, and pH of the enrichment cultures.

David quickly proved that a little tenderness worked, by isolating and characterizing a thermophilic, anaerobic bacterium that became *Bacillus infernus* (figure 1.8),[60] the only bacterium to star in a Hollywood movie.[61] Why "*infernus*"? David had taken Latin in high school and knew that *inferni* had two meanings, one referring to people residing in hell and the other to people who simply lived underground. Although his bacterium was a thermophile, he opted for the latter meaning because it was the first bacillus that was an obligate anaerobe, that is, it couldn't grow in the presence of O_2. Until Boone's discovery, all known bacillus species had been aerobes or facultative anaerobes.[62] This trait suggested that *B. infernus* must have adapted to living underground in the absence of O_2. Discovering a novel bacterium from 9,000 feet beneath the surface in pristine core nuggets was

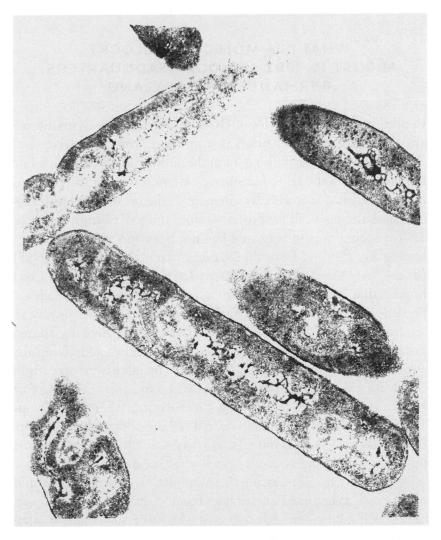

FIGURE 1.8. TEM photomicrograph of *Bacillus infernus*, 1 micrometer by 0.25 micrometer (from D. Boone).

the victory that Frank had hoped for with the Thorn Hill drilling project. But Frank was still left with the task of addressing the original question raised by Bastin and Greer sixty-five years earlier: how could they determine whether *B. infernus* or its ancestors were from the Triassic?

WHAT IS A MOLECULAR CLOCK?
AUGUST 15, 1992, U.S. DOE HEADQUARTERS,
GERMANTOWN, MARYLAND

I pulled into the parking lot of DOE headquarters in Germantown after having driven three hours that morning from New Jersey and then walked from the parking lot to the auditorium located in a side building, where the DOE Subsurface Science Program meeting was being held. I had seen an advertisement in the fax of the *Federal Register* announcing a call for proposals from the SSP to study a core of Triassic sediment, and because I had not been too successful in obtaining any funding from the National Science Foundation (NSF) for my $^{40}Ar/^{39}Ar$ (argon-40/argon-39) laser-dating laboratory, I was hoping that perhaps a grant from DOE might be the lifeline I needed. As I stepped into the stuffy auditorium I saw about fifty attendees, but thankfully none whom I knew. The meeting started promptly at 8 a.m. with one speaker after another addressing the audience about contaminant chemistry.[63] The talks went on continuously for two hours with no break. By 11 a.m., I was thinking that I had made a mistake attending this meeting when T. J. Phelps stepped up to the podium and started talking about orders of magnitude, tracers, QA/QC, and a Triassic basin, at which point I perked up. I was struggling to understand what he meant by anaerobic, saline-tolerant, metal-reducing thermophiles. Occasionally he would pause to ask a rhetorical question and then stare out at the audience, almost as if taking a mental census of who understood him enough to answer his questions. Clearly he was a microbiologist, an incredibly enthusiastic microbiologist who seemed to know a lot about drilling and a lot about geology. When he finished, people started quickly moving out of the auditorium to the cafeteria next door for lunch. I followed them to the cafeteria, grabbed a sandwich, and debated whether to hang on for another three hours. But with nothing to lose, I decided to return to the auditorium.[64] Finally, when the talks were over, a very large man took the stage and started spelling out the various goals of what were apparently his programs without any slides whatsoever. When he referred to the Triassic basin study, he started talking about developing dating methods for subsurface microbial communities, and again my ears perked up. Thinking that my patience was about to be rewarded, he then went on to say that "I have no interest in funding esoteric argon dating schemes. I am looking, however, for novel molecular approaches to constrain the age of microorganisms. We are organizing a Molecular Clocks workshop to develop these approaches." My whole purpose for attending this meeting had just been completely blown out of the water by this man. But something about the Triassic study still intrigued me. It reminded me of Michael Crichton's recent book, *Jurassic Park*. Could ancient, living microorganisms be preserved deep beneath the Earth's surface? They would be the microscopic versions of the coelacanth, a relic fish from the Cretaceous Period that had been discovered off the coast of South Africa in the late 1930s. So I decided to go up to the program manager after the meeting and at least compliment him on what I thought was a very audacious proposition. Staring down at me through his tinted glasses, he disagreed with my characterization in a way that was quite disarming. I then expressed interest in his Molecular Clocks workshop though I didn't think I had much to offer, not revealing that I was an esoteric argon dater, the very thing he had no interest in funding. He dismissed that remark and introduced me to Tommy Phelps, who was standing next to him. Then he pointed me to a woman standing next to the podium and told me that if I was interested in the Clocks Workshop that I should give Madilyn my name and address. I walked over and introduced myself to Madilyn Fletcher and expressed my interest in attending the Clocks Workshop, though I did not know what a molecular clock was. She eyed me doubtfully but dutifully took down my information.

On the long hot drive back from Germantown, I went over the talks in my head and struggled to find any application of my esoteric argon dating technique to any of the projects discussed that day with the exception of the Triassic basin study. It was halfway between Wilmington and Philadelphia that it finally struck me what the Triassic basin study really meant. Bacteria that had been deposited 230 million years ago and then buried beneath the surface were still re-producing. Even though the dinosaurs had been extinguished from the surface of our planet by a giant meteorite impact and the ice ages had come and gone, those deeply buried bacteria were still living down there. A nuclear holocaust could set our atmosphere on fire or plunge us into a nuclear winter and it would have no effect on the microorganisms 9,000 feet down in the Taylorsville Basin. It meant that life could exist beneath the surface of Mars. Aliens!! Ever since I had read Carl Sagan and I. S. Schlovskii's *Intelligent Life in the Universe* in junior high school, I had a fascination with life on other planets. I never thought in my wildest dreams as I drove to the DOE meeting that morning that it would give life to a passion that had lain dormant inside me all that time. I explained it all to my wife, Nora, over dinner that evening, and she agreed that it was an incredible discovery.

In the first week of September I received an invitation to attend the Evolutionary Clocks Workshop to be held at the end of September in Annapolis, Maryland. The letter came from Professor Sandra Nierzwicki-Bauer, a molecular biologist at Rensselaer Polytechnic Institute, and promised that more information would arrive. The next week, sure enough, from DOE headquarters came a fifty-page document on the Phase II Preliminary Plan to determine the origins of microorganisms in the deep subsurface. The week after that, a twelve-page fax arrived from Ellyn Murphy of PNL,[65] describing the preliminary agenda and providing very, very explicit instructions of what our presentations needed to include. In particular, the fax stated that it was up to each speaker to relate his or her methodology to constraining "microbial origins." The problem was that I knew nothing about microbiology, and I seriously doubted I could measure the age of a bacterial cell using argon isotopes. With only two weeks to go and classes already starting at Princeton University, I was swamped. Nevertheless, I thought it prudent to go down the hall and see if I could find a book or two about microbiology in the biology library. I started in the card catalog, which led me to a row of shelves. As I started browsing I found books on halophiles, acidophiles, methane-producing bacteria, iron bacteria, thermophiles, psychrophiles, cyanobacteria, and . . . the list went on and on (figure 1.9). An hour later I walked out with a dozen books on various types of microorganisms, returned to my office, and placed the stack on my desk. It was a monumentally impossible undertaking to understand or even grasp everything in these books in a week so as to remotely sound intelligent during a talk to the experts. This was crazy, I thought. I had a choice. I could either continue studying the diffusion of argon through mineral grains as a means of constraining the geological record, or I could search for life beneath the surface of our planet and perhaps on Mars. I thought to myself, what would Spock have done? I turned off my desktop PC and turned on my AC. I sat down and grabbed the book on top of the stack about thermophiles.

On September 30 I arrived at the Holiday Inn located near the waterfront in Annapolis at 7 a.m. after having spent three hours on the road dodging the rush-hour traffic of Philadelphia, Wilmington, and Baltimore. I walked into the restaurant to grab a quick breakfast before the meeting and came upon the DOE program manager, Frank Wobber, who was just finishing his. He invited me to join him, and while I was quickly downing Maryland-style crab cakes and eggs he explained how he was a geologist by training. As a Fulbright Fellow at the University of Cardiff he had studied sedimentology. He reminisced fondly about a workshop on contaminant transport at Princeton's ostentatious Dillon estate that he had attended seven years earlier and that had been led by the famed hydrologist George Pinder,[66] whom I knew somewhat. Back in those days, getting the SSP started on this risky path of research had not been easy. But, he explained, now that the results from cores collected in South Carolina had begun to appear in the scientific literature, the supporting evidence of the tracers combined with the novelty of the microbial isolates had begun to shift the broader scientific opinion from disbelief toward at least a recognition of the possibility that life could survive, perhaps even thrive, deep beneath the Earth's surface. The tens of millions of active heterotrophic, organic carbon–consuming bacteria per gram of sediment that his microbiologists had grown in

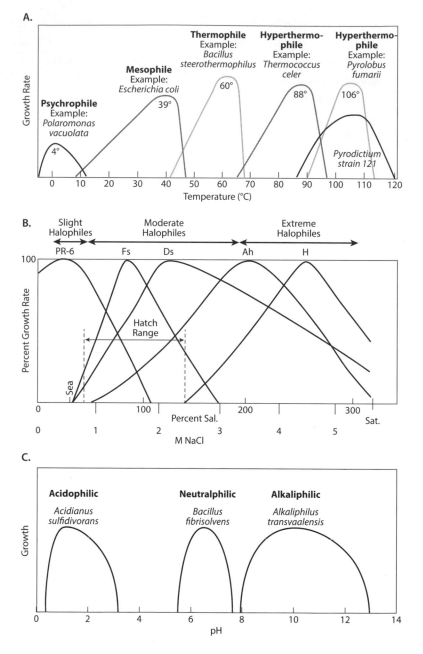

the lab meant that they had also captured a potent force for remediating polluted aquifers if the bacteria were stimulated with the right substrates.[67] Getting this far had required the concerted efforts of a few, and those few included Tommy and his colleagues in the SSP. The key requirement was not a paradigm shift per se, but the close integration of scientists from different disciplines that he had engineered. Frank reminded me of the old saying "Nobody wanted to crank the engine of a Model T, but once it was started everyone wanted to climb on for the ride!" Apparently after eight years of cranking, Frank had his Model T going, and he was taking a lot of scientists on the ride of their life. The question for me was whether I was willing to step on board. Having finished eating, Frank abruptly got up to go make certain that everything was in order for his workshop. It was probably the most important breakfast I had ever had.

The workshop was in a small, cozy meeting room with most of the seats in the back already filled by the time I arrived. One of the first speakers was a young microbiologist named Jim Fredrickson. He began his slides with a Far Side cartoon of a caveman looking down a primitive microscope and made some passing reference to Frank, to which Frank retorted jokingly from the back of the room. I clearly was not getting the joke (yet) behind this banter. Jim carried on with a remarkable forty-five-minute description of all the reports that had been published about supposedly ancient bacteria. These included

---

FIGURE 1.9. A. Growth temperature ranges of microorganisms (from Brock and Madigan 1991) modified to include the hyperthermophilic *Pyrodictum* strain 121 (from Takai et al. 2008). B. Growth salinity ranges for different halophiles (from Vreeland and Hochstein 1992). PR-6 is a cyanobacterium (*Agmenellum quadraplicatum*), Fs is a protozoan (*Fabrea salina*), Ds is an algal species (*Dunaliella salina*), Ah is a cyanobacterium (*Aphanotece halophitica*), and H is an archaeal halophile (*Halobacterium*). Definition of halophilic ranges are from Ollivier et al. 1994. C. Growth pH ranges approximately based upon *Acidianus sulfidivorans* (from Plumb et al. 2007), *Bacillus fibrisolvens* (from Rosso et al. 1995), and *Alkaphilus transvaalensis* (from Takai et al. 2001).

accounts of thermophiles trapped in fluid inclusions from Yellow-stone National Park cores,[68] halophiles trapped in fluid inclusions in salt crystals from the United Kingdom,[69] DNA (deoxyribonucleic acid) from Miocene-age black shale in Idaho,[70] living cultures of bac-teria isolated from Siberian permafrost that had been trapped and frozen for millions of years,[71] and DNA and isolates from amber de-posits tens of millions of years old.[72] He then followed with a sum-mary of what the SSP had uncovered at sites in South Carolina, Idaho, and Washington, using terms such as "eutrophic," "electron acceptors," "lithotrophy," "oligotrophy," "anaerobic," "phospholipids," and "adenosine triphosphate." None of these terms were in the hard-rock lexicon in my head yet were apparently common knowledge to everyone else in the room.

A couple of talks later, Sandra Nierzwicki-Bauer began speaking about 16S, 23S, 18S rRNA and molecular clocks. I had heard of RNA (ribonucleic acid) before but did not know what the little *r* meant or what a ribosome was. I was wondering whether the 16S rRNA was some nomenclature for stable isotopes of sulfur perhaps,[73] but the other numbers didn't match. She then talked about restriction frag-ment length polymorphisms, rRNA probes, and intergenic spacers.[74] Basically the only slide I understood was her title slide, after which I was totally lost. She was followed by speakers talking about restric-tion enzymes, recombination, neutral drift, O-antigens, etc., etc., and other aspects associated with molecular clocks (figure 1.10). My scribbled notes were now drowning in question marks.

The speakers did not escape being questioned from the back of the room by Frank. He would ask questions designed to keep the speaker focused on the goals of the workshop, prefacing his questions with, "I am certain that you will forget more than I will ever learn about [the topic of the talk], but could you explain how this relates to . . . ."

FIGURE 1.10. A. Universal phylogenetic tree (from Woese et al. 1990) rooted to a hypothetical last common ancestor and based on comparison of 16S (Eubacteria and Archaea) and 18S (Eucarya) rRNA gene sequences from nineteen species. Woese

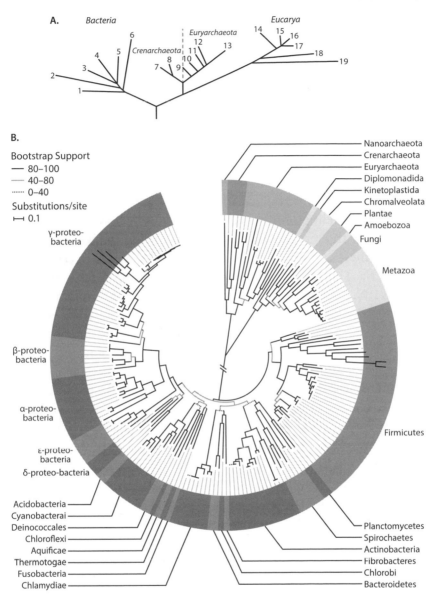

et al. (1990) used this tree to redefine life into three domains (Eucarya, Bacteria, and Archaea) and two kingdoms (Crenarchaeota, Euryarchaeota). B. Universal phylogenetic tree (from Ciccarelli et al. 2006) based on comparison of 31 homologous genes associated with the ribosomes of 191 species with complete genomes.

Finally it was noon, and I used our brief lunch break to talk to Jim Fredrickson for the first time. I explained to him how his talk started me thinking about how one might radiometrically date and microbially analyze the same sample. I showed him a sketch that I had made during the morning discussion time. After lunch, it was the geologists' turn and after a couple of talks, I stood up and proceeded to describe a project that I had recently finished on deriving the thermal history of a Mesozoic sedimentary basin using laser $^{40}Ar/^{39}Ar$ dating of potassium (K) feldspar overgrowths.[75] I went on to explain how fluid inclusions trapped in minerals were time capsules that contained ancient groundwater. I described how they could provide estimates of the temperature and salinity during the basin's history and that these conditions could be related to the growth conditions of bacteria. This was something I had learned about in my readings (figure 1.9). I ended the talk by trying to connect it to the microbial origins focus of the workshop. I postulated a narrative in which a Triassic dinosaur on a shoreline pooped its latest meal and deposited a ton of bacteria into the sediment. As this sediment was subsequently buried, the thermal history would select for thermophiles. At the end of my talk the audience made sure to correct my misunderstandings about the origins of microbial diversity in soil. I sat down afterward thinking that I had given one of the most embarrassing talks of my career, but behind me Frank looked over to Ray Wildung and nodded to him.[76] Ray knew what he had to do.

The workshop continued for another day and a half, during which I learned a lot more about osmosis, luminescent bacteria, PEX genes, transformation, transduction, conjugation, and Rec A. I also learned about the difference between a eukaryote, a eubacterium, and an archaebacterium (figure 1.10) and how to spell "archaebacteria" (not the way geologists spell "Archean"). More time was spent in discussion of the talks than in giving the talks, and we could not leave on Friday until we had drafted the conference report with the wording and punctuation approved by Frank. I drove home Friday afternoon through the various rush hours, energized. Even though I had not understood one tenth of what was presented, I was impressed with how Frank could bring together the two apparently disparate scien-

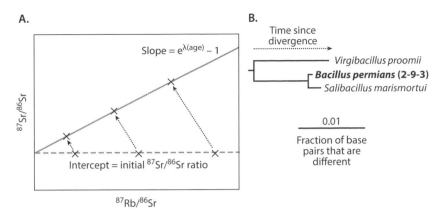

FIGURE 1.11. A. An isochron plot for $^{87}Sr/^{86}Sr$ versus $^{87}Rb/^{86}Sr$ illustrating the principles behind the radiometric clocks used by geochronologists to date rocks. The $^{87}Sr/^{86}Sr$ of each rock sample evolves from a progenitor initial ratio at different rates depending on the $^{87}Rb/^{86}Sr$ value. B. A phylogenetic tree of 16S rRNA sequences from a subsurface isolate compared with similar surface isolates illustrating the principles behind dating the time of emergence from an ancestral 16S rRNA sequence. Each species evolves from a common ancestor but at a different rate (from Vreeland et al. 2000).

tific disciplines of radiometric dating and microbial phylogeny, each speaking entirely different languages, into one room to focus on a single question as important as the evolution of life (figure 1.11).

Before Thanksgiving I received the final report on the workshop from DOE headquarters. By then I had already been invited to attend a one-day Taylorsville Basin stocktaking meeting to be held in Annapolis on December 2. After the usual three-hour commute, I stepped into the same meeting room we had used in September. The tables had already been set with copies of the transparencies to be used by each speaker. There must have been a stack of twenty talks sitting in front of me. I wasn't quite sure where I was going to put all these handouts. Tommy Phelps from ORNL, the same microbiologist I had met the previous summer in the DOE auditorium, ran the meeting. Tim Griffin began the meeting by guiding us through one transparency after another illustrating the geological history of the Taylorsville Basin. The sidewall cores had sampled muddy sedimentary rock originally deposited in a lake that looked very much like the deep, anoxic, fault-bound Lake Tanganyika that exists today along the rift zone of eastern Africa. The various speakers who followed Tim described how the cores had been acquired and processed. Tommy then gave a very long series of talks on tracers with various acronyms (PFT and IMT) and comparisons of different microbial analytical techniques with more acronyms (PLFA, CLPP, MIDI, and FAMEs). He then explained how these were applied to various types of samples, each having their own acronyms (CS, RP, PC, PP, DM, and RS). The acronyms swarmed across my notebook, refusing to be defined. All of this was presented in a tour de force of bar charts, principal component analysis charts, and various tree diagrams, all suggesting that some cores had not been contaminated with drilling mud. By the lunch break I began to realize that there really weren't thousands of feet of continuous cores of Triassic sedimentary rock available for my studies, and I began to worry. We broke for lunch, during which I met David Boone, D. C. White, Rick Colwell, and Tom Kieft, all of whom were microbiologists. After lunch David presented his results on an isolate from one of the cores, *Bacillus infernus*. With the little background reading I had done, I could at least understand David's presentation. I was wondering whether *Bacillus infernus* was as dangerous as *Bacillus anthrax*. During his talk he reiterated the estimate of 176°F for the maximum temperature (T_{max}) of the basin that had been given earlier that morning. He stated that it was consistent with the theory that the ancestor of *B. infernus* had been deposited with the lake sediments in the Triassic and had survived for 230 million years. This would be true only if the present-day environment at depth had not changed radically over the geological history of the basin. During the discussion, however, we were perplexed as to how the sediments would have entered the gas window at such a low temperature.[77] Unless, of course, the CH_4 was not thermogenic, but biogenic, and perhaps Todd's recovery of methanogens from the drilling muds was actually telling us that methanogens were also responsible for the gas detected in the Thorn Hill no. 1 borehole. Tommy was quick to rule this out because the ratio of CH_4 to the other hydrocarbons was too low. Biogenic CH_4 deposits are more than 95% CH_4. There was too much ethane and propane in the Thorn Hill no. 1 gas plays for it all to have come from methanogens.

It was left to D. C. White, however, to end this discussion with one of his colorful metaphors. He described how gram-positive bacteria, like Boone's *B. infernus*, are thick-skinned, like basketballs, and gram-negative bacteria, which are the rest of the bacterial species, are thin-skinned, like balloons. The Gram staining technique had determined their cell wall composition.[78] D. C. firmly believed that Boone's gram-positive bacteria could have survived from the Triassic even enduring the 80–120°C temperatures where oil formed from buried organic matter, the so-called oil window. "Bacteria aren't pantywaists!" he concluded the discussion emphatically. By this time it was 6 p.m. and Frank started heading back to his home in Gaithersburg. The rest of us were having beers in Tim's hotel room, where I learned from Todd that there were no samples available for my analyses. I started to get visibly and audibly annoyed. Tim reached over to calm me down and assured me that he had bags of cuttings back at their office in Richland. He could lend me all the samples I wanted.

After Christmas I flew out to Richland, Washington, where I visited Tim at his Golder Associates office and picked up about twenty bags of the coarsest-grained cuttings along with a small box of thin sections made from some of the cores. I then visited Ray Wildung, at the Environmental Science and Research Center (ESRC), and with his help I was able to submit a proposal to obtain some funding to partially support a new graduate student, Hsin-Yi Tseng. Together Hsin-Yi and I started to work on the thermal history of the Triassic samples from Taylorsville Basin. The drill cuttings that Tim Griffin had so diligently collected during those fateful days in March would prove to be the keys to unlocking the secret to the origin of Boone's *B. infernus*.

CHAPTER 2

THE TREASURE OF CERRO NEGRO

Bacteria as individuals might have infinite longevity. They might be capable of resting in suspended animation for unlimited periods of time and some might have survived in this state since the Pre-Cambrian. The evidence is highly suggestive but not conclusive.

—Claude E. ZoBell[1]

TINY POCKETS OF ANCIENT BRINE: A HOME FOR DEEP LIFE? JANUARY 1993, PRINCETON UNIVERSITY

In my first look at the thin sections of the sidewall cores from the Thorn Hill drill site, I was delighted to see late-stage framboidal pyrite (fool's gold) in the pore spaces between the detrital grains. This formation ranks as one of the sexiest mineral morphotypes you can see in a thin section—gold jewelry, if you will. As the technical name implies, it is a delicate raspberry-like cluster of goldish crystals that is spherical in shape, ranges from a few to several hundred microns in diameter, and composed of individual submicron pyrite crystals. Surely this was a smoking gun indicating that the thermophilic SRBs that David Boone had isolated had been active in these very same sediments for millions of years and could not be drilling contamination.[2] I could see that the diagenetic cements, minerals that fill the pores between the grains of detritus that settled on the ancient lake floor, contained two-phase, fluid inclusions,[3] but these could have been produced during thin-section preparation. Carefully prepared thin sections of the cuttings would be required to find fluid inclusions that could constrain the thermal history. Hsin-Yi Tseng did not need much convincing and started to work immediately.[4] I also contacted an expert on fluid inclusions, a former graduate student of our department, Bob Burruss. He was working at the U.S. Geological Survey (USGS) in Denver and was very open to the idea of Hsin-Yi visiting his lab to make the necessary measurements.

By now, I was receiving weekly faxes from DOE headquarters advising me on upcoming meetings and project reviews associated with the Deep Microbiology Subprogram. The purpose of my grant from the ESRC was to determine whether fluid inclusions could also trap and preserve ancient bacteria, much like the scenario in the Jurassic Park novel. To get started, Jim and I organized a small workshop in Salt Lake City, Utah, in March to which we invited several experts on the topic.[5] At this meeting Jim and I decided to pursue investigations of fluid inclusions in quartz crystals and salt crystals (figure 2.1). We discussed various techniques on how to extract cells and DNA from fluid inclusions. To obtain samples of the salt crystals, I contacted yet another former Princeton University graduate, Dennis Powers. Dennis was a sedimentologist working as a consultant for the Waste Isolation Pilot Plant (WIPP) site, located outside Carlsbad, New Mexico. If we could get underground access to the WIPP site, we should be able to obtain fresh salt crystals with primary fluid inclusions that were Permian in age. If living bacteria were trapped in those, they truly would represent living fossils from the late Permian Period, 250 million years ago.

Meanwhile, by carefully studying the fluid inclusions within the minerals of the Thorn Hill no. 1 sidewall cores, Hsin-Yi had discovered that at one time the maximum temperature, T_{max}, in the strata had been approximately 160–200°C, not 175°F. I later showed some of these fluid inclusions to the then director of the DOE Office of Health and Environmental Research (OHER) at a meeting that Frank organized in June.[6] This range of temperatures was consistent with a thermogenic origin of the natural gas, but it exceeded not only the maximum growth temperature of Boone's B. infernus, but also the 110°C for any known bacteria (actually Archaea) grown in a laboratory.[7] Perhaps the bacterium that David Boone had isolated was not

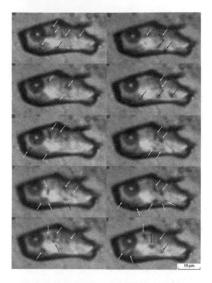

FIGURE 2.1. (Top) Time-series microscopic images of bacterial bodies moving around inside two-phase fluid inclusions in quartz (from Naumov et al. 2013). These are very much like the ones first reported by Keith Bargar (1985). (Bottom) (A) Archaeal halophiles trapped in salt crystals grown in the lab; (B) trapped in 9,600-year-old salt crystals from eight meters depth in the Badwater, Death Valley salt core; (C) trapped in 97,000-year-old crystals in the Badwater, Death Valley salt core; (D) is a fluid inclusion inside a salt crystal grown under sterile conditions (from Mormile et al. 2003).

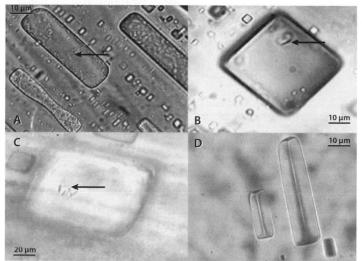

from the Triassic after all. If true, it was a shame because we had gotten quite thrilled by the possibility of a Triassic Park paper after just seeing the movie *Jurassic Park*. However, John Baross and Jody Deming, two marine microbiologists who studied deep-sea black smokers, had published a paper ten years earlier on the existence of bacteria that could grow at temperatures as high as 250°C.[8] Could *B.*

infernus have descended from such a "superthermophile"? Hsin Yi obtained her results just days before I was to present a poster at the second International Symposium on Subsurface Microbiology (ISSM) held in Bath, England, in September 1993. This meeting was largely sponsored by DOE (i.e., Frank). It was my first microbiology meeting, and I was blown away immediately by the venue and the accommodations. I wondered if all microbiology meetings were held in as extravagant settings as this one. It was a full week of complete immersion into topics I had never guessed existed and little understood. Among the twelve books that I had checked out of the library last year was the conference proceedings of the first ISSM, which had been held in Florida in 1990.[9] Many of the people who attended that meeting were also at this one, but many more scientists from the international community were participating. Frank could not attend, but he was like the Lone Ranger—everyone wanted to know, who was Frank Wobber and where was he?[10]

I sat in one session devoted to microbial transport to find out if it was feasible for bacteria to migrate from the surface 9,000 feet downward through the impermeable sediments of Taylorsville Basin.[11] But it was clear from their talks that no one had managed to demonstrate the migration of subsurface bacteria over a distance of more than about ten yards in shallow, permeable aquifers. I also attended an exciting group of talks on hyperthermophiles that occur in oil reservoirs that had been stimulated by the recent discovery by German microbiologist Karl Stetter of hyperthermophilic sulfate-reducing Archaea deep beneath the surface in an Alaskan oil field.[12] He had stated that the oil field, however, had been flooded, so they could not establish whether or not the hyperthermophile was indigenous to the reservoir. At the same session I met Steve Hinton, a microbiologist working for Exxon, who presented his work on microbial remediation of the Exxon Valdez oil spill.[13] I was surprised to learn that Steve's lab was located near my university in New Jersey. I asked Steve about Karl's discovery, what "flooding" meant, and whether anyone had tried measuring the rate of microbial transport in oil reservoirs. He did not know, but he did know that biodegradation of petroleum was restricted to temperatures of no more than 80°C. As far as he was concerned, Stetter's hyperthermophile was probably transported to the subsurface by seawater flooding.

As I sat through one talk after another, I felt as if I was an undergraduate again, completely anonymous and free to focus on learning. In my naiveté I decided to take a poll at my poster of those who stopped by, asking them two questions: Do you believe *B. infernus*'s ancestors could have survived the 160–200°C thermal maximum? Or do you believe *B. infernus* was transported 9,000 feet down from the surface following the basin's thermal maximum? The results of the poll were that 30% of the attendees believed that the bacteria could survive 160–200°C, 30% did not believe they could survive such temperatures and must have been transported, and 40% just did not know.[14]

CAN RADIATION FEED BACTERIA?
SEPTEMBER 23, 1993, DEVON COAST

The next day, the entire session was devoted to how bacteria interacted with radioactive waste. I learned that both Canada and Sweden had created underground labs to test the concepts of nuclear reactor waste storage, much like the concept of using the volcanic tuffs at Yucca Mountain in the United States. I was particularly drawn to this session because of two side-by-side papers that I had read in the proceedings of the first ISSM. The first was by Derek Lovley, a microbiologist at the USGS, showing that his ferric-iron-reducing bacterium, GS15, could reduce ferric iron using H_2. His paper was followed immediately by a paper written by Beda Hofmann, a geochemist at the University of Bern, describing reduction spheroids, which are greenish spheroids that occur in reddish, ferric-iron-rich sediments, or red beds. The classic theory on how they formed was that organic matter gets buried in red sediment and then in the subsurface at higher temperatures, the organic matter reduces the red ferric iron, producing a greenish, ferrous-iron-rich diffusion halo around it. Beda Hofmann, however, noted that the cores of the reduction spheroids were also concentrated in radioactive minerals. He showed theoretically that the radioactive minerals would produce H_2 by a process called radiolysis. The H_2 would then diffuse outward, abiologically reduce the ferric iron, and produce the spherical shape. I had already learned from David Boone and from Tommy in the previous months that both *B. infernus* and TH23, Tommy's ferric-iron-reducing isolate from Taylorsville Basin,[15] both used H_2. Was it possible that ferric-iron-reducing subsurface bacteria created these green spheroids and they did so by using the H_2 generated by radiolysis?

The following day, which was a one-day break in the middle of the meeting, Nora and I visited a beach along the Devon coast line with Tony Kemp, a geochemist from the University of Bristol, who was about to publish a paper on the Permo-Triassic reduction spheroids exposed there. Along the cliffs we could see the red sediments and mottled layers containing the reduction spheroids. We collected samples of the spheroids from rocks that had fallen on the beach. Later I explained to Tony my crazy idea, which he was willing to entertain, over a traditional Devon clotted cream. (Yum!) Upon returning to Princeton and sectioning the spheroids, I found that their centers were radioactive as well. Given my previous research on the effects of neutron irradiation on minerals, I had readily grasped the physics behind Beda's arguments, but testing this hypothesis that radiolysis could support bacteria would require subsurface microbial samples.

I had little time to pursue this thought, however, because I had just received the geophysical logs from the Wilkins borehole and Thorn Hill no. 1. Frank had requested another Taylorsville stocktaking meeting in Gaithersburg at the end of October,[16] so using these logs I quickly modeled the in situ environment of *B. infernus*.[17] It was during this meeting I had a chance to really talk with D. C. White and better understand his phospholipid fatty acid (PLFA) analyses that were so pivotal to proving that *B. infernus* was not a drilling contaminant (figure 2.2).[18]

Upon returning from the meeting, I discovered a message on my answering machine from Dennis Powers, who had come through and arranged for me to go underground at the WIPP site to collect salt samples for microbiology. Since I had never sampled anything for microbiology and had no idea what to do, I called Tom Kieft, who

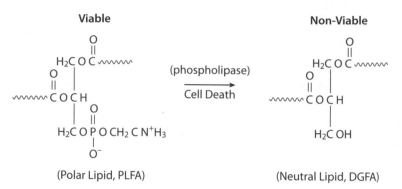

FIGURE 2.2. Phospholipid fatty acids (PLFA) and their conversion to glycolipid fatty acids upon cell death (from White and Ringelberg 1997).

was a professor of microbiology at New Mexico Tech (NMT) in Socorro. I asked him if he would be interested in helping me collect some samples from WIPP. Tom was dead keen on the idea, and we arranged for the trip over the Thanksgiving holiday.

BACK TO THE SALT MINES IN
SEARCH OF ANCIENT MICROBES:
NOVEMBER 26, 1993, CARLSBAD, NEW MEXICO

Several weeks after I talked to Tom I was visiting my mom at our home in Carlsbad. My father had recently passed away and my mother was still living there. I was going through my dad's belongings when I came across a certificate of appreciation decorated with colorful Christmas gnomes from the AEC. My dad and my uncle Bert had worked for many years in the potash mines of the Duval Sulfur and Potash Company, just south of Carlsbad. These potash mines had transformed the cattle-raising, cotton-farming community of nascent Carlsbad into the thriving county seat in the late 1950s. But Dad and Uncle Bert had quit the mining business and set up their own construction company, which was how my dad came to work as a contractor for the AEC (now the DOE) to construct the shaft and tunnel that the AEC planned to use to set off an underground atomic bomb as part of Project Gnome.[19] I was six years old at the time when Mom and I were allowed to go underground to visit the tunnel where the bomb would be placed. During the detonation, my sister Toni and I sat in the back of my dad's green '59 Chevy pickup about five miles away from ground zero, along with many other curious Carlsbadians. We felt the ground shudder, then heard a blast, and finally saw a white plume of smoke rise into the atmosphere. I was quite disappointed. It hadn't looked anything like the mushroom cloud I had seen in the science fiction movies I watched on our old black-and-white Zenith TV. Keep in mind that this was back in the days when in first grade we would have "duck-and-cover drills" in the classroom to protect ourselves from a Soviet ICBM attack.[20] So it was not surprising that we had not been overly concerned about fallout from a tiny subsurface nuclear blast like this one.

I met Tom outside the Motel Stevens in downtown Carlsbad early the next morning. I had met him only briefly at my first meeting on the Taylorsville Basin project.[21] We headed out of town on the Pecos Highway to Loving, turned onto State Highway 31, passed one of the old Potash mine headgears,[22] and turned right onto the highway to Jal. Once we left the green irrigated fields surrounding the Pecos River, we entered a land as desolate as any found in an Italian spaghetti western with nothing but caliche soil, sparse mesquite, and pump jacks stretching to the horizon.[23] It was no wonder this location was selected for dumping radioactive waste. A radioactive leak out here might harm a few cattle, but that's about it. We turned left onto the WIPP road and finally pulled into the parking lot outside of its beige-colored buildings, where we met Dennis.

After checking through security and getting our hard hats, we took the elevator down 1,800 feet to the main level. We were walking through the same formation in which the Project Gnome atomic bomb had been detonated and were just six miles away from the enormous cavity that had been excavated by the blast. The WIPP tunnels were huge, large enough for trucks to drive through. Dennis had brought along a portable, electric drill with a five-inch-diameter saw-toothed bit. He took us to one of his favorite sites, where he could show us Hopper crystals, the crystals of salt that form from evaporating seawater, in this case Permian seawater. He showed us the difference between these textures and the more massive and clear salt that was formed by the dissolution and precipitation of salt with groundwater migration that occurred long after the salt was deposited. Although this salt, which was generated where we found it, was presumably equally ancient, the salt with Hopper texture was indisputably formed as the seawater evaporated from the Permian sea.

The salt was remarkably clear. We took the lights off our hard hats, placed them against the tunnel wall, and peered inside the salt bed to see the light reflecting off the three-dimensional relict textures of a Permian seafloor. I peered into the salt beds as I scanned my headlamp around as if I were scuba diving across the Permian seabed. Everything was frozen in place, like the shoebox dinosaur dioramas that I used to build in elementary school. Could it be that within this congealed seawater, halophiles still breathed? Was ours the first light that their pigments had detected since they were buried alive 250 million years ago? It was amazing that such fine details could be preserved for so many millennia.

Dennis selected the target for drilling. With a high-pitched squeal of the drill, Dennis cored into the salt several centimeters. Tom had opened the cooler he had brought down, which carried Blue Ice, several pairs of latex gloves, and some peculiar-looking plastic bags, one of which he handed to me while he put on his gloves. I looked at it and must have had the "what do I do with this?" perplexed expression on my face. "That is a sterile Whirl-Pak," he commented as he patiently rinsed his gloves with the ethanol he had brought. He reached over, taking the bag from my hands, and deftly ripped off the top of the bag, pinched it open, and then while holding the bag with one hand, he pulled the salt disk from the corer with the other, pushed it into the bag, and with a quick flip of the wrists whirled the bag twice between his hands and twisted the wire clasp on top of it. He showed me how to do it and soon I was collecting my first microbial samples! Tom also collected some of the salty soil on the tunnel floor as a control for contamination. Samples in hand, we strode back to the cage. By now we could taste the salt on our tongues, smell the salt in our nostrils, and feel the salt collecting in the sweat of our skin, which was soaking up our moisture and forming a salty coating.

As we rode back up toward the surface, Dennis pointed out the stratigraphy of the Salado Formation and explained how the tunnels and storage rooms were excavated. Once they were filled with radioactive waste, the salt would slowly collapse around it from the weight of the overlying strata.[24] On the drive back to Carlsbad, Tom and I stopped on the side of the highway at the Laguna del Sol, the basin where the potash mines disposed of their mine water, to collect some modern-day halophiles. If the 16S rRNA genes of these modern halophiles turned out to be identical to any halophiles entombed in the salt layers of the Salado Formation beneath us, then the possibility of infiltration of the Permian salt layers with halophile-bearing brine from the surface had to be considered. After all, according to the molecular clock, 250 million years of evolution should have introduced some single point mutations into the sequences of the 16S rRNA genes of the same halophilic species. The mud on the shore of the lagoon had a salty crust, which Tom dug through with a trowel. Beneath it was a colorful mud cake of pink, purple, green, and black layers. "It's a classic Winogradsky column,"[25] he cried ecstatically as he gently lowered the salty layers into another Whirl-Pak bag. "Oh yeah," I replied, as if I knew what he was talking about, but of course, I was clueless. Satisfied with our trophies we parted ways in Carlsbad, Tom driving back to Socorro, where he would ship some samples to Jim Fredrickson. I flew back to Princeton with some of the samples to perform electron microprobe analyses and attempt to date them using my laser $^{40}Ar/^{39}Ar$ technique.

The visit to WIPP was impressive in two respects. Dennis Powers and my uncle Bert had both mentioned that occasionally the miners would tunnel or drill into huge pockets of saltwater that they believed were trapped Permian ocean water. Could these larger-scale equivalents of the brine fluid inclusions, super-sized fluid inclusions, actually host living microbial ecosystems? The other aspect of WIPP that so impressed me was that I could actually see the geology and pick my samples. For years I had been doing geology on the surface, in roadcuts, or in open pits in the tropical jungles of South America and Africa. The only time I had ever gone underground was in a leech-infested adit, or mine tunnel, in the Amazonian jungle, and then only out of sheer desperation, to find unweathered rock. Underground mines had no appeal to me, but they definitely had an advantage for microbial sampling because one wasn't completely blind. Instead of spending a million dollars drilling Texaco's dry hole, a researcher could piggyback on the mine infrastructure to choose interesting sites. Couldn't one then, for a fraction of the cost or perhaps no cost whatsoever, core a hole from the tunnel into the target, such as a super fluid inclusion? I knew little about the details of drilling, but that would change sooner than I expected.

BACTERIA AS OLD AS METHUSELAH, REALLY? DECEMBER 4, 1993, OUTSIDE OF KNOXVILLE, TENNESSEE

The week following my return from New Mexico I met with Tommy at his new home outside of Knoxville, Tennessee. Susan and he had just finished building it near a beautiful lake surrounded by woods. Susan was out of town and, thanks to a freak snowstorm, I was trapped in the house with Tommy for two days. The only food we had was celery, peanut butter, and wine. Frank had arranged for another all-hands stocktaking meeting, this one to review the Deep Microbiology Subprogram to be held two weeks from now, and Tommy was helping me prepare my ten slides on the Taylorsville Basin. Like many newcomers to the SSP, Tommy was looking out for me by making sure that my presentation would satisfy Frank's expectations. As one of Frank's PIs had once commented, "Working for Frank Wobber was like raking leaves in a mine field!" Tommy was making sure I didn't step on any land mines.

Tommy also gave me an intense two-day tutorial on microbial physiology, survival, and calculating in situ rates of metabolisms from geochemical constraints and, thereby, inferring average cell turnover times. In one of his recently submitted papers, Tommy had estimated the cell turnover time for the living bacteria they had found 600 to 1,200 feet beneath the surface in South Carolina.[26] His estimates ranged from hundreds to hundreds of thousands of years! He explained to me how prokaryotes have a circular chromosome and no telomeres but had DNA repair systems.[27] This essentially means that prokaryotes do not age, there were no limits as to the number of times they can replicate their genome, and a single cell could, in principle, live forever. We estimated the in situ rate of microbial respiration in the Thorn Hill no. 1 samples based on my analyses, and we came up with a million years for the average cell turnover time. I was blown away. Could the original *B. infernus* have been that old before it reproduced in Boone's lab?

But the most intriguing aspect of my two-day tutorial was the discussion of Tommy's recently published discovery of the high radiation resistance of subsurface isolates.[28] Tommy had expected that because the subsurface isolates had been buried beneath the surface of our planet for up to 100 million years, they would be more sensitive to UV. But he discovered that the subsurface isolates were more resistant to UV than the typical *E. coli*. This was great news for the SSP, which could count on subsurface bacteria being active in subsurface sites contaminated by radioactive waste. These bacteria had not been exposed to UV in the subsurface, but they would have been exposed to gamma rays from the decay of natural uranium, thorium, and ^{40}K. This intrigued us because at Thorn Hill no. 1 most of the isolates came from the sidewall cores into a shale zone in which the spontaneous gamma-ray counts were quite high. Using the gamma-ray logs, we could estimate the annual dosage of bacteria living within the shale, and it was clear that the dosage was high enough that being radiation resistant was an advantage. If the kill times from radiation were shorter than the cell turnover times calculated from the geochemical constraints, then fairly quickly the indigenous microbial community would decline in numbers, perhaps even becoming extinct. Now I was wrestling with two opposing hypotheses, one in which radiation supported subsurface life and the other, in which radiation extinguished it. We could make some simple model predictions, but in the end how could I test whether these processes were actually occurring in the subsurface?

HOW DO YOU DETERMINE THE ORIGIN
OF SUBSURFACE BACTERIA?
DECEMBER 16–17, 1993, ANNAPOLIS, MARYLAND

The Deep Microbiology Subprogram meeting was held in Annapolis, Maryland, in December over a three-day period and had more than fifty attendees. The request had gone out that we arrive at the meeting with fifty paper copies of our transparencies. A new crop of investigators, which included me, was joining the SSP veterans with their primary focus to be directed toward what was being called the Origins field site, and this meeting was their orientation. Early in the morning of the first day, as I was hurriedly photocopying handouts for my presentation, I hacked off the tip of my thumb with a paper cutter. Using duct tape and paper towel, I bandaged my thumb and started my regular commute down to Annapolis from Princeton at 4 a.m. Next to me on the seat were several portable plastic file boxes filled with previous workshop presentations that I had brought just in case I needed them. On the first day, SSP veterans summarized the results from the older field sites at PNL, INEL, SRP, and the NTS. On the second day the focus was on the new Origins field efforts, including our presentation on Taylorsville Basin, followed by three hours during which the new investigators, who had just received their funding, presented a summary of their proposal. The final day was focused on literally drafting the research program and making sure that the new investigators were plugged into the planned field campaign. It was at the end of this long day that Frank announced to the assembled that I would become one of the coleaders of the Origins field campaign, along with Jim Fredrickson. I was stunned, but over a quick dinner, Jim reassured me everything was going to be fine.

After dinner, we returned to the conference room, where Frank and the investigators in charge of the field implementation for the Origins site were all seated at a big round table. For the next two hours, Frank went though the time line for the field campaign and the long list of items to be done and who was going to do them. I scribbled notes furiously to keep up. The primary site was one Frank had picked out in northern New Mexico during one of his road trips. It was a 3.3-million-year-old volcanic plug named Cerro Negro, which intrudes the Cretaceous Mancos Shale Formation. The goal was to determine whether the microorganisms found in this 90- to 97-million-year-old shale formation were descendants of microbes originally deposited with these marine sediments or had been transported from the surface more recently by groundwater, specifically since the time the subvolcanic intrusion had occurred. If for some reason the logistics and access fell through for this primary site, a secondary site had been selected in the Naval Oil Shale Reserves (NOSR) near Parachute, Colorado, in one of the hydraulically tight gas reservoirs that the DOE was involved in drilling.

Frank insisted that drilling at Cerro Negro begin that June, which meant that we had five months to do the following. We had to collect and analyze rock samples from outcrops exposed at the nearby Jackpile uranium mine and groundwater samples from the monitoring wells. Gravity, magnetic, and seismic surveys had to be performed across the volcanic feature in order to produce the best possible three-dimensional structure for identifying the best locations for the first vertical well and the second angled borehole. The goal was to sample the same sedimentary stratigraphy within the thermal aureole of the volcanic intrusion as sampled outside of the thermal aureole,[29] but the size of that aureole still had not been determined. The final access rights also had to be negotiated. The details concerning the coring and the type of drilling fluid that would be used still had to be finalized, and the drilling contracts submitted, evalutated, selected, and completed.

THE DOE DESCENDS ON TINY SEBOYETA: JANUARY 21, 1994, FORTY-FIVE MILES WEST OF ALBUQUERQUE, NEW MEXICO

The SSP went into hyperdrive. As Frank said, "This is the time where the rubber meets the road."[30] We all exchanged our new emails because Frank was now becoming adept at the new technology and the reams of fax paper were no more. Any hesitancy I may have had about committing myself 100% to Frank's program went out the window because it was either all hands on deck or you were off the ship. On Friday, January 21, I flew out to Albuquerque, where I met up with Brent, Tim, Tommy, Tom, and Jim, and we drove about forty-five miles west of Albuquerque to the Cerro Negro site for a tour and sample collection. It was bitterly cold that morning when we visited the Jackpile uranium mine on the Laguna reservation. The mine had ceased production in the early 1980s, but at that time had been the largest surface uranium mine in the world. Since January 1990, the members of the Kawaik tribe had been working on environmental restoration of the 2,700 acres damaged by the mining activity. Escorted by one of the supervisors, we drove into the mine. On the walls of the mine's 300-foot-deep open pit was exposed the entire stratigraphic section that we would be drilling further north, right down to the whitish Jackpile Sandstone of the Jurassic Morrison Formation, which contained all the uranium. We pulled up to one of the mined faces and with sledgehammers, chisels, and Whirl-Pak bags and collected several kilograms of Mancos shale for distribution to the various PIs, both new and old. Then we came across a layer of beautiful gray-colored, somewhat flattened rock spheres that were up to a foot in diameter. These were carbonate concretions. I grew quite excited, because D. C. White had just handed me a manuscript he'd published with Max Coleman of the University of Reading on how Fe(III)- and sulfate-reducing subsurface bacteria produce these marvelous structures.[31] They were the perfect example of how subsurface microbial activity permanently impacts the structure and texture of rock. They were proof that anaerobic subsurface bacteria had existed in the Mancos Shale soon after deposition. I took one of the smaller concretions for fluid-inclusion analyses back to Princeton. Perhaps Cretaceous Period bacteria were still trapped inside waiting to get out.

Just four miles further up the road, within the Spanish land-grant community of Seboyeta, was Cerro Negro. We turned right onto New Mexico 334, and within a mile the twin basaltic plugs poked up 900 feet to our left. We got out of the car and proceeded to hike the

half mile up to the top of Cerro Negro to examine the contact between the basaltic intrusion and the surrounding cliffs of Gallup
Sandstone for signs of thermal overprinting.[32] The sky was hazy from
the dust, but we could easily make out the Jackpile uranium mine
with its tailings piles and the L-Bar Mine tailings pond a couple of
miles to our southeast. Stretching to the east and northeast was a
parched expanse of yellowish-gray Mancos shale partially covered
with sage grass. Further east the soil turned more reddish, and we
could just make out the western rim of the great Rio Grande rift valley. But from our northeast to our south, the plains ascended along
juniper-sprinkled slopes to the white cliffs of the Gallup Sandstone,
where they paused and then ascended further to the black lava-
covered mesas above Cerro Negro. The lavas all came from the extinct stratovolcano of Mt. Taylor ten miles to our west.[33] Its snow-
covered, 10,000-foot-high peaks poking above the mesas were clearly
visible to us. During the springtime, that snow would melt and feed
the streams running down the mountain, but some of that meltwater
would penetrate the fractures of the volcanic edifice and slowly sink
deep beneath the surface. The water would eventually follow the
dike contacts down until it intersected the permeable sandstone
aquifers and then at a snail's pace follow the aquifers down the gradient and ultimately reach Cerro Negro, fifteen hundred feet beneath
us. That flowing groundwater would herd the bacteria along their
subsurface journey toward the Cerro Negro intrusion, where they
would have flourished after it had cooled down enough for recolonization, 3.3 million years ago. In the less permeable Mancos shale,
however, the heat-sterilized zone might likely have remained sterile
to this day. If that was true, then we could be confident that the bacteria found in the Mancos shale away from the intrusion were there
before 3.3 million years ago.[34]

Below us, Brent Russell was already deploying a team of geophysicists to carry out magnetometer and low-frequency radar surveys to
get a handle on the shape of the volcanic plug and the three-
dimensional surface separating the Mancos Shale and underlying
Dakota Sandstone. In all of our minds we hoped that Cerro Negro
was a simple textbook cylindrical plug. That would greatly simplify

the thermal model, establish the distance to the target zone, and determine the angle for drilling. After sample collection, we met for
another day at the Best Western in Albuquerque to go over the steps
needed if we were to be coring by June. The go-no-go decision would
have to happen very soon. We needed geophysical certainty on the
shape of the intrusion in order to write the specifications for the
drilling contracts. In particular, we needed to angle the second borehole so that it sampled the Mancos Shale below the water table but
before it intersected the thermal aureole around Cerro Negro. From
this initial survey it looked like it would require an angle of about
30° from the horizontal. From what was known of the stratigraphy, it
looked as if just the coring alone would require about $300,000, but
we would see after the bids arrived. We would require an inventory
of supplies and equipment that would include what people could
bring to the site at no charge. Phil Long, a geologist from PNL, was
in charge of drafting the science implementation plan (SIP), with
assistance from Tim and Brent.

The sandstone and shale units were all fractured along six different
planes, so contamination from drilling mud was going to be a big
problem. To minimize the contamination we planned to use compressed air for drilling down to the water table before switching to
either water with no recirculation or mud with circulation below the
water table. The important thing about drilling the vertical control
hole first was that it would not only preview the drilling targets for
the angled borehole, but also supply the water for the angled coring.
For tracers we would use PFTs, fluorescent microspheres, and perhaps lithium bromide (LiBr). Brent, Tommy, and Jim were in charge
of the tracer QA/QC and safety plan. Brent and Tim would have to
file the document required by the National Environmental Policy
Act for the drilling to make sure there was no long-term environmental disturbance from the project.[35] They also needed to negotiate an
access agreement with the land grant governor and perform an ar-
cheological survey to make sure the drilling activities would not have
an impact on any Native American ruins. We estimated that the
number of investigators on site at any time would be eight. This in-

cluded a local from Seboyeta—the land grant governor's teenage son—whom we hired as our much-needed tech gofer. There would be an additional six drillers. Due to the relative isolation of the site, several RVs would need to be rented for accommodations.

On the way back from the field meeting, I stopped off in Denver to see Bob Burruss, to show him some of the fluid inclusion sections from Thorn Hill no. 1 cuttings. He was quickly able to find blue-fluorescing, petroleum-bearing fluid inclusions decorating fractures in clear detrital quartz, an indication that oil had migrated after T_{max}, the peak temperature in the rock. He proposed that by analyzing the composition of that oil we could determine whether it was biodegraded or water-washed.[36] Either way it was a big step toward explaining how and when Boone's *B. infernus* arrived in the Thorn Hill shale formation upon cooling from T_{max}. Now we only had to answer the question of when this groundwater migration had occurred. Bob also spotted CH_4 inclusions. This excited him. He told me we could use these to estimate the pressure and thus the depth of burial at the time of T_{max}.

By March 3, 1994, it was clear that the January survey had indicated that the location selected for the "control" vertical core of the Mancos Shale/Dakota Sandstone section was free of any subsurface basaltic intrusive. This meant we were beyond the thermal aureole of any unseen satellite dike emanating from Cerro Negro. Deployment to that location could now proceed as scheduled. But the survey was inconclusive concerning the shape of the volcanic plug itself. Golder, therefore, performed a time domain electromagnetic field (TDEF) survey in mid-March, which indicated that the plug did have a vertical contact at depth. Knowing the shape of the intrusion, we could then estimate the distance from the contact to the 80–120°C paleo-isotherm in the Mancos Shale and fix the second drill site and angle. By the end of April, Tim and Brent had completed a survey of the groundwater wells surrounding Cerro Negro that had been used to monitor groundwater pollution from the nearby L-Bar tailings pond containing the uranium mine wastes. They had also just received five bids from drilling companies that they would evaluate with Phil

Long.[37] We now had less than two months to move the drill rig and microbial processing labs to Cerro Negro, and investigators were still analyzing the outcrop samples collected at the end of January.[38]

NOT ONE, BUT TWO, DRILL SITES: APRIL 7–8, 1994, VIRDEN CENTER, LEWES, DELAWARE

As we were reaching the final stages of preparation for the Cerro Negro site, Rick and I turned our attention to the potential drill site in Piceance Basin, in the northwestern corner of Colorado. We were working with Tim Meyers, a geologist from Argonne National Laboratory, who had spent time at DOE with Frank. DOE had an ongoing campaign of "defensive drilling" to a depth of three kilometers around the perimeter of the NOSR. This drilling and well development was a way of keeping private companies, whose drill sites covered the private lands around the NOSR, from tapping into the government's gas reservoirs underlying the NOSR. This could provide an inexpensive means of obtaining deep subsurface microbiology samples. In September, DOE was going to start drilling just 8.5 miles west of Rifle, Colorado (drill hole 1-M-8 on their map). In October they would drill from the same site at an angle to the southwest into section 18 (drill hole 1-M-18 on their map).

Rick, Tim Meyers, and I had found this drilling opportunity more promising than Cerro Negro for several reasons. First of all, the physical properties and thermal history of the sedimentary formations at this location were better defined than those at Cerro Negro. More than a hundred papers had been published on them as a result of the DOE Multi-Well Experiment (MWX) site. The MWX ran from 1979 to 1988 and had arisen from the 1973 Arab oil embargo, which motivated a shift to gas production by the petroleum industry. It was also motivated by the failure of hydrofracturing to significantly improve gas flow from these very tight formations. The Department of Defense and DOE partnered and then set off a 40-kiloton atomic device in September 1969 at a depth of 8,400 feet, just ten miles southwest of the 1-M-18 site, in the same formation we were proposing to sample. This blast tripled the rate of gas flow, but only temporarily. It also resulted in a lot of tritium, or ^3H, in the CH_4, and 20% of the gas was H_2 produced by the detonation![39] Given that nuking the gas out of the sandstone was not a viable option, the MWX was designed to determine the hydraulic and gas conductivities on a field scale and which geological factors controlled them in order to better understand how to commercially produce the gas.[40]

Second, the 1-M-8/1-M-18 drill site was located hydrologically and topographically up-gradient from all the other private gas wells that had been drilled in this area and so was less likely to have been impacted or "contaminated" by other drilling activities.[41] Third, it provided an opportunity to obtain samples from the same depths as those acquired at the Taylorsville Basin but in a different geological setting and with the opportunity to obtain large-diameter, thirty-foot-long cores instead of the tiny sidewall cores. But the most appealing aspect of this drill site was the fact that the predicted bottom hole temperatures were going to be about 115°C, which was slightly greater than the known upper temperature limit of life (at that time), and the hole would bottom in an organic-rich coal seam that could provide organic substrates to any hyperthermophiles. Any recovery of hyperthermophiles from this borehole would test the limit of the subsurface biosphere. But like the Triassic sediments of Taylorsville Basin, the Tertiary-Cretaceous sediments of Piceance Basin had been buried deeply and heated to temperatures that 10 million years ago were 40°C higher. We could readily identify three target zones: (1) a shallow depth, at which the sediment had been continuously habitable with temperatures never exceeding 110°C; (2) an intermediate depth, at which the temperatures had exceeded 110°C but were now less than that and presumably habitable; and (3) a zone of existing sterilization, in which the current temperatures were greater than 110°C. The entire section had cooled from its maximum temperatures within the last 10 million years with the uplift of the Colorado Plateau. The gas-rich formations at depth were all overpressurized, which meant that we could expect pressures as high as 250 atmospheres at 6,500 feet below the surface. This would be an ideal hunting ground for barophiles.[42]

Tim Meyers, Rick, and I began assembling a draft proposal to Frank on pursuing this area as a second Origins site after we had received faxes from Ray Wildung at PNL strongly encouraging us to attend an ESRC review meeting set for April 7–8 at the Virden Center in Lewes, Delaware. We decided to make our pitch to Frank at this meeting.[43] Lewes was one of these sleepy seaside towns typical of the DelMarVa region and a frequent site for Frank's review meetings.

Phil Long spoke first in the afternoon and presented the plans for the Cerro Negro project along with some of the initial results on the surface samples and geophysical surveys. Phil had a great deal of experience running large, field-based DOE programs and had led the Geological Microbial Hydrological Experiment (GeMHEx) microbial drilling project at PNL three years earlier. He was a master of these presentations at the stocktaking meetings and was backed by the formidable resources of PNL.[44]

Following Phil, Rick stood up at the front of the room and made the case for the Parachute site, while Tim Meyers and I sat in the audience with crossed fingers. For Rick it was déjà vu. Rick and Phil had presented two competing drilling proposals before, at one of Frank's meetings in September 1989 in Idaho Falls.[45] Phil's site was the GeMHEx and had targeted 6- to 8-million-year-old paleosols and lake sediments,[46] and Rick's drill site had aimed at characterizing the Quaternary Snake River Plains basaltic aquifer beneath INEL. Geologically they were on opposite ends of the spectrum, but in terms of their merits, they ranked exactly the same. Frank had funded both! They cored Rick's site during the summer of 1991 (figure 2.3) and then cored the GeMHEx site in the fall of 1991 (figure 2.4). They were the first microbial drill sites after the C10 in South Carolina.

Four years later Rick was now presenting the case for the underdog site, again! He explained how we would use the cuttings and mud logs from the 1-M-8 drill hole to select the core points for the 1-M-18 drill hole. Fluor Daniel, Inc. was DOE's contractor for drilling the NOSR, and they would be starting the 1-M-18 in October. The schedule would be very tight. We would have to move the sampling trailer from Cerro Negro up to Parachute, which meant drilling would have to be finished at Cerro Negro in four months, and the coring at Cerro Negro had not even begun.[47] Our proposal had reminded Frank of

FIGURE 2.3. Typical gusty summer day in 1991 at the WO1 drill site located on the INEL site northwest of Idaho Falls. The cryogenic argon tank stands next to the INEL drill rig. Argon gas was used as the drilling fluid (courtesy of F. S. Colwell).

Project Gasbuggy, an earlier atomic bomb detonation in a New Mexico gas well, which had sparked his own interest in subsurface bacteria.[48] He decided to approve preparation for both Cerro Negro and what we began to call the Parachute Origins Site. Rick would take the lead in writing the SIP, and Tim Griffin would start deciding what type of coring technology we would use. Rick decided that our nom de guerre in our emails should be Les Chutres, a somewhat Francophied abbreviation for "parachuters."

HOW DO YOU DRILL FOR DEEP-DWELLING MICROBES? MAY 21, 1994, SEBOYETA, NEW MEXICO

Six weeks later we were shacked up in camper trailers parked a little way outside of Seboyeta on the gravel road to Cerro Negro. Jim and I had just flown in from a JASON meeting with Frank in San Diego.[49]

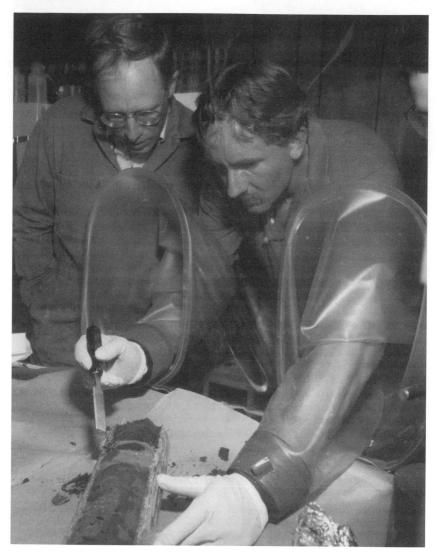

FIGURE 2.4. Jim Fredrickson paring a core in an anaerobic glove bag while Phil Long observes at GeMHEx field laboratory, fall 1991 (courtesy of PNNL).

We had all selected our bunks and stowed our personal gear when all of a sudden a rental car came roaring down the dirt road and skidded into the trailer site pulling a cloud of dust behind it. To our amazement who should step out of the car but Tommy, wearing a neck brace beneath his red beard. He had decided at the last minute to join us despite the problems he was having from his plane accident. We grabbed him and headed off to the local cantina to celebrate his arrival. The primary source of all our sustenance was Don Jose's Cocina, in Bibo, a cafe, store, and post office where we could count on high-quality beans, rice, chili, and freshly deep-fried sopapillas. The local bar was in a brown adobe building with a tin roof above the front porch. It had begun life as a Lebanese-operated trading post serving the local Kaiwaks. On the inside it looked like any mining-town tavern with the bar running the length of the main room, a footrest made of drill pipe, and drill bits standing on both ends of the bar. The pool tables were lined up along the back wall. Stuffed elk heads looked down at us from above, and, of course, the ceiling had been decorated with bullet holes, some reputedly put there by the land-grant governor himself.

The next morning we traveled about a mile further east and turned onto the gravel road that led to a thirty-foot-tall reddish-orange drill rig. It and a semitrailer full of drilling gear marked the drill site for the first vertical borehole. It was only about a hundred yards north of the gravel road. Two fifty-foot-long white trailers sat upwind of the drill rig. We met Tim Griffin in one of the trailers, which served as an office and the sample-processing lab. Tim was like the sheriff of Seboyeta, and Bruce Hallet his deputy, but without any badges. We stood there with a half dozen other investigators as he went through the morning safety discussion and the goals of the day. Each of us had a designated job. Jim McKinley and Shirley Rawson, two scientists from PNL, were in charge of the tracers, another couple of people were in charge of prepping the core barrel for microbial coring, another was in charge of geological logging of the cores, and another group was in charge of processing the cores.

We were going to be coring down 600 feet and I had assumed that all the cores were collected for microbiology. That was the first surprise for me. I began to learn that it took a good long day to process a five-foot-long "microbial" core in the glove bag, and the drillers could produce fifteen to twenty such cores in twenty-four hours. The drillers could have cored at a much faster rate, but Tommy relentlessly badgered them to slow down in order to improve the core recovery. At one point Tommy almost came to blows before the chief

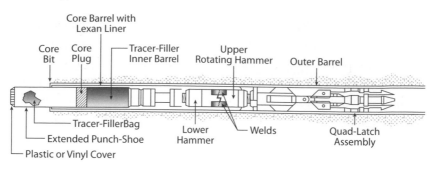

FIGURE 2.5. Triple-barrel wireline coring technique used at C10 in 1988 and variations thereof at WO1 in 1991 and Cerro Negro in 1994 (from Fredrickson and Phelps 1997).

driller finally relented. Usually the cores came out of the wireline barrel in its shiny aluminum sampler. We could mark it along its axis with a red and green permanent marker to indicate which direction was top and which was bottom, photograph the core, write a brief description, and put it in the box. But the microbial core got special treatment. First, the core barrel itself was steam cleaned. Then a clear polycarbonate tube that had been bleached was placed inside the core barrel. With his ethanol-sterilized, latex-gloved hands, Jim Fredrickson showed me how to gently push a tiny Whirl-Pak bag containing a greenish solution of fluorescent microbeads into the shoe of the core barrel (figure 2.5). The entire assembly was carried over to the drillers, who slid it down the inside of their drill stem using the wireline.

Jim McKinley showed Shirley Rawson and me how the PFT was injected into the high-pressure air hose. Compressed air with its PFT was blown down the drill rods as the drillers cranked up the diesel engine to rotate the drill rods. He also showed me how he added the LiBr tracer using a bucket to lower the concentrated solution and dump it into the bottom of the hole just before the coring barrel was slid down to begin the coring. When the compressed air was switched on, it pushed the LiBr solution through the formation as it was being cored. A high-pressure hose carried the fine dust and cuttings from the top of the drill hole to a cyclone separator, which dropped the cuttings on the ground while sending a plume of fine dust into the air to be blown downwind from the lab trailers (figure 2.6).

To do the groundwater sampling from the more permeable sandstone layers after the coring was complete, Jim McKinley had a stainless steel bladder pump they were going to put down the borehole. That was just for starters. He also mentioned how they were going to put in a system called the MLS (multilevel samplers), which they had tested at GeMHEx to measure the groundwater compositional gradients between the sandstones layers and the shale layers.[50] With the MLS they could measure the dissolved H_2 gas concentrations between the shale layer where H_2 was produced by microorganisms and the sandstone layer where the microorganisms were consuming H_2. I was thoroughly impressed—but wait, there was even more. They were then going to take some of the sandstone cores, autoclave them, and dip them into various nutrient solutions and then insert them back into the borehole at their original stratigraphic position to incubate them. After a few weeks they would retrieve the samples and determine which types of bacteria had colonized and grown within the sandstone cores. After taking notes on all of this, I was convinced that if you gave Jim McKinley a cased well, he could come up with about fifty different ways to measure microbial activity in it.

Our first microbial core had just arrived at the surface, and we quickly moved inside the trailer to see how it was cut down to size and its polycarbonate liner removed. I had never seen a glove bag before. This one must have been fifteen feet long and made of vinyl plastic. It had two pairs of arms sticking out from both sides, literally, because of the positive pressure of argon gas on the inside. At one end was a long, aluminum, cylindrical air lock to which was attached a vacuum pump and strict instructions on how to use it. We would place the drill core with its polycarbonate liner inside the air lock and evacuate it three times, refilling with argon gas before opening the inner air lock door and sliding the tube inside, along with all the flame sterilized processing tools. We would remove the liner, and push it back out of the air lock, and then photograph the core. At the other end of the glove bag sat a big red hydraulic rock splitter that we would use to slice the core into "cookies" once we had decided how

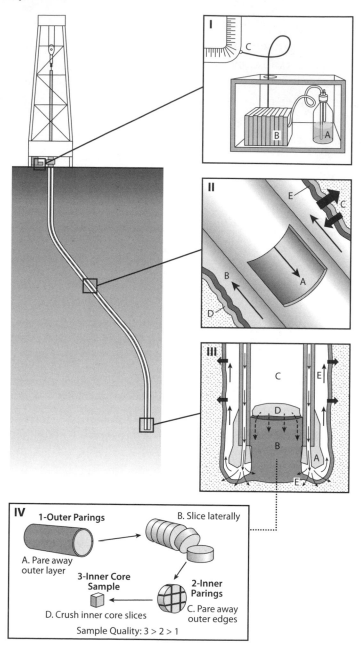

IV

1-Outer Parings

A. Pare away outer layer

B. Slice laterally

3-Inner Core Sample

D. Crush inner core slices

2-Inner Parings

C. Pare away outer edges

Sample Quality: 3 > 2 > 1

to dissect the core (figure 2.6). The outer surfaces of the cookies were "pared" with the splitter by hand, and the pared fragments were collected for tracer analysis. This process left behind an internal nugget, which we then bashed into pea-sized chips for general distribution to the microbiologists. We then scooped the pea-sized fragments into double Whirl-Pak bags and passed them through another aluminum cylinder into a second smaller glove bag. There we packed the bags into mason jars containing a pellet of palladium catalyst, and we then filled the jars with Ar and H_2 gases (the palladium would catalyze a reaction between H_2 and O_2 to make H_2O, thereby removing any O_2 in the jars). We packed the jars into Styrofoam coolers containing frozen Blue Ice and shipped them to microbiologists working on the isolation of anaerobes. Core plugs were taken from the core for permeability measurements and mercury porosimetry.[51] All the samples were then passed out through a second, much shorter aluminum cylinder air lock with its own pump.

FIGURE 2.6. Application of tracers to microbial rotary coring. I. Tracer injection into drill mud. A is the bottle of PFC tracer and B is the HPLC pump. II. Positive circulation of drilling mud flow as it is pumped down the drill string and through the core barrel and up between the core barrel and the rock carrying the cuttings to the surface where the cuttings settle in a drill mud tank and the drill mud is returned back down the drill string. A. The drilling mud is pumped down the drill string. B. The drilling mud with tracer returns to the surface outside the drill string, carrying rock cuttings up with it. C. The drilling mud, tracer, and drilling-mud bacteria penetrate the formation. Formation fluids and indigenous microorganisms diffuse into the drilling mud of the borehole. D and E are the mud-cake layers. III. The coring bit cutting into the rock formation. A. The drilling mud leaving the bit faces, where it cools the bit and removes the rock cuttings. The drilling fluid passes between the outer coring barrel and the inner polycarbonate liner, thereby not contaminating the core. B is the rock core penetrating the polycarbonate liner, C. D is the bag of fluorescent microspheres that have burst and is releasing microspheres around the outer edge of the core. E. The drilling fluid penetrates the outer rim of the core as it is being cut. IV. How the core A is pared to remove the outer layer, B and C, to reveal an inner nugget, D (courtesy of Tim Griffin).

I was pretty useless in the field at Cerro Negro, having none of the required skills in microbiology or geochemistry. All I could really do was help process the cores, take lots of notes, and learn. Every investigator who arrived on site, however, brought some equipment to share. Tom had brought his fluorescence, phase-contrast micro-scope from his lab in Socorro. I brought along a magnetic suscepti-bility bridge that could measure the paramagnetic Fe content of the cores, which might be related to the bioavailable Fe. I had also brought Vee-Tech. His real name was Witold J. Grzymala-Busse, and he was an undergraduate summer student, but he simply went by Vee-Tech. He stood six feet five, weighed 250 pounds of solid muscle, and could carry two high-pressure gas cylinders from the truck to the trailer, each one tucked under an arm that was bigger around than my thigh.

For recreation we played beach volleyball, actually more like gravel volleyball, at the deli on the road to Seboyeta. It was a good place to resolve tensions between the various "collaborating" research groups. But the levels of play and dedication to inflict dermal abrasion on oneself were highly variable. So we mixed and matched the teams to level the field. It soon became clear, however, that Vee-Tech could block any spike or serve coming across the net. One by one his team members would desert him for the opposing team, leaving him alone on the opposite side of the net. Only then did the rest of us have a chance of a fair game.

Bob Griffiths, a microbiologist from Oregon State University, had a graduate student living at the site all summer as well. He was taking crushed samples from the microbial cores and measuring the enzy-matic activities for phosphate and glucose uptake. In our lab trailer he had set up a fluorometric device. He tried to explain to me how this worked in conjunction with a fluorescent dye. With too little chemistry and biology background, I pretty much did not follow him. Then Tom came to my rescue and loaned me his copy of the sixth edition of Thomas Brock's microbiology text,[52] and I found a summary of enzyme assays in one chapter. Tom had done me a big favor in finding me an excellent primer for geomicrobiology. As I flipped through the pages of this book, I came across wonderful color diagrams of the inner workings of bacterial cells and then upon a color photograph of a Winogradsky column.[53] Tom had been right about the sediments of Laguna del Sol. "Oh yeah," I repeated to myself as I read about the late, great Winogradsky.

Thanks to Tom's loan, I finally began to understand how the DNA-to-RNA-to-protein machinery of a cell worked, and with further study, much to my delight, I could now understand whole sections of *Scientific American* magazine that previously had been completely unintelligible to me. In spare moments we spent time discussing the future of the Origins program and SSP. Frank had informed all of us that the SSP would be terminated after FY95–96 and that we all should prepare for it. We bemoaned the irony of this situation, in which SSP was finally getting to the point of answering the important questions about subsurface life only to have its plug pulled before completing the analyses of the samples we were collecting that summer. We decided that perhaps some favorable popular press was called for, and Tom decided to see if he could get National Public Radio interested, and I decided to contact Walter Sullivan of the *New York Times*.[54]

Tom had also brought an electric field detector that his wife, Sandy, had borrowed from the NMT atmospheric physicists. It provided a warning that a lightning strike was imminent. The entire area around Mt. Taylor was a focal point for the summertime, monsoon-season, hair-raising (literally) storms coming up from the Pacific, and the thirty-foot-high drill tower was a lightning rod for these storms (figure 2.7A). The electric field detector provided the drillers with the advance warning they needed to clear the rig area. But coring progressed smoothly through the summer without incident or accident. We penetrated the three top sandstone formations—the Two Wells Sandstone, the Paguate Sandstone, and the Cubero Sandstone—and stopped when the driller penetrated the Jackpile Sandstone at 760 feet depth. The well was then cased, with intervals perforated at each sandstone aquifer. Various research groups rotated on and off the site during the summer, with the exception of Phil Long and Tim Griffin, who pretty much remained encamped at Cerro Negro for the duration.

A.

B.

FIGURE 2.7. A. The CNV (vertical borehole) drill site at twilight, with the sun capturing the Cerro Negro peaks and the white cliffs of the Gallup Sandstone, during the summer of 1994 (courtesy of F. J. Wobber). B. The CNAR (angled borehole) drill site in fall of 1994 taken from the top of the Gallup Sandstone cliff looking south toward the Jackpile uranium mine. A spherical cryogenic N_2 tank sits between supply trailer and the field lab trailers (courtesy of T. Griffin).

By the end of August, the rig had completed coring of a second vertical borehole about a hundred meters from the first in order to satisfy the Heterogeneity Subprogram of the Origins project. The rig was moved from its location a mile away from Cerro Negro, up a dirt road to within 450 yards of it, and nestled between the white talus blocks of the Gallup Sandstone. The reddish-orange rig was tilted about 34° from the horizontal and pointed directly at Cerro Negro with the goal of intersecting the intrusion at depth (figure 2.7B). We began by using N_2 gas to minimize potentially deleterious impacts of O_2 on anaerobic microorganisms residing in the shale and sandstone.[55] We switched to a drilling mud system as we penetrated the water table and the borehole angled downward. Inclination data clearly indicated that the drill bit was directly underneath the southern plug of Cerro Negro, but so far we had intersected only a few thin dikes. Back at the vertical hole, Jim McKinley's group was lowering and raising various MLSs. By the end of September, we had penetrated the entire Cretaceous section and had bottomed out in the Jurassic Jackpile sandstone after having drilled and cored more than five hundred yards. Still we had not penetrated the Cerro Negro intrusion itself. The vertical contacts inferred from the TDEF survey appeared not to be correct, and we might have found a more carrot-shaped intrusion. It was approaching late October, and the sampling equipment had to be deployed to Parachute immediately. So we began demobilizing the Cerro Negro drill camp. Over the course of four months, however, seventy-six microbial cores had been collected and distributed to twenty labs across the country. A new record for the SSP!

LES CHUTRES GET THEIR TURN AT DEEP CORING: LATE OCTOBER 1994, NINE MILES NORTHEAST OF PARACHUTE, COLORADO

On October 4, William Broad had published an article on DOE's "juggernaut" deep microbiology exploration program in the Science section of the *New York Times*. In his usual punchy, exciting style he stressed the idea of the potential antiquity of deep subsurface micro-organisms and raised the question of whether we might be opening a Pandora's box of virulent dinosaur pathogens. Richard Preston's *The Hot Zone* had just been published and was a popular read at the Cerro Negro drill site. We wondered whether we would be drilling into our own hot zone at Parachute. If so, were the protocols we had developed for processing the core within the glove bag and trans-ferring them into air-tight mason jars enough to protect us from whatever was down there? Boone, however, was quick to dismiss the suggestion that his *Bacillus infernus* represented a threat to the metro-politan northeast United States. But it was a fear that I would soon face again.[56] We were doubly honored by a second article, by Walter Sullivan, but he questioned the authenticity of our claims despite our rigorous tracer techniques. It all reminded him of the discovery of living microorganisms in a meteorite that had fallen in France in the mid-1860s.[57] That discovery had quickly been endorsed by eminent scientists and Nobel laureates but was soon dismissed when it was discovered that the meteoritic samples contained ragweed pollen. It was a sobering commentary that shook my confidence in SSP to the bone, and it was a portent of coming events.

In late October, Tim Griffin and Rick drove one of the lab trailers from Cerro Negro up I-25, turned onto I-70 heading toward Denver, and crossed the Rockies to the Parachute drill pad. It was located two miles north of I-70 and accessed by a graded dirt road that crossed a river at the bottom of a canyon, which at that time of year was dry. The road then climbed 200 feet up a 30° slope with two switchbacks. Tim drafted a site bulldozer and operator to pull the lab trailer up the road, but negotiating the trailer around those hairpin turns without sliding off and down into the canyon was heart pounding nonetheless. Back at Princeton I had just received the first cuttings from the 1-M-8 borehole,[58] and we quickly made slides and exam-ined the fluid inclusions of the deeper layers. We were looking for two-phase inclusions and getting T_{max} estimates to start picking the target depths.

Two weeks later we all flew out to Grand Rapids, Colorado, and drove to the site, arriving with our steel-toed boots and coveralls

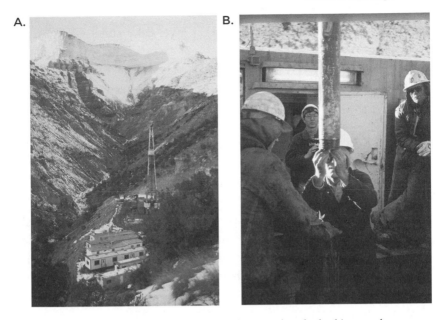

FIGURE 2.8. A. 1M-18 drill site near Parachute, Colorado, looking up the canyon to the Roan cliffs of the Green River Shale in November 1994 (courtesy of T. Griffin). B. Mark Lehman packing a plastic bag of fluorescent microspheres into the bottom of the Gel Core, while Chuanlun Zhang looks on in November 1994 (courtesy of F. S. Colwell).

ready to get to work.[59] Tim had everything set up in the trailer, with bleached tile floors and a fully inflated anaerobic glove bag. The twelve-story-tall drill rig with all its extraneous generators, pumps, vehicles, and crew quarters was crammed precariously onto a thirty-by one-hundred-yard shelf carved into the canyon wall. We donned our hard hats and gloves and stepped outside for the tour. Despite the chill in the air, the insulated coveralls kept us toasty. I had thought that the rig at Cerro Negro was big, but it was a pygmy compared with this monster. It was a shiny silver tower with some teal-green cabins and storage units around its base (figure 2.8A). The roughnecks were busy moving thirty-foot rods around the platform as the bit churned away some 2,250 feet beneath our feet. It was approaching the first core point that we had selected based on the drill logs of the 1-M-8 borehole. Exposed along the canyon walls surrounding us was the Wasatch Formation. It was a layer-cake stack of light-tan sandstones interbedded with pinkish-red layers that represented ancient paleosols. They were much like the paleosols of the Ringold Formation cored at GeMHEx site on PNL, only five times older and ten times deeper. Could the ancestors of the microorganisms that resided in those soils when they formed at the surface 60 million years ago be alive today after having been buried a mile beneath the surface? Our core target was even deeper, a porous, fluvial sandstone 2,820 feet beneath us in the Molina Member of the Wasatch Formation. As was done at Thorn Hill no. 1, we would core a shale/sandstone transition in the hope that the nutrient fluxes from the shale into the sandstone would have sustained deep life for millions of years at thermophilic temperatures. Peering down on our drill site from 3,000 feet above us were the white cliffs formed by the Eocene Green River Shale. Up in those cliffs the DOE had supported an oil-shale mining operation a decade earlier.[60]

Microbial contamination of the Parachute cores was going to be a serious issue since we were using drilling mud. We had also wanted to collect the core at the in situ pressure because of our concern that during normal coring operations the pressure in the core decreases rapidly upon approaching the surface and this might kill any barophilic microorganisms. Equipment did exist for preserving the in situ formation pressure in a core, but the cost had proven prohibitive. Tim suggested we try something the SSP had never used before, the gel coring system from Baker Hughes. This system was commonly used in the oil patch (that is, the petroleum industry) and employed a viscous gel of polypropylene glycol that sat inside the core barrel, protecting the core barrel from drilling mud as the core barrel slid down the borehole. As the core bit began to grind into the rock, the rock surface pressed against a spring at the base of the CoreGel, which opened holes through which the gel oozed to coat the rock core and seal it from drilling mud. This system was designed by Baker Hughes to obtain cores with their internal formation water and hydrocarbons intact, without any contamination from the drilling fluid. Tim reasoned that it should also prevent microbial contamination from the drilling mud filtrate and better preserve the indigenous microbial community.

Tim showed us his CoreGel,[61] which basically looked like a 30-foot-long aluminum cylinder. Unlike with the Cerro Negro coring, this tube would slide into a solid coring tube with its own bit. As Tim explained to all of us standing around the CoreGel, once the drillers reached our target depth, they would have to pull all the drill pipes out of the borehole along with the bit. Then they would mount the CoreGel/core barrel, attach a coring bit, and start pushing it back down the hole. Given that they were currently advancing the hole about 54 feet a day, we estimated that we had twenty-four hours before they would attach the CoreGel. After Tim's briefing, I visited the Fluor Daniel geologist on site, who was in one of the cabs examining the drill cuttings collected from the mud and scanning them with an ultraviolet (UV) light to look for hydrocarbons. Most of the cuttings had the granular, oxidized sand texture of the Wasatch Formation sediments that had been deposited in a river about 60 million years ago but looked as if they had been deposited yesterday. I was happy to see the reddish-colored ferric oxides because they would definitely provide a good electron acceptor for any subsurface thermophilic, iron-reducing bacteria. I was amazed at how the geologist could ascertain the exact stratigraphic position based upon scrutinizing the cuttings like he was reading tea leaves.

I returned to our trailer and sat down with Tim and the rest to go over the core logs. We estimated that we had about eight hours to go before the drillers stopped drilling and started pulling up the rods. That would take them about fourteen hours if there were no problems. We decided that we would turn on the pump for the PFT right before they attached the CoreGel. I asked Jim McKinley to show me the PFT setup and explain how it was going to work this time. I followed him back outside to behind the mud tanks, and he showed me the HPLC pump, which was encapsulated in a Plexiglas cubicle with a tiny tube attached to a humongous pipe. It was hard to imagine how such a tiny vial could yield a detectable signal in such a huge volume of mud. Afterward we drove back down to Parachute to check in at an abandoned Exxon facility on Battlement Mesa that Tim had leased for cheap accommodations.

The phone call came in the middle of the night when the drillers were just about finished pulling out the drill rods and we needed to be ready with our CoreGel. We carefully drove up to the site in the dark. Jim initiated the PFT tracer while we watched the roughnecks pull the aluminum tube in its coring barrel up above the rig deck. Then they all stepped back to let the scientists do their thing. Mark Lehman carried a bag full of green fluorescent microspheres out to the suspended core barrel and ever so delicately wrapped the wires of the WhirlPak bag into the foot spring at the base of the CoreGel (figure 2.8B). The scientists then retreated as the roughnecks grabbed the core bit, screwed it onto the end of the barrel, and tightened it with what I thought must be the world's biggest channellocks. They were so big it took two men to operate each one. We then descended the stairs as the roughnecks wrestled thirty-foot pipes to the deck, screwed them one by one onto the drill string, and lowered them into the borehole. Our gel coring system was now inching its way down to the target. It would be another day and a half before it would appear again.

We went back to Battlement Mesa to get some sleep, then woke up for a late breakfast and spent the day visiting a local school to give a talk to the students about our work. It was still dark when early the next morning we received word from the rig that our CoreGel was coming up to the surface, so we donned our coveralls and drove up to greet it. When we arrived, the roughnecks had just laid the long aluminum core barrel down on the wooden platform. Tim Myers walked over to the core barrel and tried to determine just how much of it was filled with core by banging on it with his rock hammer. Thunk, thunk, thunk, and then twang! The upper part of the barrel echoed like a bell. We did not have a complete core, but only about half of one. I started running red and green magic markers down the length of the barrel, which was still warm to the touch. Tim Griffin and I then began to mark it into three-foot-long lengths. It looked like we had six three-foot cores and a bit. A roughneck then came over with the biggest rotary saw I had ever seen. It had a two-foot circular blade that sliced through the aluminum tube, showering us with hot sparks. He quickly cut our core into three-foot slices as if he were cutting up a loaf of sourdough bread. We weren't too concerned about contamination from the blade because, with all the sparks, the friction must have cauterized the end surfaces of the cut rock cores. Mark Lehman had been busy back in the trailer sterilizing polycarbonate tube liners that we were going to use to store the cores until we could process them in the glove bag.

Inside the trailer we removed the aluminum tube by sliding out the rock cores, along with all the gel, which was a black, sticky goo. Removing the gel from the core's surface before we could process the samples turned out to be like wrestling Uncle Remus's tar baby. We had to rub the gel off the core with sterile rags, and although we could get it off the core, we could not get it off the rags. Soon it was on our fingers, and from there on our coveralls and then on our faces. The only way we could get it off of ourselves was by using kerosene to remove it. We then slid the three-foot-long cores into the sterilized polycarbonate tubes and stored them in the tubes under N_2 gas while we processed each core separately in the anaerobic glove bag.

The cores appeared dark gray and shiny, but it was clear that we had captured a beautiful sandstone-to-shale transition. Mineralized fractures cut across the core, which revealed open cavities, some a half a millimeter wide. That seemed huge to me. It was astonishing that such open space could exist at such great depth and under such great pressure. For the bacteria navigating their way through the rock, it was the equivalent of the St. Lawrence Seaway. We could not have had better luck capturing that transition. But the real surprise came when we cut the cores of silty sandstones with our hydraulic rock splitter into two-inch-thick, five-inch-diameter disks. As we used the hydraulic splitter to pare the disks down to cubical nuggets we could see that the sandstone was bone dry on the inside. I guess we should not have been surprised since we were drilling in a gas reservoir, but my preconception was that at 2,820 feet depth, all the pores would be filled with a fair amount of water. We expected to see some moisture. We were all disappointed because we really needed to have some way of measuring the fluid chemistry, and the dryness of the core would seem inimical to microbial life. On the good side, the CoreGel clearly had kept the drilling-mud filtrate out of the core.[62]

To keep track of which investigator got what, we had a whiteboard hanging on the wall of the trailer opposite the glove bag so we could see everyone's specifications for their samples. Beneath the bright ceiling lights of the trailer, we pared the five-inch cores down to nuggets using the hydraulic splitter, smashed them to smithereens, then scooped up the "homogenized" chips into Whirl-Pak bags and placed into the mason jars. We then packed the jars into Styrofoam coolers with frozen Blue Ice to keep the samples cold during transit to the receiving labs. We took some of the samples and placed them into a small titanium cylinder, which we then pressurized with argon gas to 2500 psi. These we would send to Art Yayanos, a microbiologist at Scripps Institute of Oceanography who specialized in culturing barophiles. The odd thing was that one of the recipients, a microbiologist from the University of Oklahoma by the name of Joe Suflita, did not want us to send him any homogenized samples. Instead, he wanted us to pick out intact cores several centimeters in length and place them inside ammo cans that he had shipped out to us along with desiccant and a palladium catalyst designed to scrub out any O_2 leaking into the can. We were all joking about Suflita's ammo-can samples. What was this guy thinking? Unbeknown to us he had a very clever idea, indeed. When the first core from the Wasatch was completely processed, Rick and I drove the boxes full of mason jars down to Grand Junction for overnight shipment to all the labs. It was during our drive to and from Grand Junction that I learned how Rick Colwell got his start in Frank's program.

SSP'S FIRST CORING OF VOLCANIC ROCKS: SUMMER 1990, IDAHO NATIONAL ENGINEERING LABORATORY

When Rick started his postdoc at INEL in 1988, he was a highly trained aquatic microbiologist who had studied how microbes in rivers formed epilithons, the biofilms that form on rock surfaces in aquatic habitats. For his work at INEL, he had been researching how the bacterium *Thiobacillus ferrooxidans* oxidized sulfides, as part of a U.S. Bureau of Mines project.[63] At that time, his understanding of subsurface life had been like that of everyone else—nil. He had known about pathogenic microbes and protozoans that reached the subsurface as a result of poor waste disposal practices and could thereby contaminate aquifers. He had also realized that wherever a well was drilled, surface microorganisms were likely to inoculate the subsurface. Then one spring morning in 1988, LaMar Johnson, who was the manager at the INEL, and Brent Russell, of Golder Associates in Richland, Washington, had walked into Rick's office and told him that he should be doing subsurface microbiology with the Wobber group. Rick thought they were crazy. He had just wanted a quiet place to work in a lab, maybe do some fieldwork, but not any of this hyper-disciplinary stuff in which you were working with scientists and engineers from all sorts of disciplines who spoke different languages. But in a few short months Rick was leading the INEL subsurface microbiology drilling program, the first project to tackle volcanic rock aquifers, a big step from the sandy, porous Cretaceous aquifers of South Carolina, with a whole set of new problems with respect to drilling contamination. It was also clear from the outset that Rick would have to drop his other project and devote himself full time to SSP, if for no other reason than to keep up with Frank's almost incessant queries, demands, and schedules. Rick had taken advantage of some coring that was occurring during the summer of 1988 at the Radioactive Waste Management Complex (RWMC) to collect his first subsurface microbial samples.[64] The drillers there had used HEPA-filtered air from a compressor to lift the cuttings out of the hole as they cored down to 590 feet, just penetrating the water table of the Eastern Snake River Plain Aquifer, the source of all the irrigation water for Idaho potato farms. The cores were mostly from 75,000- to 600,000-year-old basalt, very young volcanic rocks by geological standards. He focused on the thin sedimentary layers in between the basalts and was able to find a fairly high number of microbial cells, not as high as those found in the cores from the P series wells of the SRP but still high enough to look like a real signal. Although he had not used tracers, it was enough background information to formulate his first drilling proposal to Frank. It also didn't hurt that Tommy had taken him under his wing, and with Tommy's help he drafted a successful proposal to Frank in January 1990, to drill the young volcanic aquifer beneath INEL.

When the funds arrived, coring into the Snake Rive Plains basalt aquifer began in the summer of 1990. But before the drilling at INEL began, Brent had to modify the air rotary system that Rick had used earlier at the RWMC site, because the O_2-rich air would certainly deal death to any anaerobes inhabiting the sedimentary layers between the basalts. They also couldn't use the pink drilling mud and the Denison core barrel with extended shoe that had been used so successfully at C10. Tommy and Brent had the idea of using argon gas as the drilling fluid. The gas could be filtered with 0.2 micrometer filters to remove any microbial contaminants. Now the only contaminating bacteria would be those coming from the drill rods, which would be steam cleaned. Argon was perfect for the PFC tracers, which were volatile. Because argon was denser than air, it kept the borehole and cores anaerobic. Because gas is a much less effective coolant than water, however, they had to tape temperature strips to the inside of the coring barrel to monitor the rock-core temperature. They had to make certain that frictional heating from the core bit was not raising the core temperatures to the point of sterility. Using this approach they made certain to core slowly enough when collecting the microbial cores so that the temperatures never exceeded 40°C. Since there was no drilling water for the bromide tracer, it was put into the same Whirl-Pak bags that held a trillion fluorescent beads and was strapped to the shoe of the core barrel. When the rock core penetrated the core tube, it broke the bag, and the beads and bromide washed over the five-foot-long microbial cores.

Their drill site was located in the most remote and desolate location within the INEL reservation. They were two and a half miles away from any facilities and had to drive along a single-track dirt road through a parched landscape of sagebrush and volcanic mounds to get to the drill pad. The verdant green, irrigated potato fields with their massive circular sprinklers and water pumps that surrounded

INEL were fifteen miles away to the east and south. The team was as far away from potential contamination as they could get. Nothing on the black rocky flat where the rig sat protected them from the blustery winds that roared down from the southern reaches of the Bitterroot Mountains fifteen miles to the northwest. So remote and bleak was the site that none of INEL's mechanics wanted to work out there (figure 2.3). If the drill rig broke down, major bribery was required to coax repairmen from their warm, cozy, coffee-machine-filled workshops. They could not use the Blue Bird MMEL, because if it was contaminated with radionuclides, Tommy would never be able to get it back. So Brent and Rick set up a core processing lab in a trailer upwind from the drill pad, which they named NPR-WO1 (in which "NPR" stands for the New Production Reactor site, where they were located, and "WO" stands for "Wobber"). The anaerobic glove bag had grown to sixteen feet long with an aluminum torpedo tube to transfer the six-foot-long core into the workspace. They now wanted to look at the microbiology of the volcanic rocks, not just the soft sediment between the lava flows, so they had to figure out how to subsample igneous rock. Tommy and Brent designed a hydraulic core splitter to cut and pare the core and a steel Plattner mortar to crush basalt down to pea-sized chunks before placing the crushed rock into double Whirl-Pak bags. Keeping the wind-blown dust from drifting into the trailer was a thankless battle, but the rock samples never felt the wind because Rick's team carefully passed them through another air lock into a packaging glove bag where the samples were weighed, labeled, and placed in mason jars. As the wind buffeted the trailer, the jars were passed out of the second air lock and packed on Blue Ice in Styrofoam coolers. Tommy would then set new land-speed records driving the coolers down the winding dirt track to the INEL shipment facility. The dual-bag arrangement sped up processing, but processing one core still would take five people five hours. One step in the core processing required the delicate surgery of scoring the polycarbonate core liner with a battery-operated circular saw. This was accomplished inside the glove bag with the operator wearing a pair of gloves on top of those in the glove bag sleeve. Amazingly enough, no fingers had been lost.

The coring campaign at NPR-WO1 finally stopped in the fall of 1990 at 675 feet after having collected about fifty microbial cores that summer. It was a new record for the SSP. The tracers had not worked perfectly, but they had done well enough to estimate the potential microbial contamination to be no more than one cell in fifty kilograms of basalt.[65] This had been a good result given that during the subsequent months the investigators found that the basaltic rock core contained very few viable microorganisms. In some cases no microbial activity or growth had been detected, which, in and of itself, had been a testament that the samples had not been contaminated during drilling, recovery, or processing. The PFC tracer had again yielded miraculous results, showing that the argon gas had penetrated into the basalt formation ahead of the drill bit.[66]

LIFE AS A DEEP-DWELLING BUG

After we returned from Grand Junction, Rick sent an email to Les Chutres informing them of the samples coming their way and their tracking numbers. On the plane ride back to Princeton the following day, I imagined how all this coring and sample processing would seem to one of our little deep-dwelling friends living 2,820 feet beneath the drill site.

It had been quite happy in *its* tiny universe of water and mineral grains, eking out a meager existence in the dark. Occasionally *it* would have enough energy to divide, and sex was way too infrequent,[67] but *it* was always able to find around every corner of *its* home a new fresh surface of iron that *it* could rest on and reenergize. Suddenly *it* was vibrating through an earthquake unlike any *it* had experienced in its thousand years of existence. Its proteinaceous fimbriae attached to the iron mineral surface were being ripped out of its cell membranes,[68] and finally *it* was hanging on with just its flagella. *It* was being slung around its mineral house. Then as suddenly as the shaking had begun, the shaking stopped. *It* only felt the random motion of the fluid and an intermittent tug of gravity pulling it down in the floor of its home. *It* detected discernable cooling as time passed because its membrane began to stiffen. Some of *its* chemoreceptors were detecting nutrients that *it* had not detected for hundreds of years, and *it* detached itself from the floor and started swirling its flagella in search of the food that literally must have been shaken out of inaccessible pores onto the floor of its home. Just as things seemed to settle down while *it* was consuming its new feast and electrons were coursing through *its* membrane, the earthquakes were back. This time they occurred as intermittent, high-magnitude, pounding spikes followed by a pause. The thin film of water that had surrounded *it* and had made *it* so happy started to retreat, leaving *it* attached to the mineral grains by the slenderest layer of moisture, just enough to sustain its internal pressure. Even stranger, *its* genes for photoreceptors, which had long ceased activity and had been dormant, were suddenly being activated. *Its* defenses against radical oxidants were also producing messenger RNA. The portion of *its* genome that *it* had inherited from its ancestors was suddenly taking over its cellular machinery. *It* was getting drier and drier, colder and colder, and *it* started making itself into a spore, but *it* was slowing down. Ice crystals began to form in *its* cytoplasmic membrane as *it* tried to frantically finish a spore coating, but it was too late. *It* stopped.

As I landed at Newark Airport, I could only hope that *it* would survive a similar trip to the labs in time for the microbiologists to resuscitate *it*.

LES CHUTRES' HUNT FOR HOT BUGS TURNS COLD: THANKSGIVING HOLIDAY, NINE MILES NORTHEAST OF PARACHUTE, COLORADO

We returned two weeks later, over Thanksgiving, for the second sampling depth only to find the drill site covered with snow. The intermittent melting had turned the road into an icy slalom course. The core run at 6,600 feet arrived at the surface at midnight, and we could tell something was wrong just by the way the roughnecks had removed the shoe and were peering up into the CoreGel. For some reason, the rock core had not penetrated the CoreGel and only 70 centimeters of rock were captured in the shoe. The drilling mud had probably contaminated the core; nonetheless, we processed it as a "contamination witness." John Lorenz performed the CoreGel "autopsy" to find out why the core jam had occurred. He deduced that they had unfortunately cored right into a vertical fracture running down the length of the core. It seemed plausible that when the drilling fluid at 3,000 pounds per square inch hit the fracture, it caused it to expand until both sides were rubbing against the core shoe with so much friction that the core bit could no longer cut any deeper. John had seen the vertical fractures at the MWX boreholes down in the valley not two and a half miles from us. A bitterly cold front moved in as the drillers started tripping down the rods to drill again.[69] We now were on the horns of a dilemma.

Over beers at Jay's Steakhouse in Rifle, Colorado, Rick, the two Tims, Mark, Chuanlun, John, and I discussed our two choices. Either we immediately return to the drill site and make a second attempt at collecting core from this interval, or we stick to the original plan and take a third core in the coal seam at 8,800 feet and 115°C. We had no other options. No sidewall coring tools like those Rick and Tim had used at Taylorsville were available. The drillers were now advancing the hole at about 60 feet per day and in 36 days would be at the target depth. I advocated going for the deeper, hotter core, a risky option that might come up with nothing. Earlier in the year, the microbiologists working with Tommy Gold had isolated a thermophile from his borehole in the Siljan Ring Complex in Sweden that came from a depth of at least 9,800 feet,[70] if not 17,000 feet, setting a new depth record for the biosphere. I wanted to go deeper. But Rick countered that it would not fulfill the primary goal of the Parachute drill campaign, which was to test recolonization of the sandstone in the current habitable zone, where the chances were better that we would find bacteria. After a heated debate chased by many beers, we decided to go for a second core run at 7,000 feet. Jim Fredrickson had flown in to help us by now, having completed the job at Cerro Negro. The drillers tripped out the drill rods one last time for us and attached our last CoreGel. As Mark Lehman stuffed in the bag of microspheres, we all prayed to the god of drilling for a perfect core. The roughnecks screwed on the bit, and down the borehole the core barrel went. Later in the day, when the core made contact with the borehole bottom I decided to visit the drill house, just to see how it was done and perhaps learn why the previous core had jammed. One of the drillers was at a panel with knobs and meters, looking pretty serious. He turned and grinned in that kind of Clint Eastwood way where you couldn't tell if he was happy to see you or to shoot you. As I hesitated at the door, the freezing air swept into his relatively cozy compartment. He waved me to come in quickly with the motion of his gloved hand. I started asking him questions about what that knob did and what this dial does. "Do you want to do the drilling?" he said. I was shocked, because I thought it would be breaking a bazillion rules. He turned down one dial and the rods outside stopped turning. He then flicked two switches and stood aside. I moved over to the panel and pulled off my mitts. "First thing you want to do is to release the lock on the rods by flicking that switch. Now your rotator is already locked on the rods and you slowly turn up that knob until you get to about there." He pointed at the dial. I followed his directions and looked out to see the drill rods turning. "Am I coring yet?" I asked. "No, you now need to release the brake lock with that switch," he replied, which I did with a confused look on my face. "Now how do I push down on the rods?" I asked. "You don't. The weight of all of that steel is more than enough push. In fact you are pulling up on the rods all the time so you don't crush the bit or jam it." He showed me how to ease off the brake and adjust it to about the rate of penetration that he was using for these rocks. I got it. I had a lot of experience coring short rock cores in my former paleomagnetic career using a portable chainsaw-powered drill. Back then I would be ecstatic if I could get a one-foot core. Now I was turning a 7,000-foot-long core barrel with an engine the size of a locomotive and aiming to get a 30-foot-long core. This was amazing. I decided I had better step away from the drill controls and let the expert handle our last shot before I jinxed everything. But this time, the CoreGel worked as advertised. It was way below zero when the drillers finally pulled the CoreGel out of the borehole early in the morning. We put electrically heated blankets on top of the CoreGel to keep the polypropylene glycol gel from stiffening before we could get the cores out.

In the trailer we pulled out sixteen feet of massive cross-bedded sandstone from the core barrel. The cores had peculiar organic-rich seams, or stylolites, in which the quartz had been dissolved during very deep burial and heating. It was as if some acidity from the organics had acted to dissolve the quartz while it was being compressed under great stress. Rick and I drove off with the processed samples in the back of our carryall, this time fishtailing down the hairpin slalom like we were in a Warren Miller extreme ski movie.[71] We crossed the frozen gorge below and zoomed down the dirt road to the highway and on to the Grand Junction FedEx office.

The next day Jim Fredrickson and I flew to Richland, Washington, while the rest of the group went fossil hunting in the Anvil Points oil shale mine, high up on the cliffs overlooking our drill site. Jim and I were still trying to make some progress on the original project that got me started, investigating fluid inclusions as a tomb for bacteria. Our department's machinist at Princeton, George Rose, had constructed a crushing cell based on the design that Jim and I cooked up in San Diego in June. Keith Bargar had sent me some thin sections, and I identified the specific location on these sections containing his bugs. We spent the evening trying to crack the quartz and then flushing it with distilled sterile water to extract any fragment that Jim might target with polymerase chain reaction (PCR). But we ended up crushing the crusher and not so much the quartz. We had to devise a different approach and apply it to softer minerals, like the salt samples that Tom Kieft and I had collected the previous Thanksgiving.

I finally returned home in time for Christmas 1994. Fluor Daniels would finish drilling 1-M-18 by mid-January. We had obtained the most pristine microbial samples ever obtained from 7,000 feet depth, along with gas and formation fluids to tell us the origin of any bacteria that might exist in those cores. Right before Christmas, however, Jim and I received an email from Frank telling us that we must organize an all-hands meeting in six weeks for an update on the analyses of the precious cores, the treasure from Cerro Negro and Parachute.

CHAPTER 3

BIKERS, BOMBS, AND THE DEATH-O-METER

"Isn't the earth's radius about 3,926 miles?" [Axel]

"It's 3,926⅔" [Professor Lidenbrock]

"Let's round it up to 4,000. Out of a journey of 4,000 miles, we've gone 40?" [Axel]

"That's right" [Professor Lidenbrock]

"And this at a cost of some 210 miles diagonally?" [Axel]

"Exactly." [Professor Lidenbrock]

"In about twenty days?" [Axel]

"Twenty days." [Professor Lidenbrock]

"Now then, 40 miles are 1/100 of the earth's radius. So if we keep this up, we'll take 2,000 days, or nearly 5½ years, for our descent alone!" [Axel]

"To blazes with your calculations!" my uncle countered with an angry gesture.

—*Jules Verne, Journey to the Center of the Earth*[1]

THE LONG SUBSURFACE MICROBIAL JOURNEY TO CERRO NEGRO: FEBRUARY 25–28, 1995, MARRIOTT CENTER, SALT LAKE CITY, UTAH

Frank had warned us that the end of the deep subsurface microbiology program was a little over a year away. We spent what little time we had remaining trying to finish analyses of the samples collected from Cerro Negro and Parachute and publishing the results. Frank

scheduled stocktaking meetings throughout 1995, beginning February 25–28 at the Marriott Center in Salt Lake City. It was clear from
the outset of the meeting that both sites had presented new challenges to microbial sampling and that our time and resources were
too limited to address them all.

At Cerro Negro the microbiologists had found indigenous bacterial populations of 100,000 bacteria per gram in the sedimentary
strata both far away from and close to the basaltic intrusion. Most of
the microbial analyses pointed toward no apparent differences between the microbial content of the sandstone and shale within the
proposed thermal aureole of Cerro Negro and that of the sandstone
and shale outside the thermal aureole. But the microbial results also
had issues. To begin with, the air used to drill the vertical borehole
drove the PFT tracers into the undersaturated cores, raising the spec-
ter of contamination. To make matters worse, the negative control
cores seemed to yield signs of growth even though they had been
fired at 400°C. It was Todd Stevens who discovered that when he
flamed the outside surface of the pared nuggets, they did not yield
any growth. Apparently, because these core samples had been relatively dry, they generated electrostatically charged dust when they
were cut and pared inside the glove bag. The particles of dust from
the microbially contaminated outside surfaces of the cores were then
jumping onto the surfaces of the pared internal nuggets. This cross-
contamination may have been responsible for the disparities between the different types of microbial assays and was a focus of
heated discussion way into the night. It left an indelible impression
in my mind about the need to track and to sterilize the outer surface
of all rocks and chips in the glove bag during paring.[2]

But some of the microbial evidence for indigenous inhabitants
was extremely compelling. On the first day, Joe Suflita showed us
autoradiographic images of local hot spots of sulfate reduction activity inside the sandstone cores, especially for those whole cores in
contact with the shale.[3] Joe and his postdoc Lee Krumholz had been
developing photographic images of in situ sulfate-reducing microbial activity. Because their samples were not pared, they had not suf-

fered from the issues of cross-contamination. This is what they had been doing with those ammo-can cores. Even though they did not isolate any SRBs, they could show that bacterial activity increased when crushed shale was brought into contact with the sandstone and that the bacteria were mesophilic.[4] The sulfate-reducing activity was greatest at the interface between the shale and the sandstone. The sulfate-reducing activity was negligible in the shale because the size of the pores was much less than 0.2 micrometers, which was the minimal width of a typical bacterium. In the sandstone, however, the pores were much larger. The data also indicated that bacteria could easily migrate through the spacious sandstone aquifers, but migration through the Mancos Shale was restricted to fractures larger than 0.2 micrometers in width.[5] The SRBs were not reducing sulfate left over from Cretaceous seawater sulfate. The stable sulfur isotopic data on the sulfate indicated that it originated from oxidation of Cretaceous-age pyrite by aerobic groundwater in the recharge zones of the sandstone aquifers. The age of the groundwater moving through the sandstone aquifers and intersecting Cerro Negro ranged from a few thousand to 25,000 years.[6] This was much, much younger than the 3.3-million-year age of Cerro Negro. The rapid migration of water from Mt. Taylor to Cerro Negro had been facilitated by the numerous fractures that provided conduits for the bacteria to migrate even into portions of the Mancos Shale (figure 3.1A). Theoretical models of the groundwater flow yielded seepage velocities on the order of 0.2 meters, or about 6 inches, per year. This sounds very slow, but given 3.3 million years, groundwater would migrate about 470 miles! This is much further than the distance across the sterilization zone of Cerro Negro.

The 1,500-foot-long shallow-angle borehole at Cerro Negro was supposed to have sampled the sterilization zone of the basaltic plug's thermal aureole, but for some reason it seemed to have missed the main intrusion and intersected only a few narrow dikes (figure 3.1B). The uncertainty in the shape of the intrusion meant that predictions of the radius of sterilization around the intrusion based upon thermal models were equally uncertain. Identifying where that steriliza-

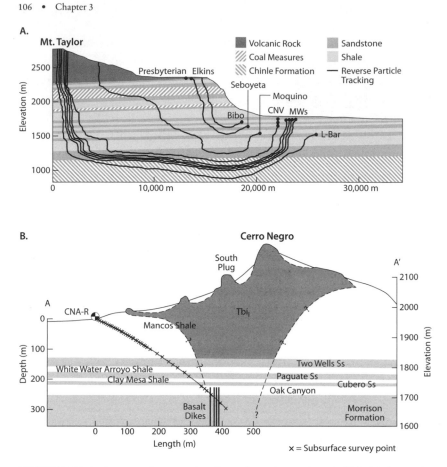

FIGURE 3.1. A. Cross-section running from the extinct volcano of Mt. Taylor to the Cerro Negro vertical borehole (CNV) with the names of other monitoring and water wells indicated (from Walvoord et al. 1999). The reverse particle tracks represent the predicted pathways that particles of water would take to arrive at different wells based upon the porosity and permeability of the different formations. In the case of Cerro Negro, the water originates from the top of Mt. Taylor. B. Cross-section of Cerro Negro showing the location of the 1,500-foot-long angle borehole (CNA) designed to sample the sterilized thermal aureole of Cerro Negro. The numerous question marks indicate uncertainty in the shape of the intrusion despite numerous geophysical surveys performed in advance of the drilling.

tion zone existed had to rely heavily upon input from the members of the research team who were organic geochemists and geochronologists who used geothermometers and thermochronometers.[7]

Gas chromatography–mass spectrometry (GC-MS) analyses of the extractable organic matter from the Mancos Shale indicated that it had not experienced temperatures higher than 90°C.[8] This was consistent with our fluid inclusion results on the carbonate cements in the shale, which indicated a maximum temperature of approximately 65–90°C. This finding, combined with the vitrinite reflectance data,[9] indicated that the shale and sandstone units experienced a T_{max} of about 75°C from 75 to 15 million years before they cooled down to their current temperatures of about 20°C between 15 and 5 million years ago. K-Ar dates on the potassium-bearing, illite clay[10] from volcanic tuffs in the shale ranged from 38 to 48 million years.[11] These ages were much younger than the age of the tuffs when they were deposited, but much older than the age of Cerro Negro, and these ages coincided with the age range of T_{max}. Vitrinite reflectance data suggested that a short thermal pulse sufficient to raise the temperature of the shale to approximately 100–125°C, the temperature deemed sufficient for sterilization, had occurred within about 150 feet of the contact with the basalt plug. The vitrinite reflectance data also indicated that hot, hydrothermal water had penetrated the shale formation along fractures in the angle borehole and that cooler groundwater had penetrated the thermal aureole of the intrusion in the more permeable sandstones. Given that the sterilization zone was only about 150 feet and the groundwater migrated six inches per year, water would have traversed the sterilization zone during the 300 years after the intrusion had cooled down.

Microbiologists had detected microorganisms very close to the intrusive contact, so clearly they had recolonized the sterilized zone in less than 3.3 million years. How fast do bacteria travel underground compared with the rate of water flow? A lot of factors control the rate of subsurface microbial transport, and Frank had funded many PIs who focused on answering this question. Microbes will collide with the surface of mineral grains and adhere to them for some time before detaching themselves to rejoin the flowing groundwater. This process is called retardation. The tendency to attach to a grain depends upon a lot of detailed physics and chemistry. Microbes are typically negatively charged, but mineral grains can be either positively or negatively charged, depending on the pH. As a result, a negatively charged bacterium would adhere strongly to a positively charged mineral grain and hang around there for quite a long time.

But under different circumstances, microorganisms can actually move faster than the bulk of the water. Imagine a microbe is a barge flowing down a river and as that river flows down a relatively straight channel, the raft will move more slowly than the river. But when the river slows down and enters a delta, like the Mississippi Delta, the water begins flowing through lots of little side channels, or disperses, and its average forward velocity toward the Gulf of Mexico slows even further. But the barge is too big to enter these little tributaries and sticks to the main channel, which is flowing faster than the side channels, and thus the barge's velocity is greater than the average velocity of the river water in the Mississippi Delta. In the Mancos Shale there were lots of little channels that were too small for bacteria to enter, but there were a fair number of larger-diameter fractures through which the water and bacteria were moving at a faster rate. The ^{14}C dates, however, measured the average velocity of the water, not just the water following the fast paths. As a result the bacteria may have penetrated the 150-foot-wide sterilization zone and reached Cerro Negro in less than 300 years.

NEUTRON BOMBS, TECTOGENESIS, AND DEEP MICROBIAL MIGRATIONS: SPRING 1995, PRINCETON UNIVERSITY

Using other processes, microorganisms can also travel faster than the groundwater migrates. Many microbes are motile, having flagella that rotate like propellers on a barge. Flagellar rotation would require some metabolic energy on the part of the microbe, adenosine 5′-triphosphate (ATP) in this case, or diesel fuel in the case of the barge. But on a long journey, both would quickly run out of fuel and both would have to stop and refuel. From all of the previous SSP work, the subsurface biosphere appeared to be a very energy-poor environment because the microbes had already eaten all of the easy-to-get food. But Tommy hypothesized that the sterilization zone of Cerro Negro had provided a feast for the newly arriving, colonizing microbes. Tommy had likened the process to a neutron bomb that eradicated life without destroying the food. They could have fed upon the remains of the dead microbes (i.e., the amino acids, peptides, nucleic acids, and pieces of DNA and lipids) and the substrates generated from heating the shale (i.e., carboxylic acids) and the H_2 produced by the fresh basaltic minerals. Not only did they have sufficient energy to swim faster than the water, they could also reproduce! Reproduction counters the effects of dispersion and also increases the speed of migration. Now you have a whole fleet of barges migrating through the Mississippi Delta. Last, microbes can be chemotactic. In other words they contain proteins that will react to the presence of a certain metabolic substrate that they require. In combination with their swimming motion, the microbes can detect the direction from which the substrate appears and swim directly toward it. Now you have a captain on your barge looking at the signs for the fuel depots and charting a course for them so as to never run out of fuel.

I found Tommy's hypothesis absolutely fascinating because it was based upon experimental observation of the behavior in the lab of many different types of bacteria. Tommy was just extrapolating that behavior to a geological environment and adding geological time. Eight years earlier, Francis Chapelle had worked with a microbiologist from the University of Maryland who explained to him that the difference between geology and microbiology is the difference between the wrong way of doing science and the right way of doing science, that is, inductive versus deductive reasoning.[12] To a large degree, this difference in philosophical approach was why the geologists and microbiologists had seemed unable to accept each other's ideas about the importance of microorganisms in the geological cycle and on a geological scale. But Tommy embraced both of these opposed philosophical approaches with no visible signs of indigestion as far as I could tell. As F. Scott Fitzgerald once said, "The test of a first-rate intelligence is the ability to hold two opposing ideas in the mind at the same time and still retain the ability to function."[13]

In the SSP, Frank had performed a shotgun marriage of these equal opposites that was pivotal to its success. One example of this merger was the heavy reliance on the culturing and isolation of subsurface microorganisms in order to determine whether their physiological characteristics—their tolerance to temperature, salinity, pH, radiation, nutrient deprivation—were consistent with the environment and the geological history inferred by the geologists. The optimum growth temperatures of the isolates were particularly important, because the geologists had developed a wide assortment of geothermometers and thermochronometers sensitive to the temperature range of 60–125°C.[14]

The thermochronometers and geothermometers had indicated that the Mancos Shale attained a T_{max} of about 75°C and had probably stayed at that temperature for 60 million years. Given the uncertainties in the geothermometers, this estimate was indistinguishable from the 80°C that Steve Hinton had told me was the temperature limit of microbial activity detected in oil reservoirs. Only hyperthermophiles could have survived those temperatures within the shale since deposition in the Cretaceous, but had there been enough resources to sustain them for 60 millions years? The mesophilic bacteria the microbiologists had isolated, therefore, must have been immigrants transported by the groundwater during the uplift and erosion of the San Juan Basin beginning 15 million years ago that resulted in the formation of the Colorado Plateau and Colorado River.

At the Parachute Origins Site, we had succeeded in obtaining large, high-quality, low-contamination cores and water and gas samples from a depth of 6,000 feet with ambient temperatures higher than 80°C and ambient pressures of 260 atmospheres (figure 3.2A).[15] We had definitely taken a significant step beyond Taylorsville Basin. As at Taylorsville Basin, thermophilic, ferric-iron-reducing bacteria were enriched and isolated, one by Rick's group. It was from the sandstone body in the Wasatch Group at a depth of 2,580 feet. Amazingly, the

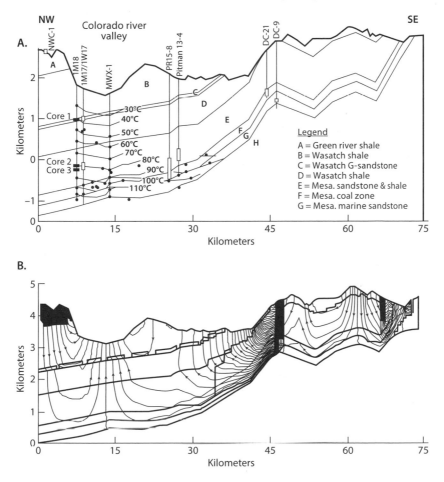

FIGURE 3.2. A. Cross-section of Piceance Basin from northwest to southeast, including the 1M18 drilling site. The depth and temperature of the three Parachute drill cores are shown (from Colwell et al. 2003). B. Same cross-section as in A but showing the direction of groundwater flow based on the porosity and permeability of the sedimentary units of the basin (from Colwell et al. 2003). In this case the bacteria isolated from Core 1 in A would have originated from the Roan Plateau overlooking the Parachute drill site. The particle travel time from the Roan Plateau to core 1 in the Wasatch G sandstone was 0.5 million years, whereas the particular travel time to cores 2 and 3 was 1.2 million years.

16S rRNA gene sequence of this bacterium was similar to that of *Desulfotomaculum*, an SRB! This finding was initially surprising, but the paper by Max Coleman on concretions had isolated an SRB that would switch from sulfate reduction at high H_2 concentrations to ferric iron reduction at low H_2 concentrations,[16] and this Parachute isolate had been grown without additional H_2. Intriguingly, John Lorenz identified centimeter-scale $FeCO_3$, or siderite, concretions in the same core.[17] Siderite is the mineral that ferric-iron-reducing bacteria produce in the presence of a bicarbonate-rich media. Were these siderite concretions produced by Rick's isolate? Tommy had also isolated from the same core a ferric-iron-reducing thermophile that produced massive quantities of magnetite.[18] Or as Tommy put it, "It sh@#!ed magnetite at an industrial scale." But in the presence of CO_2 it also produced prodigious siderite at 60°C! In outcrops of the same sandstone formation, we later identified cannon-ball carbonate concretions up to five feet in diameter that would have required millions of years to form.[19] It boggled my mind that given a few million years, spherical structures the size of a human could be produced by microorganisms that were one quintillion-trillionth the size of a human.

But the biomass concentrations of the second and third sandstone cores taken from the Mesa Verde Formation at 6,088 and 6,285 feet depth, respectively, were very close to the limits of detection, and they yielded just one microbial enrichment. Unlike Cerro Negro, the deeper cores at Parachute were definitely sampling sterilized sandstone layers in spite of their large pore size (up to 1 millimeter) and porosity (5–7%).

To unravel the mystery surrounding the presence of bacteria in the shallower Parachute cores and their absence in the deepest cores, Qing-jun Yao, a postdoctoral associate of mine, determined the thermal history of the rock strata using fluid inclusions and vitrinite reflectance data. Both sets of data indicated that core 1 had been heated to a temperature of 120°C and that cores 2 and 3 had been heated to 145–150°C, and the geothermal gradient had been a sweltering 50°C per kilometer of depth, double the typical continental geothermal gradient. The CH_4 had formed during maximum burial of the formations, between 35 and 5 million years ago. Qing-jun's data clearly indicated that the sedimentary deposits we had cored at Parachute had been rendered lifeless by the extreme heat. Fission track data on apatite crystals also indicated that the formations had only started cooling below these temperatures in the last 5 million years, when the Colorado River had begun cutting its way down through the Piceance Basin strata.[20] The fluid inclusions contained freshwater, so Qing-jun developed a fluid flow model for groundwater motion through the fractures in the Piceance Basin to understand why indigenous bacteria were present in core 1 but not in cores 2 and 3 (figure 3.2B). Water collected from the same Wasatch sandstone a mile away yielded an age of 1.6 million years since it had been at the surface.[21] According to Qing-jun's model, the Wasatch sandstone had cooled to 75°C at that time. The maximum growth temperature of the ferric-iron-reducing thermophile that Tommy had isolated was 75°C. Was this just a coincidence, or had the ancestor to Tommy's bacteria migrated more than a mile into the Earth during the Pliocene Epoch?

Qing-jun's model showed that the high topographic relief generated by the erosion of the Colorado River drew fresh surface water from the Roan Plateau deep into the tight gas reservoirs (figure 3.2B).[22] In spite of the high topographic relief the low permeability of the formations surrounding the gas-rich sandstones meant that the water seepage velocity was only two-tenths of an inch per year. At that rate, Tommy's bacterial ancestor would have had to travel for at least a half million years to cover the two mile flow path from the surface to the Wasatch sandstone where it or its ancestor was retrieved 1.6 million years later by us![23]

While we were frantically finishing the research at Cerro Negro and Parachute in preparation for a second stocktaking meeting in early July, Hsin-Yi had performed an exhaustive analysis of the thermal and fluid flow histories of Taylorsville Basin, including modeling fission track dates.[24] She discovered that the microbially sampled strata had cooled to the present-day temperatures about 140 million years ago.[25] The sediments deposited in the Triassic lake that eventually became Taylorsville Basin undoubtedly contained microorganisms similar to those discovered in the cores. But with the continued burial of sediments during the late Triassic, the temperatures eventually increased to the point that the entire rock formation was sterilized. With the aid of a numerical fluid-flow model, she showed that when the Taylorsville Basin was uplifted in the Jurassic during a compressional tectonic episode, meteoric groundwater would have penetrated the formerly sterilized zones of rock at a velocity of one-half inch per year, accounting for the water-washed composition of the oil in the fluid inclusions. The observations from both Parachute and Thorn Hill no. 1 pointed to a tectogenetic origin for the deep subsurface bacteria; that is, they had occupied their current niche during last major tectonic event and had been isolated since then.[26]

If our tectogenetic theory was correct, then the ancestors of *B. infernus* should have infiltrated the Thorn Hill no. 1 lake sediments during the Jurassic Period. The evidence from the Parachute drill site indicated that it was plausible. If so, the journey down the 25-mile-long flow path would have taken a minimum of 3 million years. When we analyzed the pore structure in the sandstone cores from Thorn Hill no. 1, however, we discovered that the maximum pore diameter was only 0.03 micrometers! This was far too small for even the tiniest bacteria or spore to squeeze through. But when did this pore structure form? Fortunately, the pores were filled with clay minerals, including the potassium-rich illite clay. We had just developed a technique for determining the age of microgram quantities of illite using our laser $^{40}Ar/^{39}Ar$ microprobe.[27] The analyses indicated that the illite clay crystallized about 80 millions years ago, implying that clay mineralization sealed the rock 60 million years after the strata had become cool enough to be microbially inhabited. It also indicated that the bacteria had been entombed within their mineral cage since the Cretaceous! The results of DNA analyses of *B. infernus*, which were completed the same week as the illite dating, revealed that the bacteria appeared to be a new species, one never before identified on the surface of the Earth!

By performing detailed electron microscopy and stable isotope analyses, we were able to determine how much calcite and pyrite had resulted from eons of bacterial respiration in the Thorn Hill no. 1 cores. Using the equation Tommy had given me, we calculated that the average doubling time of the observed bacterial population was on the order of thousands to millions of years! This growth rate was so slow, 10^{15} times slower than that of bacteria present in sewage sludge, it suggested that the bacteria were surviving but not reproducing. If microorganisms can evolve only by reproduction of their DNA, then these inferences suggested that the bacteria represented species that were present in the basin at least 80 million years ago and had not changed significantly since that time. The bacteria were for all intents and purposes living microfossils, much like the monsters that inhabited Jules Verne's Lidenbrock Sea, except they were much, much tinier!

LIFE IN THE SLOW LANE FOR THE MICROBES LIVING DEEP BENEATH THE NEVADA TEST SITE

Using geological constraints, isotopic dating of groundwater, and hydrological modeling, we seemed to have determined that microorganisms had existed in some deep subsurface environments for tens of millions of years! How did they manage to survive for so long? In the GeMHEx core collected at the Hanford site, soil bacteria had been trapped in the reddish paleosol and subsequently buried beneath lake deposits 6 to 8 million years ago.[28] Sedimentary rocks frequently contain organic carbon that was produced by photosynthetic organisms on the surface of soil or lakes and subsequently buried during deposition at low temperatures. As long as this organic source remained available, heterotrophic microorganisms within the pores of the sediments could continue to survive. These subsurface microorganisms obtained energy by oxidizing this ancient organic matter and reducing chemical species, such as sulfate or ferric iron, after the O_2 in the water had long since been depleted. Ferric iron was abundant in the oxidized Fe-rich volcanic detritus in the Ringold Formation and the reddish paleosols. As the sediments became more deeply buried over geological time, as at Cerro Negro, Piceance Basin, and Taylorsville Basin, they became compacted and cemented, decreasing the permeability and porosity and thus reducing the supply of nutrients. As a consequence, the overall microbial metabolic rates in the sedimentary rock gradually declined, except in microhabitats directly surrounding nutrient sources. The distribution of microorganisms should thus have become "patchy" in environments where the supply of water and nutrients was low. Microcolonies or individual cells collected in these "microbial oases," which were separated by large volumes of sediment lacking viable microorganisms. In the Hanford Site sediments, including the GeMHEx cores, microbial activity was almost always detected in 100 grams of sediment, frequently detected in 10 grams of sediment, and rarely detected in 0.1–1 gram of sediment.[29]

The biological characteristics of the subsurface microbial isolates appeared to indicate that they were indeed starving today. As do higher organisms, bacteria rely on metabolic reserves during periods of long-term starvation. They can metabolize polyhydroxybutyrate, an energy-storage compound that they manufacture and store when times are good. Eventually the bacteria diminish in size as they deplete their reserves, shrinking from a healthy size of a few micrometers down to as small as 0.2 micrometers in diameter, or 100 to 1,000 times smaller in mass. These tiny, starved bacteria are called ultramicro-, or dwarf, bacteria. They predominate in nutrient-deprived environments and have been observed frequently in the deep subsurface as well as in the oceans. Subsurface bacteria appear to be exceptionally talented at maintaining the viability of their shrunken mass at little or no metabolic cost.[30]

The percentage of cells that were cultured on laboratory media also generally decreased with depth or as the flux of water and nutrients in the subsurface decreased. Total microbial population concentrations were enumerated by microscopic counts (figure 3.3) or by extracting and measuring the concentration of PLFA in the cores. More than 10% of the total count of microbes in the sandy sediments beneath the Savannah River Plant, where the flux of water and nutrients was relatively high, could be grown in the lab and thus deemed "viable."[31] The exceptions to this rule were the sands above the water table and the clay-rich aquitards[32] that were surprisingly undersaturated with respect to water.[33]

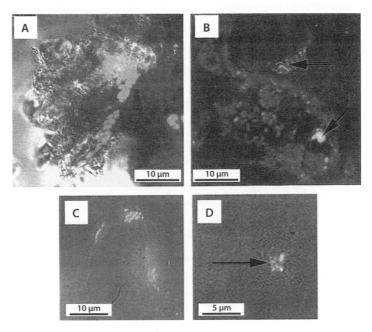

FIGURE 3.3. A, B. Laser confocal and C, D. fluorescent microscopic images of propidium iodide–stained cells on the fractured surfaces of Snake River Plains basalt from 118 meters below the surface. Images in A are of diatoms, and B–D of rod-shaped bacteria distributed at clusters (from Tobin et al. 2000).

Less than 0.1% of the total cells could be cultured from unsaturated sediments in the 600-foot-deep Ringold Formation at the Hanford Site, where the flux of water was low to nonexistent. Rick Colwell had observed the same discrepancy in the sedimentary interbeds of the Snake River Plain basalts.[34] In many instances the results from the various culture-based, microbial assays from different laboratories did not agree with each other, as if there were small spots in the rock from which microbes could be grown and other spots where nothing could be grown even though a fairly large number of cells could be quantified under the microscope using fluorescent DNA stains, such as acridine orange.[35] It seemed more likely that inability to culture the subsurface microorganisms was related to the paucity of nutrients in their environments and that they had suffered too much damage over time to be able to repair and replicate even when fed nutrients in the lab. These dead or dying cells may decompose very slowly in the subsurface, but may still be detected by acridine orange direct counts, PLFA, or propidium iodide.[36] Alternatively, the organisms could be alive but might have lost the ability to replicate due to DNA damage from stresses under extended, severe starvation.

The same paucity of viable microorganisms was also discovered at a third field site in the west located at the NTS.[37] DOE was using Alpine miners[38] to tunnel into the Miocene Epoch–age rhyolitic welded ash flow and ashfall tuffs and ignimbrites of Rainier Mesa located on the northernmost sector of the NTS, which was the site of the first underground nuclear test in 1957. The four tunnels constituting the U-tunnel complex penetrated the mesa to depths ranging from 150 to 1350 feet and totaled 15 miles in length, providing opportunities to study the subsurface microbial spatial distribution. Frank recognized that obtaining fresh uncontaminated tunnel samples, however, with an Alpine miner was challenging. Nonetheless, a young associate professor by the name of Penny Amy, who was located close by in the Department of Biological Sciences at the University of Nevada–Las Vegas, took up that challenge.

As deep as these tunnels were, the water table was still hundreds of feet deeper, so most of the samples represented the deepest vadose zone ever sampled for microbiology.[39] Some water was present that had descended quite recently from the surface along fault zones cutting the rhyolitic tuffs. This water still contained cosmogenic ^{3}H and was, therefore, less than 50 years old.[40] But the migration of water from these fault zones through the matrix of the tuff was much slower, requiring perhaps 250,000 years.[41] The migration of bacteria would be still slower because of the very low water content, or matric potential,[42] where the rock porosity is filled mostly with air and the water confined to a film that is thinner than the diameter of a bacterium. So the colonization of the rhyolytic tuffs could date back to 12 to 14 million years ago, when the tuff had been water saturated and subsequent alteration of the tuff's minerals to zeolite consumed the pore water.[43] As was the case for the Hanford Site sediments, Penny found that the viable counts ranged from less than 1 to 10^5 viable, or cultivatable, cells per gram of tuff, far less than the total cell counts, which were 10^{6-7} cells per gram.[44] No obvious pattern was observed in the distribution of the viable cells. The researchers also observed no correlation with the chemistry other than the fact that viable nitrate-reducing bacteria were found in tuff that contained nitrate,[45] suggesting that these microorganisms were surviving in the tuff.[46]

THE DEATH-O-METER, SLIME, AND THE END OF THE DOE SSP: JULY 1995, PORTLAND, OREGON

Microbiologists argued about whether to attribute this discrepancy between total counts and viable counts to not knowing how to grow all of the different species of bacteria, and about calling them viable but not cultivatable. Tommy, however, wondered whether the non-cultivatable bacteria were analogous to "rusting (automobile) hulks in the desert"—their exteriors were still well preserved but their engines no longer functioned. In addition, the total cell counts provided by Tom Kieft were sometimes greater than the fluorescent in situ hybridization (FISH) counts performed by Nierzwicki-Bauer's group, suggesting that a much smaller fraction of the total population was active. Similarly the estimates of the number of cells from the PLFA concentration were also less than the total cell counts. From these observations Tommy constructed his "Death-o-Meter," a blue dial with a skull and crossbones going from procreation on the right to mineralization on the left (figure 3.4). He described how the cultivatable microorganisms they were finding in their cores were like him when he was younger. As bacteria begin to starve, they do not grow or reproduce, similar to what happens when we all become middle-aged. When a bacterium begins to shrink under chronic starvation, it consumes itself, similar to what happens to us when we enter our senior years. A dead bacterium will still have chemically reactive cellular remains, like the rusting hulks of junked cars. Tommy's description was a bit more colorful stating that "they are just like a dead fisherman who still smells for many, many years afterwards." On his Death-o-Meter he could plot each of the subsurface sites sampled during the previous ten years.

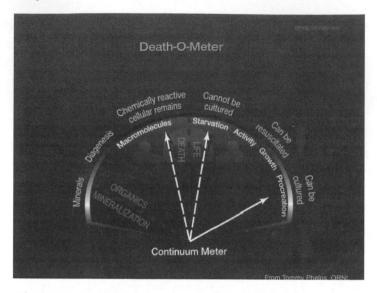

FIGURE 3.4. The Death-O-Meter (courtesy of T. J. Phelps, summer 1995, Portland, Oregon).

In order to understand the Death-o-Meter, you have to understand that temperature is not the only limitation imposed upon microorganisms in the deep subsurface. The capacity of the geochemical environment to support life, the flux of water, and the availability of pore space are also important. The subsurface environment must provide carbon, nitrogen, phosphorous, and trace metals, so that microorganisms can synthesize cellular constituents, including DNA and proteins. Most important, some form of fuel must also be present so that microorganisms can produce ATP, the energy currency required for biosynthesis (figure 3.5). When there is a dearth of energy or ATP, which is what we assumed was the normal state for the subsurface, the microbes would be on the left side of the Death-o-Meter. What controls the boundary between life and death for the microbes is the balance between the energy flux to the microbe and the energy it requires to maintain its metabolic and anabolic processes. An autotroph requires more energy per unit time to maintain itself than does a heterotroph, and a thermophile requires more energy per unit time to maintain itself than does a mesophile. Subsur-

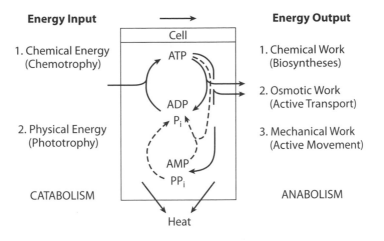

FIGURE 3.5. Bioenergetics and the relationship between catabolism, the synthesis of adenosine 5′-triphosphate (ATP) and anabolism (from fig. 1 of Thauer et al. 1977).

face microorganisms were adapted to survive in an energy-poor environment.

All taken together, the data from Taylorsville Basin, Parachute, and Cerro Negro suggested that in tectonically active environments like Parachute and Cerro Negro, the bacterial migration rate and microbial activity were high compared to more quiescent geological terranes,[47] such as Taylorsville Basin. There the subsurface microbial migration had long since ceased and the bacteria had presumably become relatively dormant. Could it be that the age of deep subsurface microbial communities corresponded to the last significant tectonic event in which high topographic gradients and syntectonic fractures and faults produced the greatest groundwater flow?[48] It was Tommy who speculated that the flow would transport bacteria into geologically sterilized rock, where they could rapidly grow on nutrients without having to compete with already-established microbial inhabitants. This was Tommy's biker-bar/neutron-bomb hypothesis, as he described it, at the July 1995 meeting in Portland.[49]

Imagine you're a stranger in town and you go into a biker bar for a beer. Your chances of quenching your thirst depends upon how successful you are in muscling your way past the big bikers to the bar before getting tossed out on your head. That's how difficult it is for a colonizing bacterium from the surface to successfully invade a subsurface microbial community. Compare this to the situation where you enter a town that has just been hit by a neutron bomb. All the people are dead, including the bikers, but the steaks are still sizzling on the grill and the beer kegs are still flowing. Subsurface environments that have been geologically heated to the point of sterilization, like Parachute and Cerro Negro, have been neutron bombed. When they are cool enough for microbial habitation, the first pioneering microbes get to eat all the steaks, drink all the beer and then they become the new bikers.[50]

Toward the end of the same meeting, Todd Stevens unveiled the discovery that Jim McKinley and he had been toiling on for the past two years, a new source of H_2 that could provide energy to subsurface chemolithoautotrophic microbial ecosystems,[51] which they called SLiME. At the end of his talk, Todd responded to a question from Frank, who, as usual, was sitting in the back of the conference room, "I wanted to tell you how I was spending your money."

Todd and Jim's discovery, like many scientific discoveries in an unexplored field, came about by a sequence of serendipitous circumstances beginning with the GeMHEx project. In November 1990, after the drilling had been completed at NPR-WO1 at INEL, the drilling at the Yakima Barricade Borehole at the Hanford site had begun. This site was located on the west side of the Hanford Nuclear Reservation, where a relatively thick sequence (about 675 feet) of lake sediments and a reddish-colored 6-million-year-old paleosol of the Ringold Formation overlie the Columbia River Basalt.[52] The proposed drill site in the Hanford Nuclear Reservation was hydrologically up gradient from the contaminated aquifers underlying the radioactive waste tanks that contained the waste that had been generated by plutonium processing. Given the 300-foot-deep water table and the high porosity and softness of the sediment,[53] Phil Long had selected an old-fashioned cable percussion tool for coring to avoid the drilling mud contamination that had been the biggest source of microbial contamination in the cores from SRP and C10. Fluorescent micro-spheres and bromide solution were used as tracers, instead of PFCs. The hole was slowly advancing beneath the water table in early 1991 when an explosion occurred at the drill site. During one of the pro-cedures, the drillers had taken a lunch break and covered the hole with a thick steel plate while they left the well idle. When they re-turned and restarted the percussion hammer, the borehole exploded, shooting the thick piece of steel on top of the well into the air. No injuries occurred but the GeMHEx drilling was halted as an investi-gation on the origin of the explosion was undertaken. The initial in-vestigation carried out by Jim McKinley had pointed to the anaero-bic groundwater interacting with the steel casing creating H_2 gas that built up in the well while it was covered with the steel plate. The spark generated by restarting the percussion tool had created the ex-plosion. But even this setback was enlightening. These results sug-gested to the PNL group that the subsurface microorganisms might be consuming the H_2 produced during drilling and installation of the casing.[54] Since all deep commercial drilling, whether for water or for oil, uses steel, it became apparent to Frank and the SSP group that this very human invasion of the subsurface world was feeding it H_2. According to Tommy, microbes eat H_2 like humans eat junk food.[55]

Todd, however, wondered whether a similar type of reaction in-volving the Fe-bearing minerals in basalt would also produce H_2. If so, could this source of H_2 sustain anaerobic communities in the con-fined Columbia River basaltic aquifers deep beneath the Hanford Reservation? McKinley had worked on the geochemistry of the groundwater that came from wells intersecting that aquifer as part of his Ph.D. thesis. They began sampling the wells. The deepest well they sampled, DC-6, was located just outside the remains of the pre-war town of Hanford.[56]

Todd developed a sterilized sediment trap that McKinley and he then attached to the wells to collect the planktonic and, hopefully, sessile microorganisms.[57] From his "bug" traps, Todd was able to cul-ture H_2-utilizing methanogens and acetogens that fixed dissolved in-organic carbon (DIC) to make their biomass and did not use dis-solved organic carbon (DOC).[58] McKinley subsequently discovered that the $\delta^{13}C$ of the DIC was very positive and was inversely correlated with the DIC concentration. This looked very much like the stable isotopic signature of preferential $^{12}CO_2$ uptake and utilization by autotrophs like methanogens and acetogens. Finally, the groundwater contained very high concentrations of H_2, much higher than had been recorded in the South Carolina coastal plain sediments. Todd performed sterilized microcosm experiments with basalt chips from rock cores, some of which were from Cerro Negro, and showed that the H_2 gas was produced by the reaction of O_2-poor water with the Fe-bearing minerals in the basalt at a rate sufficient to support the methanogens and acetogens. They hypothesized that the CH_4 and acetate supported an entire subsurface ecosystem without any supply of DOC from the surface.

At the end of the Portland meeting, David Boone, Tommy, Tom, Jim, and I were having breakfast, scratching our heads and wringing our hands over what we could do now. During the past ten years, the SSP had collected almost three hundred subsurface samples for microbial analyses using tracer techniques that its researchers had developed, isolated thousands of microbial species, and characterized hundreds of isolates. Most of the samples were from sediments, but basalt and rhyolitic ash had also been sampled. The ages of the rocks sampled ranged from a hundred thousand years to more than 200 million years, and the groundwater sampled ranged in age from less than a thousand years to over a million years. We had determined the rates of cellular turnover for different environments that ranged from hundreds to millions of years and identified new sources of energy to sustain those rates. We had answered Bastin's question from 1926. Finally, we could estimate the total microbial biomass present in the terrestrial subsurface down to 1,500 feet. That estimate proved to be almost as large as the total biomass of life on the surface of the Earth.[59] To present this synopsis of the achievements of the SSP to the public, a small group of us were just finishing an article for *Scientific American*.[60]

But below 3,000 feet, we had been able to collect only a few high-quality core samples, and of those, only the samples from Thorn Hill no. 1 had yielded what we considered to be indigenous microorganisms. At the Parachute site we had an environment at 6,000 feet depth that seemed perfectly capable of sustaining hyperthermophilic life yet none had been detected. Was it because it had only become habitable about a million years ago and groundwater had yet to penetrate that depth to inoculate it? Or did life not exist at depths at which the temperatures exceeded 80°C? Any conscientious discussion of the deep subsurface biosphere seemed almost fruitless given the few pinpricks that we had made so far below 3,000 feet. Over coffee and pancakes we debated whether it was possible that life was originating again within those sandstones and, given enough time isolated from the surface DNA world, whether subsurface life would emerge? Tommy Gold speculated that the deep crust could have been where life originated on our planet in the Hadean Eon.[61] To answer this question we needed to collect cores from deep subsurface sites that had been thermally sterilized in the past but had cooled down and remained isolated from surface groundwater for tens of millions of years.

But how could we get really deep subsurface samples without the substantial DOE support and the connections that made it possible? Although the International Continental Drilling Program had incorporated Frank's deep microbiology input into their new plan in 1992, obtaining support for collecting a single core could take a decade, and we would have to get in line behind the geologists. I offered that one possible place where we could obtain samples was in the ultradeep gold mines of South Africa. I explained to them that for the past year I had been in contact with a former graduate student of mine who was working for Anglo Gold/DeBeers in South Africa. He might be able to help us obtain some samples.

CHAPTER 4

MICROBES IN METEORITES! WHERE DID THEY COME FROM? HOW DID THEY GET THERE? WHAT DID THEY WANT?

> "I'm not mistaken," I said. "We've gone past Cape Portland, and those 125 miles to the southeast put us in the open sea."
>
> "*Under* the open sea," my uncle countered, rubbing his hands.
>
> "Which means," I exclaimed, "that the ocean's spreading overhead."
>
> This state of affairs may have been perfectly acceptable for the professor, but the thought of my strolling around under a huge mass of water didn't fail to concern me. And yet it generally made little difference whether the plains and mountains of Iceland were hanging over our heads or the waves of the Atlantic; all that counted was that the framework of the granite held firm.
>
> —*Jules Verne, Journey to the Center of the Earth*[1]

SLIME ON MARS AND FROM MARS:
FALL 1995 TO SUMMER 1996

Stevens and McKinley's SLiME paper was published in *Science* in the fall of 1995. The acronym immediately grabbed the attention of NASA. Mix water with basalt and you have energy for life. This could occur in the subsurface of any planet. It seemed fantastical, but Chris McKay, an exobiologist at NASA's Ames Research Center, commented that the finding supported their earlier model of potential microbial ecosystem beneath the surface of Mars.[2] NASA's Mars program began to shift its exploration strategy. Instead of looking just for fossilized remains of ancient surface life on Mars, perhaps they should be looking for signs of an extant Martian biosphere present beneath its surface. NASA began organizing workshops on how to explore for these living subsurface Martians. Stevens and McKinley's paper also had significant implications for life on Earth. Like the Columbia River basalts, the oceanic crust, which covers 70% of the Earth's crust, is composed of basaltic lava and dikes. Could Todd and Jim's SLiMEs form the base of an oceanic subsurface biosphere? At this point the only inferences that had been drawn about an oceanic crust biosphere were those gleaned from the black smoker vents at the ridge.

Stimulated by the SLiME *Science* paper, articles began to appear in the popular press almost every other month throughout 1996. Our "Hell's Cells" article, in *Scientific American*, appeared soon after Todd and Jim's *Science* paper. The "Biosphere Below" was published in *Earth* magazine and featured the Parachute drilling campaign. Even the late, great Stephen J. Gould wrote an article in *Natural History*, "Microcosmos."[3] Unfortunately, the management at DOE was not impressed, and despite its many achievements, the SSP was officially terminated and our access to the really deep subsurface with its thermophiles was effectively cut off.

Or was it? I had been doing geological fieldwork in South Africa before 1987, prior to swearing I would never return to that country again. It was simply too hard to overlook the apartheid system that would herd most of its citizens into shanty towns hidden from sight by earthen berms. But the government had just changed in 1994 with the election of Nelson Mandela as president. I had been impressed by how easy it was to access high-quality microbial samples at the underground WIPP site, and I recalled the ease with which I had been driven into one of the platinum mines in the Transvaal Province, north of Johannesburg, many years earlier. One could directly sample the layered mafic rock at the same depth that Todd and Jim had sampled the water from basalts in their Hanford wells.[4] But it was the deep gold mines that interested me the most, because heat-flow studies indicated that at their deepest depths, the rock temperatures were about 50–70°C, the range at which thermophiles grow. These mines were located within the 2.9-billion-year-old Witwatersrand Basin. The richest gold ore in the world was being extracted from these coarse-grained alluvial sandstones that contained rounded detrital pyrite, but also framboidal pyrite! Perhaps the latter was produced by living SRBs. The deepest mines with the highest temperatures occurred near Johannesburg. The few papers available on the gold-bearing layers in these mines described it as thin organic-rich seams, almost like coal. They were also uranium rich, which seemed ideal for testing my radiolysis hypothesis. Intriguingly, some papers suggested that the radiation was responsible for abiogenic formation and alteration of the organic carbon.[5] David Phillips, my former grad student, had told me that he had dated the micas in these deposits at his $^{40}Ar/^{39}Ar$ lab, and they typically yielded ages of two billion years. This recorded the approximately 300°C hydrothermal metamorphism associated with the formation of the Vredefort Impact Crater, a 300-kilometer-diameter crater that sits in the middle of the Witwatersrand Basin. So at least two billion years had passed since the formation was sterilized. Given our results from the Parachute project, this was very likely long enough for a subsurface microbial ecosystem to be established, or perhaps even a second genesis of life. Other than one Hollywood movie depicting a historical flood of one of the deep mines,[6] I could find no information describing the water down there. I was completely clueless as to whether the mines could yield samples suitable for microbial analyses.

Unbeknown to me at the time, what I was proposing to do was nothing new by any means. The first exploration for subsurface life occurred in mines. In 1793, Alexander von Humboldt, a recent graduate of the Freiberg School of Mines, studied the various species of fungi and algae that he came across in the gold mines of the Fichtel Mountains during his stint as the inspector of the Department of Mines for the Prussian government.[7] Much later, 1910–11, European microbiologists started growing bacteria from coal, trying to determine whether the CH_4 gas commonly encountered with disastrous consequences in the coal mines was, in fact, created by bacteria, since bacteria that produced CH_4 had already been isolated.[8] Two German microbiologists reported the presence of gram-positive, spore-forming bacteria in coal mined from a depth of 3,600 feet (1.1 kilometers).[9] In two papers, one from 1928 and one from 1931, C. B. Lipman, a soil microbiologist from the University of California, Berkeley, also reported growing bacteria from anthracite coal collected at 1,800 feet beneath the surface in a Pennsylvania coal mine.[10] He paid meticulous attention to eliminating laboratory sources of contamination. He performed numerous heating and inoculation experiments during which he discovered that he was able to resuscitate bacteria from coal chunks heated for two days at 160–170°C, but not from coal chunks heated for six days at 160–170°C. Because of this remarkable thermal tolerance and because he believed in the remarkable persistence of life in a spore state,[11] he proposed that the bacteria were originally deposited with the peat layer 250 million years ago and that a few spores survived the coalification process until the removal of their habitat by the mine and its arrival in Lipman's lab. Michael Farrell, a microbiologist from Lehigh University in Bethlehem, Pennsylvania, quickly disputed Lipman's findings. He went back to the same mine, but to the shallower level of 328 feet, and performed repeated sampling of fractured and unfractured coal seams. He found that the fractured coal contained bacteria similar to those found in great abundance in the mine water, suggesting that the bacteria found by Lipman either were brought in with the mining or had penetrated the coal from groundwater migration from the surface. Farrell's findings were consistent with the conclusions of an earlier German study.[12] Lipman moved on from coal to oil,[13] and then eventually to meteorites, in which he also found bacteria, which he claimed in 1935 were of extraterrestrial origin. The dismissal of coal bacteria as being mining contaminants closed the chapter on the search for ancient bacteria in mines.

Investigations of the subsurface biosphere in mines were not attempted again until the late 1950s, at the Higashiyama oil mine in Japan. The Higashiyama mine had been established during World War II but was about to shut down. Drs. Izuka and Komagata, microbiologists from the University of Tokyo, saw an opportunity to investigate subsurface microbiology and quickly collected samples of oil-saturated Tertiary Period sand, water-saturated sand, brine seeps, and fresh oil from drilled boreholes along the 7.5 miles of underground galleries, which ranged in depth from 600 to 1,000 feet. From both aerobic and anaerobic cultures, they discovered wide diversity in cultivatable hydrocarbon-utilizing bacteria and fungi.[14] They proposed that the microbiota were indigenous to the oil sands but could not definitively prove it.

I was blissfully ignorant, however, of all the controversies surrounding mining samples for microbial analyses. After a year of email pleading from his former thesis advisor (who had clearly gone a bit batty over bacteria), David convinced the Anglo Gold management to help me collect samples from Western Deep Levels.[15] I had to be in South Africa in early September 1996 to collect microbial samples, which Jim Fredrickson had agreed to analyze. The timing was ideal, since I would be attending the third ISSM meeting in Davos, Switzerland, where we would be holding a final celebration in honor of Frank. I could fly down to South Africa from Zurich afterward. I gathered random sampling supplies, not knowing exactly what I would need, and constructed an air-tight aluminum canister for bringing back rock samples anaerobically.

Just before I was to get on the plane in August, David McKay of NASA's Johnson Space Center stunned the world with his announcement that Martian meteorite ALH84001 contained the fossilized remains of ancient Martian microorganisms (figure 4.1).[16] The possible indications of ancient life in the near subsurface of Mars four billion years ago would almost guarantee the presence of life in the deep subsurface of Mars today.[17] Furthermore, it meant that some rock samples collected on the surface of Mars and returned to Earth might also contain fossils of ancient microorganisms.[18]

SWEDISH SLIME BENEATH THE BALTIC SEA: SEPTEMBER 1996, ISSM '96, DAVOS, SWITZERLAND

Unlike the previous two ISSM meetings, ISSM '96 in Davos focused less on the techniques used to obtain subsurface microbial samples

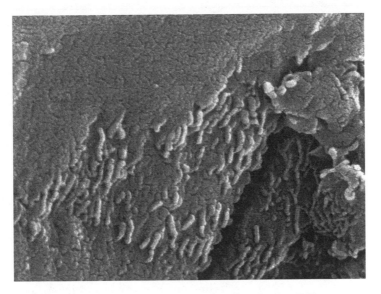

FIGURE 4.1. SEM image of fracture mineral surface in the ALH84001 meteorite (from McKay et al. 1996).

and much more on what was being learned from the analyses of the last three years. But the coffee-break conversations were all about ALH84001 and the prospect of life on or beneath Mars. This prompted D. C. White to stand up in front of the audience of a couple of hundred microbiologists and declare that the Vatican should be supporting the first trip to Mars to search for subsurface life, "To determine whether the finger of God had touched Mars before the Earth." Many other SSPers were at the conference in Davos presenting their research. I presented Hsin-Yi's latest results from the Taylorsville Basin, and Rick presented our results from Parachute, but the talk that stood out in my mind the most was that of Karsten Pedersen's comparison of 16S rRNA data with the geochemistry of the different fracture fluids from his underground laboratory located near Äspö, Sweden. As I was drooling over his data set, he mentioned during his conclusions that the H_2 in his "geogas" was derived from radiolysis and that this was supporting the autotrophs. As he pointed out to the audience, he didn't have a Rocky Mountain orogeny like Parachute to drive his microbial system and keep it from running out of gas for billions of years, he needed another source. The best candidate for that other source in his opinion was radiolysis.[19]

I was a bit crestfallen that Karsten was way ahead of me on my idea about radiolysis, but not too surprised given his background. The continental crust covering 30% of our planet is largely composed of the ancient granitic igneous rocks containing little or no organic carbon. This was the environment that Karsten had been studying. Could autotrophic microorganisms like those found in the Columbia River Basalt also dominate in his felsic igneous rock environment?[20] This question was the focus of research by Karsten's group at the University of Göteborg, Sweden, which had first detected microorganisms in water flowing through fractured granite at the Stripa iron mine. The reason such a question was of any concern in a country like Sweden is that a large fraction of Sweden's electricity is generated by nuclear power, and the waste from the nuclear industry would have to be stored in deep underground vaults excavated in Precambrian granite. The activity of the microorganisms would conceivably impact the long-term safety of these waste depositories. Fred Karlsson, the project manager of SKB AB (Swedish Nuclear Fuel and Waste Management Company) knew this and needed a microbiologist who was interested in studying the topic. Enter Karsten Pedersen, who had recently received his Ph.D. in microbiology from the University of Göteborg and was hired immediately on the faculty of its Department of General and Marine Microbiology. His research began just as the International Stripa Project was winding down, but Fred played the role, much like Frank Wobber in the United States, of maintaining sufficient long-term funding for Karsten to get him started. Until 1976, Stripa had been a functioning iron mine since its beginnings in 1448 and comprised an underground network of tunnels extending to depths of 1,480 feet and boreholes that tapped water at depths of 2,700 feet. From 1976 until 1992 the Stripa mine had been the site of a multidisciplinary, multinational study of groundwater flow through deep, fractured rock in order to better understand the possibility of the escape of radioactive waste from depositories located in such environments. In 1986 SKB on its own undertook to construct a pilot depository on the southeastern Baltic coast island of Äspö, located 140 miles south-southwest of Stockholm. The site provided access to a major shear zone, as well as to unsheared 1.7-billion-year-old granite, so both ideal and nonideal rock types could be compared. Between 1990 and 1994 the SKB started with a vertical elevator shaft that went down 1,480 feet (450 meters), which they then followed with a drivable, helical tunnel down to the same depth (figure 4.2). The entrance to the tunnel was on the mainland and passed 500 feet beneath an estuary to the Baltic Sea. The fantastic thing about the helical tunnel was that it provided access to a much larger three-dimensional volume of the rock for study than did the single elevator shaft.

In the early 1990s at Äspö and Stripa, Karsten's team was finding a lower diversity of microorganisms in their fracture water than that being found in aquifers beneath the Savannah River Plant. These microorganisms also seemed to be utilizing simpler organic substrates, for example, lactate, and also CO_2; that is, they were mixotrophic. Only at Äspö were the researchers able to isolate anaerobic SRBs and methanogens. These bacteria must have been carried from the surface down into the fractures by groundwater flow some time after the granites had cooled 1.8 billion years ago. It was Karsten's reports that captured the attention of Tommy Gold, who was quite busy in Sweden at that time with the Siljan Ring drill site, and germinated his idea of a deep, hot biosphere.

A typical commute to work for Karsten involved driving a couple of hours from Gothenburg on the west coast, through the forest-covered countryside heading east through Jönköping and down to the scenic port town of Oskarshamn. Then he turned left and drove about twenty kilometers north along E22, past more fir-tree forests and grassy pastures, and after a few twists and turns, he was on a side road heading north for the estuary, where he suddenly descended into a tunnel as if he were driving into the Bat Cave. He would pass beneath the Baltic Sea before reaching the island and winding his way down 450 meters. He could see to his left the fracture zone where Baltic Sea water was seeping into the tunnel. Further down and to his right he could see the fractures where the meteoric water of the island was seeping into the tunnel and the copious Fe-rust biofilms.[21] De-

A.

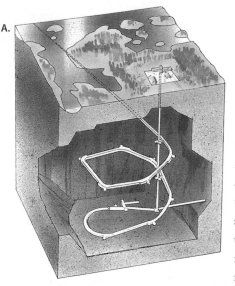

FIGURE 4.2. A. The underground research laboratory at Äspö Island, Sweden. B. Ant-farm view of Äspö underground research laboratory showing various experiments and the elevator access (courtesy of Swedish Nuclear Fuel and Waste Management Co, Illustrator Jan Rojmar).

B.

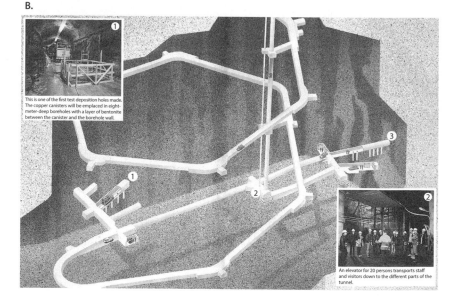

This is one of the first test deposition holes made. The copper canisters will be emplaced in eight-meter-deep boreholes with a layer of bentonite between the canister and the borehole wall.

An elevator for 20 persons transports staff and visitors down to the different parts of the tunnel.

scending further and to his right he could see fractures containing water dating from the time of the ice ages. Finally, at 1,400 feet beneath the surface of the island, he parked next to his laboratory, which on the outside looks like a green shipping container but on the inside is any microbiologist's dream field laboratory. Next to his lab he had three boreholes drilled into fluid-filled fracture zones with high-pressure valves sealing them. Thin stainless steel tubing snaked from these valves to the lab where he could draw in the fracture water with no worries of contamination and perform microbial experiments on the fluids, passing them through rock cartridges made from the cores of the borehole, and observe the microbial activities as if he were watching it take place in the fractures themselves.

On some days Karsten would drive across the bridge to Äspö to the surface facility where he keeps a full-scale microbiology laboratory and then just take the elevator down to do his fieldwork. Of course, microbial experiments were not the only thing happening at Äspö. Several experimental vaults had been excavated for testing the proto-type radioactive waste containers. The Swedish concept of radioactive containment involved three levels, beginning with the copper canister, which is resistant to corrosion, surrounded by a layer of bentonite clay to absorb water, and ending with the rock itself. It is in the rock where the microorganisms may play a key role in impeding the migration of the radioactive waste that manages to get past the canisters and clay and into the fractures. Inside the vaults, the copper canisters, which are about five meters long and one meter in diameter, were placed, like giant sarcophagi, inside tombs of solid granite and lined with bentonite clay bricks. The tunnels of each copper-canister crypt were then backfilled with mixtures of clay and rock. The copper canisters did not contain radioactive materials but did hold electrical heaters designed to simulate the heat production of the radioactive waste. Over a period of multiple years, the canisters would be excavated one by one so see how corroded they were and whether groundwater had penetrated them. Karsten's team would then analyze the samples to see if they contained different types of microorganisms and their abundance.

The research infrastructure and experimental program at Äspö clearly demonstrated the advantage of underground research laboratories (URLs) in addressing the fundamental questions concerning the deep biosphere. The problem for the ex-SSPers was that the United States did not have any deep underground laboratories in which similar facilities could be constructed; it had only radioactive depositories, like WIPP.

At the last evening banquet of the third ISSM conference, Tom Kieft leaned over to ask me, "What ever happened to that South Africa idea you told us about in Portland over a year ago?" "I fly there tomorrow," I replied, still uncertain as to what was going to happen when I landed in Johannesburg.

It was a miracle that I managed to return to the United States with the samples I had gathered at Western Deep Levels. The Anglo Gold geologist who had so kindly fetched me very early that morning from the hotel in Rosebank and taken me to the mine picked me up after I returned to the surface. He drove me back to David's $^{40}Ar/^{39}Ar$ laboratory at Anglo Gold's headquarters, where I used his argon gas to flush the air out of my aluminum canister after placing my two, thirty-pound rock samples inside it. I used David's Geiger counter to verify that the rock containing the thin dark seam of blackish "carbon leader," the thin band of organic carbon containing the highest concentrations of gold and uranium, was indeed radioactive, and the counts were above background even outside the aluminum canister. But at the airports, neither South African Airlines security nor U.S. Customs had any interest in this slightly radioactive, metal container, which I pushed along in my trolley. Back in my lab I pared the rock and put internal chunks into Whirl-Pak bags and mason jars flushed with argon and shipped it all, along with some of the precious water sample, to Jim at the former PNL (recently renamed the Pacific Northwest National Laboratory, PNNL) for microbial analyses. In Princeton, we began looking at the geochemistry of the water and the rock. Overall, the entire trip to South Africa and the shipping had only put me back personally five thousand dollars, just within the limit of my credit card.

ANCIENT MARINE BACTERIA TRAPPED IN SALT ON EARTH AND MARS? SPRING–SUMMER 1997

Like many of my fellow SSP colleagues, I was scrambling for support. Jim and I wondered whether NASA's exobiology program might be interested in the work we had started on the salt fluid inclusions. Jim had been developing direct DNA extraction techniques in his laboratory, and his group had successfully extracted PCR-amplified 16S rRNA gene product from the Laguna del Sol salt that Tom and I had collected, but not from the WIPP salt. When the WIPP salt was inoculated with a culture of an archaeal halophile, it did yield a DNA product, proving that Jim's result was not due to his technique. The ancient salt contained potassium-bearing phases between the crystals that I dated with my laser ^{40}Ar/^{39}Ar probe and obtained results slightly younger than the age of deposition, suggesting that the samples had been sealed to groundwater flow soon after deposition. The geological constraints placed the T_{max} at about 50°C, which is very similar to the maximum temperature that the Laguna del Sol salt marsh experiences on a hot summer day.[22] There was no geological reason to doubt that the WIPP salt would not contain halophiles.

The sterilization and handling procedures that Jim had been using, which apparently were working quite well, followed those of a German microbiologist named Dombrowski from the University of Freiberg, who in the late 1950s and early '60s published half a dozen papers on the isolation of living bacteria from salt crystals that had been collected from evaporites ranging in age from the Permian to the lower Cambrian from mines or cores spanning depths from 750 feet (230 meters) to 14,000 feet (4.3 kilometers).[23] Dombrowski's group practiced all the sterilization and control processes that later epitomized the SSP approach. When they heated salt crystals to 200°C, they could never grow any bacteria, very much like the baked control cores of the Thorn Hill coring campaign. They could not grow any bacteria from the 14,000-foot-deep core sample, a depth at which the ambient temperature was 130°C, a result that again suggested that their processing protocols eliminated drilling contamination. During the Mesozoic Era this salt layer would have been another 3,300 feet deeper in the crust and at a temperature of 160°C, well beyond the accepted limit of life back then (and now). The poster child of Dombrowski's experiments was *Pseudomonas halocrenaea*, which he had isolated from the Zechstein salt deposit and the Bad Nauheim saline hot springs in Germany. He described it as a living fossil from the Permian Period. Dombrowski had also pa-

tiently performed five-year-long desiccation experiments on his isolates to test their survival compared with other species of *Pseudomonas*. In other words, Dombrowski combined QA/QC and microbial
and geological arguments to test the veracity of his claim that the
bacteria were indigenous to the salt. This integration of inductive
and deductive reasoning was reinvented thirty years later by the SSP.

So why hadn't other microbiologists followed Dombrowski's lead
and pursued the isolation of ancient microbes entombed in salt? Because microbiologists from Belgium had discovered that *Pseudomonas halocrenaea* was physiologically so similar to *Pseudomonas aeru-
ginosa*, a bacterium commonly found in soil and groundwater, that it
appeared to be the same species, they had challenged Dombrowski
to demonstrate that the salt-mine deposit had not been contaminated with or inoculated by groundwater.[24] Dombrowski could not
do so with the tools available at that time, and belief in salt as a
subsurface repository for ancient bacteria soon evaporated (pun intended)—so much so, that in 1981 a textbook described mined rock
salt as free of any bacteria and, therefore, the perfect environment for
storing radioactive waste.[25]

While Jim and I had been focused on the Cerro Negro and Parachute drilling campaigns, Cynthia Norton, a microbiologist from the
University of Maine, who specialized in archaeal halophiles, teamed
up with Bill Grant and Terry McGenity, both microbiologists from
the University of Leicester, to successfully isolate archaeal halophiles
from Permian salt deposits as deep as 4,000 feet (1.2 kilometers).[26]
They had shown earlier that during the crystallization of salt, these
archaeal halophiles congregated inside incipient fluid inclusions of
salt with a seemingly purposeful intent.[27] Unlike the bacteria isolated
by Dombrowski, archaeal halophiles are obligate halophiles, which
means that they are incapable of surviving in fresh groundwater.
They are aerobic heterotrophs, generating their biomass from dissolved organic matter.[28] They grow to immense concentrations in
salt ponds where the temperatures reach 45–60°C and the O_2 concentrations are quite low because of the high temperature and salinity.
The key to their growth rate is the photon-powered bacteriorhodopsin ion pumps used to create a proton gradient that produces ATP,[29]

which in turn is used to power pumps that move Cl^- and K^+ into the cell to maintain osmotic pressure. These halophiles are even photo-tactic, swimming in the salt ponds toward the surface to obtain as much near-infrared light as possible to power their ion pumps, but not too close to the surface, to protect themselves from UV exposure by the overlying water. The archaeal halophiles found in the dark salt mines by this trio of intrepid microbiologists, however, did not have bacteriorhodopsin, so their only source of energy was from the oxida-tion of organic matter by O_2. Although the scientists could not prove without a doubt that the archaeal halophiles they isolated from salt had not come from groundwater infiltration into the salt deposits from nearby aquifers, it was by far the best evidence that existed for long-term isolation of living microorganisms in the subsurface.

In light of these recent discoveries from the U.K., our negative re-sult from the WIPP salt did not satisfy a minimum criterion for a scientific publication. If anything, our results indicated that bacteria (and probably Archaea as well) could not survive within fluid inclu-sions in a moribund state for 250 million years, and thus were on the death side of Tommy's Death-o-Meter, because amino acids spon-taneously decompose in 10^4 to 10^6 years.[30] We needed to look at younger salt deposits. Fortunately for us, Tim Lowenstein, a geologist at SUNY Albany and an expert on salt deposits, had recently col-lected a 525-foot-long (160 meters) core from salt deposits in Death Valley for the purpose of climate studies.[31] He had dated the salt lay-ers using a uranium decay scheme known as disequilibrium dating and found that they ranged back in time to 200,000 years before the present. So all we had to do was to look for primary fluid inclusions in the salt crystals of different ages, identify those containing micro-organisms, and extract the latter for DNA analyses. As long as we knew the age of the salt and that the fluid inclusions were primary, we would know the age of the entombed microbes. Tim was inter-ested enough to send us his best salt crystals containing primary inclusions.

All we needed now was some funding to perform the experiment. I had attended a NASA exobiology workshop in 1994 on salt as a re-pository for potential life on Mars that had been run by Michael

Meyer.[32] I suggested to Jim that we approach Michael about our project at a workshop at NASA Ames at nearby San Jose that Michael would be attending and to which we had been invited. The goal of this NASA workshop was to begin to forge a road map to develop a mission to Mars that would include a coring apparatus capable of reaching the watery subsurface believed to exist beneath the Martian surface. NASA had had plans for a Mars sample return (MSR) mission on its books for many years. But David McKay's discovery and NASA administrator Daniel Goldin's faster-better-cheaper philosophy was potent motivation for NASA Ames to reexamine the possible outcome of an MSR mission. In 1976, NASA's Viking missions to Mars had searched for and failed to find convincing evidence of life. Many NASA scientists were convinced that the failure of Viking to discover life had killed NASA's Mars exploration program.[33] But SLiME and ALH84001 had placed NASA on track to search for potential life on Mars again and to possibly bring it back to Earth.

The workshop was run by Geoff Briggs, a planetary scientist at NASA Ames with a penchant for drilling technologies. It provided an opportunity for Tommy, Todd, Jim Fredrickson, and Jim McKinley to present the coring techniques they had developed during DOE's SSP program. LANL also contributed a novel drilling technology based upon a cable plasma drilling tool that promised hundreds of meters of penetration through the Martian regolith.[34] Stephen Clifford, a geologist from the Lunar Planetary Science Institute, educated us on the structure of the Martian crust and just how many miles we would have to drill through frozen rock in order to attain water. As great as this challenge was, a conceited air of optimism pervaded the meeting. Almost everyone agreed that a drilling expedition could be done, could be funded, and would be imminent—so much so, that we all voted to send Tommy Phelps along to make sure that it would work. Tommy graciously accepted.[35] Once the Martian samples were back on Earth, concerns that subsurface Martian microbes might accidentally escape NASA labs and unleash an epidemic on the Earth that would wipe out all life arose, despite the reassurances by microbiologists that this was highly improbable. To assuage this concern, the unmanned MSR mission plan incorporated a fail-safe point at which the returning vehicle would fly into the Sun if the sample container had been breached during the flight![36] In talking to Michael Meyer at the end of the NASA Ames meeting, it seemed that our research into the salt inclusions was relevant to his program, so Jim and I submitted a proposal and were able to get a small grant from NASA's Exobiology Program to finish the work on the fluid inclusions.

I used the funds to construct a microdrill to sample fluid inclusions in salt. Essentially it used small diamond-tipped bits as thin as a hair, an inverted microscope, and a micropipette with glass tips to suck the fluid from the fluid inclusions. This procedure worked extremely well because the salt was relatively soft, and Justin Pavlovitch, a Princeton undergraduate, and I could readily see dozens of halophiles that he had grown, trapped inside single fluid inclusions, and we could see many halophiles from the modern salt crystals from Death Valley, although they looked skinny and spindle-shaped. We could then extract the tiny fluid volume, freeze it, and send it to Jim, or more specifically to Melanie Mormile, a postdoc in Jim's lab. Not only was she able to obtain DNA extracts from the salt crystals and then the fluids, but also she eventually grew and isolated one archaeal halophile from inclusion fluid from a 97,000-year-old salt![37]

As was typically the case in our research, this discovery raised more questions. If this single archaeal halophile was trapped in fluid inclusions within the salt for a hundred thousand years at 85 meters depth in the dark, how did it keep producing the ATP it required to maintain its osmotic pressure? If from the metabolism of dissolved organics, whence came the O_2 it required to oxidize said organics? Could it be that just enough O_2 was being supplied by radiolysis of the fluid-inclusion water by the decay of ^{40}K? Fluid inclusions do migrate through salt, albeit at a slow rate. Could this migration provide a renewable source of organic matter for halophile metabolism, or was it just recycling its own biomass? Could these processes sustain a few archaeal halophiles per kilogram of salt for 250 million years or longer?

While Jim and I had been analyzing the Death Valley salt cores, Tom Kieft was spending his sabbatical leave in Jim's lab at PNNL.

When my South African samples arrived that fall, Tom began using his green thumb for isolating thermophiles and tried various microbial media to see what kind of thermophiles he could find. The ferric-iron-reducing media yielded positive results immediately. Jim sequenced the 16S rRNA of the isolate and, surprisingly, it turned out to be a *Thermus*, *Thermus* SA. *Thermus* was known to be an aerobe and a denitrifier but not a ferric iron reducer under anaerobic conditions. The first water sample I had collected during my underground expedition in Western Deep Levels had yielded a novelty. I was jubilant. Jim also had some success isolating the DNA of heterotrophic bacteria from the carbon leader.

We quickly prepared the South African results and our salt data for conference papers at an upcoming meeting in July of the International Society for Optics and Photonics (SPIE) in San Diego.[38] A NASA scientist, Richard Hoover, had organized a special session there on the search for life in our solar system and had invited us to attend, along with more prominent, if not controversial, scientists. Tommy Gold spoke at the session on his Deep Hot Biosphere, and for the first time I got to meet the legendary Gold face-to-face and to talk to him about his 1992 paper.[39] Richard was very dedicated to searching for life in meteorites and a big fan of the ALH84001 discovery. So it came as no surprise that David McKay was a keynote speaker for the session. The conference was abuzz with new results coming in on ALH84001 from other scientists. To its credit, NASA had done due diligence in funding other non-NASA labs to analyze ALH84001 and test the veracity of McKay's conclusions. But almost all the studies had found reasons to doubt the multiple lines of evidence presented by McKay's group that the meteorite contained fossils of microscopic Martian life. Still, David was very upbeat, and over beers that afternoon talked to us about a new initiative started by NASA called the Astrobiology Institute. He said proposals would be due soon and he was looking for some microbiologists. Would we be interested? Tom Kieft jumped at the chance, while I declined the offer so I could focus on South Africa.

CHAPTER 5

LIFE IN DEEPEST, DARKEST AFRICA

"Now," my uncle said after finishing up these preparations, "let's see to our baggage; we'll divide it into three bundles and each man will strap one of them on his back; I'm talking about just the breakable items." Apparently the daredevil professor didn't include our persons in this last category. "Hans," he went on, "will look after the tools and some of the provisions, you, Axel, another third of the provisions plus the weapons, and I myself the rest of the provisions along with the delicate instruments." "But," I said, "who'll look after carrying down our clothes and all these ladders and ropes?" "They'll go down by themselves." "How?" I asked. "You'll see." My uncle was partial to extreme measures and resorted to them without hesitation. Under his direction Hans combined all the unbreakable items into one package, tied it tightly, and simply dropped it into the chasm. I heard the noisy whoosh that comes from layers of air being thrust aside. Leaning out and looking into the depths, my uncle watched the descent of his baggage with approving eyes, straightening up only after it was out of sight. "Good," he put in, "Now for us." I ask you as a right-thinking person—could anybody hear such words without getting the shivers? The professor strapped the package of instruments on his back; Hans took the one with the tools, and I the one with the weapons. Our descent got under way in the following order: Hans, my uncle, and I. It took place in absolute silence, broken only by bits of rocky rubble rushing down into the depths.

—*Jules Verne, Journey to the Center of the Earth*[1]

BROTHER, CAN YOU SPARE A MINE?
JANUARY 1998, PRINCETON UNIVERSITY

Dr. Rita Colwell, a prominent ocean microbiologist, had become the new director at the NSF and had decided that NSF should join forces with NASA to support a research project known as Life in Extreme Environments (LExEn). This was the first time that an interagency project had been funded. Microbiologists, who normally had a very difficult time finding a division in NSF that would support their work, flocked to the new program. Although we had some promising results on the South African samples, it seemed foolhardy to propose returning to South Africa with a team of microbiologists to test whether the deep mines could provide access to the subsurface biosphere and to test our idea that radiolysis within the carbon leader could support chemolithoautotrophic microorganisms. Our proposal would be a long shot compared with those submitted by experienced microbiologists working at well-established sites, such as Yellowstone National Park. Nonetheless after the SPIE meeting, Jim and I submitted a proposal based upon the data we had, doubting that the agencies would consider the deep subsurface a locale for extreme life. But we were wrong! We received NSF funding in January 1998, my first NSF grant for microbial work!

The bad news was that the Anglo Gold geologist who had overseen my first trip had not returned my emails for six months, which surprised me given that I had diligently reported to him all our geochemical and microbial results from those two rock samples and the water sample that we had collected at Western Deep Levels. What I hadn't realized is that the somewhat distant similarity between one of Jim's carbon-leader-clone 16S rDNA sequences and that of *Legionella* had raised alarm bells at Anglo Gold. According to David Phillips, the last thing they needed were scientists, foreign scientists no less, reporting potential health threats to the safety of the miners. In their words, what was in it for the mine? Great! With NSF funding now in hand, we had counted on returning to Western Deep Levels, which had been renamed Mponeng. Still, over a thousand operating mine sites existed in South Africa, a country slightly smaller than the state of Alaska. We just had to find out which ones were deep enough for our project and then go begging, hat in hand. Rob Hargraves rode to my rescue by contacting one of his old buddies from the days when he had worked in the gold mines, Morris Viljoen, a renowned geologist at the University of the Witwatersrand. Morris suggested that I meet with Rudy Boer. Rudy was a young, bright economic geologist, trained at University of Michigan but who now focused on environmental consultancy and, most important, he actually responded to emails.

With a contact in hand, Jim suggested that I should look for an experienced field microbiologist to oversee that aspect of the project. Jim had decided to use his NSF funds to hire a very talented and energetic microbiologist from Japan by the name of Ken Takai. I circulated an ad in *ASM News*, "WANTED: Microbiologist to participate in field studies in South African ultra-deep gold mines ...." A promising young microbiologist, Duane Moser, had just received his Ph.D. working under Ken Nealson, a well-established microbiologist. Duane had studied sea ice, the dry valley lakes of Antarctica, and the microbiota of mammoth bones; organized a series of cruises to study the microbial community structure of pristine deep-water Great Lakes sediments; and participated whenever he could in caving trips to West Virginia and Kentucky. Now in possession of a freshly minted Ph.D., he responded to my advertisement.

I could read from Dr. Moser's CV that he enjoyed fieldwork under conditions that tested not only the ability to do good work but also basic human survival. Accompanying his letter and CV was a photograph of a pair of boots sticking out of an incredibly small hole in the side of a small cliff of carbonate rock. The boots were presumably attached to the rest of Dr. Moser. This looked like a person who wouldn't just survive in the ultradeep mines, but relish the experience. This ex-farm kid from Wisconsin arrived at stately Princeton, all worldly possessions in the back of (or strapped onto) his battered Mazda pickup, which was hitting on three cylinders and blowing blue oil smoke through the fourth.

I recognized Duane from his boots as he walked into my lab. We appraised each other as I showed him around the department and campus. I gave Duane a few days to unpack, and then with nothing more than our hopes and some overheads, we sardined ourselves into the coach section of a South African Airways 747 for the seventeen-hour, forty-minute overnight flight from JFK to Johannesburg in search of Rudy Boer. Flying across Botswana in the early morning hours, we glimpsed our first views of Johannesburg through the pall of smoke that hung in the air from the seasonal Veldt fires we saw neat red-roofed row homes edged by squatter squalor interspersed between massive tailings piles. Not your obvious tourist destination.

We found Rudy Boer at Pulles, Howard and De Lange (PHD), a consulting company comfortably situated in an upscale neighborhood of central Johannesburg. Rudy could best be described as an aspiring hydrogeologist who was moving from hydrothermal geology into environmental remediation, a lucrative field for geologists in South Africa given the toxic legacy of a hundred years of heavy-metal mining. Rudy was very interested in remediation of acid mine drainage sites and had initiated a field-scale project at a mine in Swaziland. He was keenly interested in our deep microbes research and could readily see the potential biotechnological spin-offs. He chauffeured us to the grand old University of the Witwatersrand, where we lunched with Maurice Viljoen at the faculty club. Maurice was a jovial Afrikaner who positively entertained us with stories about "young" Hargraves and their good ol' days when they were "big pals."

Rudy then arranged a meeting with his contacts at Gold Fields, one of Anglo Gold's competitors and a major mining firm in South Africa, and Biomine, a company that specialized in the development of microbial ore-leaching processes. As with any professional facility in Joburg, Gold Fields was surrounded by gates and guards to negotiate. Once in, we entered a very spacious but almost empty modern office building. A short trip up the elevator and we were escorted to a boardroom dominated by a massive oval wood table and high-backed chairs. The wall carried the usual allotment of black-and-white photos of former CEOs visiting various new mining operations, and the occasional shelf holding an extra-large quartz vein or museum-quality pyrite. Even though the room could accommodate thirty people, only six were present, polarized on opposite sides of that massive table. On one side, the three men in suits from the mine were seen-it-all tough and, not surprisingly, suspicious of our motives. The room was hot due to a problem with the air conditioning. Cigarette smoke hung in the air and seemed to penetrate one's very skin. Outside, the golden, low-angle winter sun of late afternoon dimly lit the smoky room, and combined with the heat, cushy chairs, stale air, and jet-lag, sapped our alertness to those show-stopping questions that would inevitably come flying across the table from the other side.

I gave my short presentation at the table's head using overhead transparencies and hoping to hook them enough to provoke questions. My experience with Anglo Gold had made me very wary, and I had accurately anticipated about half the potentially show-stopping questions. But most of the discussion after the talk was surprisingly upbeat, and we quickly changed the topic to the terms of the study. It was clear that Rudy, Duane, and I were in the company of risk takers, the most forceful of whom was the director of Research and Special Projects, Rob Wilson. We both sensed that he was warming up to us as the meeting moved on. Eventually, the other two mine executives excused themselves, and Rob invited us to his office for a further chat about how the mines work, a chat that lasted late into the evening. He sketched on a notepad how the mine tunnels and stopes were engineered, where the air and water moved, how they were backfilled, and also how this study might be carried out (figure 5.1). The technical description was accented with one story after another. With Rob clearly in our camp, all seemed suddenly much more hopeful. He would find us a mine that suited our needs and set up an agreement between Gold Fields and Princeton.

The following day Rudy put us in contact with Professor Jennifer Alexander, the head of the Department of Microbiology at the University of the Witwatersrand. She was a prominent figure in the microbiology community in South Africa and held a controversial belief regarding the origin of HIV in Africa.[2] By the end of the meeting, we had obtained access to their ordering agent, Margaret Smith, and even a lab bench in their honors teaching lab with full privileges. It was an amazing breakthrough and we seemed to be gaining traction.

FIGURE 5.1. Cross-section of shafts number 2 (now Tau Tona) and 3 (now Savuka) of Western Deep Levels. Horizontal tunnels extending from the vertical shafts to the dipping ore zone represent each "level" of the mine. The tunnels following the strike of the ore zone are "cross cuts" that access the dipping "stopes" (from Oxley 1989).

That evening over dinner Rudy, Duane, and I discussed our meeting with Jennifer and pondered the state of the environmental business in South Africa. Although South African companies had patented bioleaching technologies based upon thermophiles, the environmental clean-up industry was just getting going and very little bioremediation expertise existed within South Africa at the time. From my conversations with a microbiologist, Mary DeFlaun, I could see an enormous potential for education and training in environmental biotechnology.[3] Microbiology departments in South Africa focused on human-, animal-, and plant-hosted bacterial and viral diseases. It occurred to me that perhaps this project could have a much greater impact on South Africa if it could be expanded to educate the next generation of environmental scientists on the capabilities of microorganisms in remediating toxic metals and even performing in situ mining. Rudy, however, was skeptical. He felt that the gulf was too great to cross. South Africa was in the midst of a massive education experiment in an attempt to transform the overwhelmingly poor, grossly uneducated majority into an educated middle class. Basic math skills were difficult enough, let alone more esoteric topics like environmental microbiology.

The next day Duane and I flew down to the University of Cape Town to meet with an old colleague of mine, Chris Hartnady, whom I had not seen since 1987 and to my surprise was no longer associated with the geology department but instead had become a hydrologist at a consulting firm. The morning after our arrival, we drove up to the base of Table Mountain, where the University of Cape Town's campus resides. During the days of apartheid, the University of Cape Town was South Africa's flagship university, being situated in the more liberal corner of South Africa, its southern California equivalent. It boasted a stellar faculty with international reputations and a world-renowned medical center. Some of those stars were the biochemists and microbiologists specializing in industrial microbiology, with whom we met that morning and all of whom expressed interest, although none could really commit to any form of participation in far-off Joburg. They were very much into their own projects, and research funding was in very short supply. The University of Cape

Town at the time was in desperate financial straits. The reception at the geology department was surprisingly more frigid than my numerous previous visits in the 1980s. Apparently news of our arrival had somehow preceded us because within the first two minutes into a meeting with one of the foremost geochemists working on the Witwatersrand Basin in the world, we were accused of coming to South Africa to make our fortune in gold-precipitating bacteria and were ushered out of his office. The next day, I reassured Duane that everything was fine, before he went off on a tour of the Cape of Good Hope with one of the microbiologists whom we had met at the university. I stayed back at our B and B catching up on email, but the nagging feeling that something untoward was in the air in the South African mining industry kept creeping into my mind, something that would be a showstopper for us.

IT IS WITH GREAT REGRET:
JULY 1998, PRINCETON UNIVERSITY

Once back in the States, it seemed that things were dragging on interminably in South Africa. May stretched into June and then July and still we heard nothing from Rob Wilson. Then a disturbing email arrived that pulled the rug out from under our fledgling project once again.

> It is with great regret, but not much surprise that I tell you that I, together with a lot of my colleagues, was made redundant effective today. I now fear for the project.... RBW

I was on an equestrian holiday in Ireland with Nora at the time, but I called Duane and reassured him, again, that we'd sort this out upon my return. I hung up the phone and then called Rob immediately for the details. A massive retrenchment was underway at Gold Fields as a result of poor performance on some international properties. The retrenchment was a little unusual even for the capricious metal industry, in which fluctuations in the price of gold combined with changes in the U.S. dollar–rand exchange rate (South African mining companies get paid in U.S. dollars and they pay their staff in rands) can make or break a mining company. The drill/blast technology used in the ultradeep mines was also very labor intensive, and the capital costs of sinking a mine shaft and the maintenance expenses were steep. The mining companies were unsure whether the new government would nationalize all the mines. But Rob told me to sit tight because he was working on a proposal for us.

Back at Princeton Duane was wringing his hands and thinking he'd made a serious misstep in his career. He proposed that we start looking at Homestake gold mine in South Dakota, the deepest mine in the United States, because it might provide a more stable, if less exotic, option. "Sure. Check it out," I said, at least as much to keep him preoccupied. A couple of weeks later, an email arrived from Rob. Rob and several of his colleagues, also recently retrenched, had formed a consulting company called Turgis Technology, which catered to the very company (among others) from which he had just been fired. Rob was back in the saddle. He was able to convince Gold Fields to allow us underground and had a contract for us by early August. Rob was up front with us. Our relationship had changed and now his services came at a price. But this was to be expected even if it hadn't been in the original NSF budget. After all, the alternative was unthinkable.

So with that, we hired Turgis Technology to handle our logistical matters, and Rob recommended that we work at East Driefontein Consolidated. They had recently sunk a shaft called E5, which reached a depth of 11,500 feet (3.5 kilometers). It was just beginning to go into production and offered the best chance of obtaining uncontaminated rock samples.

After several weeks of negotiations between Princeton University's lawyers, Turgis Technology's lawyers, and East Driefontein Consolidated's lawyers on indemnity and intellectual property rights, we finally secured underground access, and the only thing left to do was to pack for the trip. In one hectic week, the field gear was assembled, and along with a volunteer student from the University of Pennsylvania, Joost Hoek, Duane was off to South Africa. I followed two weeks later with more gear. Upon their arrival, despite great efforts to have the paperwork in order, all our gear was immediately confiscated by South African customs. This was the first, but certainly not the last, unpleasant and confusing encounter we were to face in dealing with customs issues between our two countries.

When I arrived, they picked me up in a red VW-like bus, locally called a Combi, and negotiated our way through Joburg rush hour traffic and down a highway and into one of the many towns built by the mining corporations around the deep shafts. There we met Rob at a Spur, one of South Africa's many meat-and-beer chains that propel South Africans toward their future. This one sported a big sign of an American Plains Indian chief's head complete with eagle-feather headdress. A little later with stomachs full and having made plans for the next day, we left Rob and, driving to the East Driefontein mine, passed through the "boom" gate into the "singles quarters" village to our new "home."

The next day we all went for a brisk jog around the village, but the thin air at 6,500 feet altitude dropped Duane and me to a not-so-brisk walk, and Joost, a seasoned cross-country runner, left us in the dust. Joost was definitely our party's Hans.[4] Around us the Transvaal dolomite and banded iron formations constituted red-stained "kopjes," and standing perched on top of one of them was the immense E5 shaft headgear. It must have been at least twenty stories high, if not more, and would have dwarfed the headgears of the potash mines back in Carlsbad. Across the road from the singles quarter's cafeteria, the no. 4 shaft headgear was spinning, and in the distance other headgears could be seen through the haze, all undoubtedly at work. After a Weetabix breakfast at Mrs. Sackie's singles mess and a pipe shower at the house, I inspected our house/field lab. Stacks of packing crates and ice chests lined the living room. A path worn into the rug ran from one bathroom with a toilet and shower into another bathroom with a bath. Typical of many South African homes, there was no heating and no air conditioning. Outside, splendid roses, marigolds, pansies, and lilies—almost every kind of flower imaginable—bordered the red dirt lawns, and the smell of jasmine filled the air. So did the songs of the many birds that proliferated around the mine village as spring encroached upon the high veldt.

We met Rob at East Driefontein's meticulously landscaped administrative building, where we shook the hands of a countless stream of administrators, geologists, and safety managers, all introduced by Rob and all looking at us very suspiciously. Then we trundled down the road to the grade plant to see our proto-lab. The mine had gone to great effort to convert what had been nothing but an attic into a lab by installing a floor above the mine's ore conveyor belts and adding walls, sinks, a lockable door, and even a large window. The room was spacious, coated with a thin layer of rock dust, and devoid of any of our equipment, which apparently was still in the hands of customs, but Duane had purchased a refrigerator. Downstairs I met the manager of the grade plant housing our lab, Peter Pretorius, a jovial, huge Afrikaner with a bone-crushing handshake. His duties done, Rob returned to Joburg while we headed toward Carletonville, the city built by Gold Fields to service the deep mines, to purchase food at the local Pick n Pay.

The next day Duane and Joost dropped me off at East Driefontein's hospital while they went to the supply house to pick up our mine kits, the standard mine-issued gumboots, gloves, belts, helmet, socks, earplugs, coverall, and bright orange vest with glow-in-the-dark stripes. The hospital sat adjacent to the dormitories that housed all the miners coming from Mozambique, Zimbabwe, Lesotho, and other countries. The mine hospital's halls were filled with queues of new workers filling out forms and being shepherded from one room to the next by a small army of nurses. As I filled out the form I wondered whether I should lie about my diabetes. Would that fact stop me from going underground? I decide to err on the side of honesty, which was a good thing because the first exam was a urine sample. After all the water that I had drunk that morning filling the urine bottle was easy. The nurse dunked in a strip and immediately looked up at me. "Diabetic?" she asked. I nodded. "Are you taking insulin?" I showed her my blood-test kit and explained to her my routine. "No problem." She stamped my slip and pointed me down the hall to another room where I was administered an eye exam that tested peripheral vision. No problem. Another note, another door, behind which awaited a hearing exam. Inside a booth with earphones, I focused on the slightest tones that chirped on and off at all sorts of frequencies in both ears. Afterward the nurse administering the exam told me that I had a deficit at high frequencies in my left ear, but that my right ear was perfect. That's what I get for using that nail gun without hearing protection ten years ago. No problem. Another slip of paper, and this time I was in line behind thirty black guys with no shirts for a chest X ray. So I removed my shirt. The only white guy in a room full of black miner wannabes chattering in Fanagalo.[5] Were they saying, "Look at that scrawny, old white guy. Am I going to be working with him?" After the X ray I finally got to see the doctor in charge, Leslie Williams. I asked her the reason for the X rays. "AIDS and tuberculosis. The miner's unions will not permit us to test for HIV AIDS. So we screen for tuberculosis. Given the close quarters in the dorms and the rampant prostitution, the spread of AIDS in the mines is a very serious concern." She gave me my ticket for "steppies" (the acclimatization test for heat tolerance that the mines require of all who go underground) and off I went, to be picked up outside by Duane and Joost.

The greatest challenge with the steppies is boredom. We drove over to the safety administration complex and entered the one-door, barn-shaped building with fans. Inside, we stripped down to our boxers and handed all our valuables over to the black manager. His assistant then stuck a thermometer into each of our mouths and recorded our temperatures. We then stepped into a cavernous sauna designed to simulate the high-humidity conditions underground and lined up with forty other miners opposite two long concrete benches about two feet tall. The thermometer guy demonstrated what we must do. A large circular light high at the end of the room flashed on red with a resounding "ding!" and the manager stepped on top of the bench. As soon as he was up another light flashed on green accompanied by a loud "dong!" Everyone was either nodding or shaking their heads, and all were wondering what the point was, but five seconds later there was a "ding!" and forty men rose in unison on top of the benches like a robot disco. Up and down, up and down. I tried to vary the pace, the lead foot, anything to offset the boredom and not stare at anyone else to see how they were doing. It seemed easy enough, but at twenty-five minutes I found my heart pounding like a trip hammer at a rate much faster than the dongs and dings, which abruptly stopped. In rushed the manager with his slightly diminutive assistant carrying a bucket full of thermometers dunked in alcohol. As they stuffed a thermometer in each of our mouths, I was thinking about what the doctor had told me earlier that morning concerning the lack of AIDS screening. A minute later the manager returned to measure our temperatures. Joost, the über-athlete, had a temperature close to his starting temperature. Duane, who was a bit on the heavy side, also had surprisingly low core temperature. I, on the other hand, had just barely passed.

It was approaching noon, and famished from our steppies, we drove to Carletonville, where we spotted a Kentucky Fried Chicken. During the sanction years, when American companies could not do business in South Africa, KFC had to pull up stakes. But they were back now, and although I wasn't a big fan, I was curious if South African KFC was different from stateside KFC. It was. This time the white people were all queued up and being served by black attendants. Through the open door of the back room, I spotted on the top shelf an unopened box on which was written "KFC—South African Starter Kit." Fast-food franchises clearly were one of the ways for fledgling South African black businessmen or women to start their rise to the middle class. Too hungry to wait, we popped down Chalcedony Street to a shop that served portable curried-chicken potpies—King Pies—a more traditional fast food that I decided should be imported to the States, because I quickly became hooked on them.

Foremost among the many things that were beginning to drive us all mad was our inability to get a phone line into the house and the fact that the phone line at the lab was almost always nonfunctional. Email was impossible. How can you run an international expedition if you cannot contact anyone internationally? Every time the lab phone went dead, we'd start blaming each other, and the tensions between Duane and Joost were getting particularly high. We were also getting impatient. Our glove bag was still stuck at customs, so we couldn't sample any rocks. We wanted to collect underground water samples, but no one knew of any water other than that used in mining and that comes from the surface. It was hard to imagine the mines being dry. I had seen a paper published a few months ago about the "fissure" water in South African gold mines, the only paper published on this topic.[6] Finally in a meeting with Nico Spoelstra, the chief ventilation officer, we broke through the communications barrier about sampling water. He recalled a weeping borehole on 46 level that was old but still flowing. He could take us down there to collect samples on Tuesday! We had four days to get ready, but a lot of our sampling gear still had not been unpacked at the house, and the H_2, N_2, and CO_2 gas cylinders that we required for preparing our anaerobic enrichments still had not been delivered to the mine supply house. We worked frantically through the weekend getting our mine kits ready, unpacking our supplies, and then packing our back packs with what we thought we might need even though we really hadn't a clue.

On Monday morning we negotiated the maelstrom of Joburg rush hour into Wits University, a quiet oasis among the hubbub of the Joburg business district. At the microbiology department, Margaret Smith, who was pushing a trolley full of reagents and glassware down a hallway, greeted us. She had bent over backward to fulfill Duane's late-Friday request for the chemicals we needed to prepare microbial media. We thanked her profusely while she escorted us to the autoclave past a lab full of undergrads working on their honors projects. She asked if we might be interested in a shaker-table incubator. The motor had died, but if we paid to repair it we could use it at the mine. Trying not to look like a naïve Yank being sold a lemon, I told her we'd think about it. We trundled off to the student mess, where we agreed to pay Margaret $500 for the incubator. When we returned, Margaret was all smiles when Duane gave her the good news. While Duane had her distracted, I slipped into Jenn Alexander's office to try calling home using her phone. The operator summarily informed me that you have to book the calls in advance and hung up. I eyed the fax machine on the table, picked up the receiver, and dialed home. "Hello, Princess?" "Farmy!" came the reply. I had beaten the evil phone police of Wits. I'm nothing if not persistent. After catching up on the home-front news, I hung up the fax machine and checked in with the boys, who were on their second round of autoclaving. I then drove across Joburg to the offices of Turgis Technology to meet Rob. He was in a meeting, so I borrowed his phone and left a half dozen absolutely urgent, "can you please ship such and such to Turgis Technology" messages across the United States. Rob strode in as I finished. "I hope you don't mind, we're seriously phone-challenged at the mine," I sheepishly offered. "Not at all. You must tell me what you need and I'll arrange it, chop-chop," he said emphatically shaking his head. "I need a freezer," I replied. "No problem, let's go to the Westgate Mall." We hopped into the van and he navigated me to the shopping mall to pick up a freezer for the lab so we'd have some place to store the samples we were going to collect the next day. On the way Rob, almost as an afterthought, informed me that the glove bag would finally be delivered tomorrow. I dropped Rob off at Turgis after thanking him profusely and flew back to Wits University to beat the afternoon traffic on the N1 to find that the boys had finished pouring the media plates and were just about to pull them out of the autoclave. Not only had they completed all the media preparations, but Duane had also sweet-talked Margaret into loaning us a second stand-up incubator and a portable autoclave! So we packed as much as we could into the Combi, along with the freezer and ourselves, and departed.

THE FIRST DESCENT: SEPTEMBER 1, 1998, EAST DRIEFONTEIN MINE

The next morning Duane's alarm went off way too early. We dragged ourselves out of bed, packed our knapsacks with water and snacks, and assembled our gear. We then drove to the no. 2 shaft, where the changing room was located, and met Nico, who slapped his card against the card reader to unlock the turnstile to let us through. Turnstiles are everywhere in every mine and are designed to regulate the flow of human traffic to and from the cages to a steady trickle. The short, horizontal bars in the turnstiles made them look more like seven-foot-tall sausage grinders turned on end, and negotiating one of these with a backpack of any size is an ordeal. We had been assigned lockers in the miners' changing room, and we quickly stripped down to our shorts and pulled on our coveralls and our three-foot-long wool socks, followed by our not-so-comfortable rubber boots, belted up, slid our gloves and earplugs into the belt, and donned our helmets. Well, at least we looked the part, but we were running behind schedule, so we dashed back through the "sausage grinder" and followed Nico in our van to the pyramid of spotlights at the top of the ridge that was the no. 5 shaft. Nico "carded" us through another turnstile, and we collected our miner's lamp, which we slung over our shoulders, and its battery pack, which we slid onto our belts, while balancing our backpacks and limping toward the cage.

When planning to go underground, everything revolves around catching the "cage," which is the triple-decker elevator car that hauls men and supplies to depth and back. A cage can handle forty men per deck and is basically a large steel box restrained by rollers much like those on a roller coaster and guided by a system of vertical steel rails. In a single shaft, there are up to six cages that move up and down independently of one another. Cages carrying miners start down very early in the morning, and cages carrying miners out about 11 a.m. We had a 7:30 a.m. cage—a late, managers', cage.

Shafts of sunlight pierced the starlit twilight as we stepped outside, and I had my first close-up look at the immense and brightly lit headgear overhead. It was every bit as tall as the drill rig at Piceance Basin but ten times the girth. I noticed for the first time how cold it was as a brisk wind was blowing through my thin cotton coverall. Then I realized it was the air being sucked down the shaft that created the wind. Not only was the shaft the portal through which all men, equipment, and about a megaliter of water per day for cooling and dust suppression entered and left, but it was also the window through which a huge volume of air, almost a hundred megaliters per minute, entered to ventilate the mine. The cage was swarming with a serpentine mesh of hoses and cables of various sizes and colors. Steel cables the width of a man's thigh were attached to the top of the waiting cage via a three-foot turnbuckle. The cables snaked up and over the

top of its twenty-story steel-gray superstructure and then off to a
"bank" where kilometers of cable were wound on massive steel
drums, all under the watchful eye of the bank man. As the cages
moved up and down, the cables whipped and stretched at a frighten-
ing speed, but all was under control. At Mponeng (formerly Western
Deep Levels), all the grisly details were completely covered with tan
tin siding, making it look like an extra-tall grain silo. Driefontein no.
5 didn't waste money on such frivolity. It was a "new" shaft that penetrated virgin rock and employed the latest in winding technology.
Actually, shaft sinking for no. 5 had started in 1982, and the mine was
just now coming on line sixteen years later. It had simply taken that
long to blast a shaft two miles deep, especially when it had to punch
through a major regional karstic aquifer. But given the richness of the
ore, Gold Fields had estimated that the investment would be recovered in the first couple of years of operation. On a good day in East
Driefontein, you could be on shaft bottom at more than 11,000 feet
depth in a mere thirty minutes after leaving the surface. We were actually taking an elevator down to a depth greater than the bottom
hole depth of Thorn Hill no. 1! The cages were not just for men. The
yard was strewn with railroad cars, some with full loads of rocks, and
others carrying conduit, drilling machines, and other implements of
mass excavation, and the track on which they sat ran right up to the
cage. A mine manager had even taken his pickup truck down with
him by dangling it beneath the cage's bottom.

You could see the stairways leading up to the three levels of the
cage with guys pulling back the security gates at each level and sliding open the doors. A few miners popped out quickly. We were part
of the more sullen mass of miners at the bottom now shuffling toward those doors. We stuck to Nico like glue as we passed through
the crowd. Black miners dressed in blue coveralls with yellow helmets, usually short, muscular, and stoic of manner, composed most
of the group. Managers were dressed mostly in white coveralls with a
variety of helmet colors, which apparently signified some pecking
order or specialized function. The white workers were a more bois-
terous lot and engaged in good-natured horseplay while yelling back
and forth above the ambient din in Afrikaans. We stepped through

the doors and shuffled to the back, where we leaned against the cage wall and sandwiched our backpacks between our legs as the door was shut and locked from the outside.

From outside, a squawking of loud metallic chirps announced the departure of this train to "Subterranium" and like a train leaving the station and entering a tunnel, our cage began falling from the collar and entered the barrel.[7] Our helmets were pelted with rock fragments and grit as a strong wind, the mine's breath, wicked over the edge and plunged into the void behind us. The last residues of surface twilight were extinguished immediately as the cage plummeted beneath the shaft collar. Some headlamps flickered on and we nervously switched on ours. The cage quickly gained speed, tugging at its greased rail restraints as it slid along in a vaguely controlled free fall. We all rocked in unison to the sway and jerk of the cage, like passengers on a subway commute to New York City, only quieter until an occasional clang or scrape from outside rudely woke us to the fact that we were a *falling* subway. I could feel the pressure building on my eardrums as we accelerated, and I had to clear my eustachian tubes as if I were scuba diving, so we must have been moving at a decent speed. Then I heard the sound of distant dribbling water far below, and in a heartbeat, a flash of light and the spattering of water on the roof the cage and then nothing. I realized we had just passed a tunnel, a cage stop.

Just as I started feeling the air temperature rise, the cage began to slow down and a couple of cage stops slid past. I felt the increased g-force as we pulled into our station, and just when I thought the cage had gone too far, the floor picked me up and then dropped me down as if I was tied to a big yo-yo. Above me, several miners had let out a collective involuntary "OOH" as the floor dropped out from under them after the first bounce. The wire gate outside swung open and someone outside started fumbling with the cage door. The biggest of the mine managers started swearing at him because the people on the floors above us were already moving out of the cage. Finally, the door slid back and everyone charged out. We stepped out onto a brightly lit concrete platform in the middle of railroad tracks where a sign overhead announced we were on 22 level. Four minutes had elapsed since we had traveled a mile and a quarter, more than four times the ride from the top of one of the World Trade Center Twin Towers. Taking into account the acceleration, deceleration, and delays, our peak velocity must have been over twenty-five miles per hour. We were now below sea level and as deep as the deepest mine in North America, yet this was only the halfway point. As we picked up our backpacks, Duane and I joked how the mines could be turned into amusement parks after the gold was gone. "Yeah, that cage ride was like Disney's Tower of Terror on steroids," I shouted.

Once out of the cage, we and the hundred-strong army of miners scrambled through more gates and turnstiles, turning into shotcreted tunnels again and again, until we came to another cage that looked identical to the one we had just left.[8] Here in the secondary, or sub-vertical, staging area, the air temperature was a muggy 28°C (82°F), and Duane and I heard a real cricket chirping notably faster than your average cricket. "The frequency of the chirp correlates with the ambient temperature," Duane, my naturalist, informed me during our hurry-up to wait. While we were waiting for the cage to arrive, Nico explained how the steel cable stretches such that the whole affair bounces several feet on its tether, accounting for the yo-yoing. The tensile strength of the cable is the only reason one cannot go directly down to three and a half kilometers depth with a single shaft. At a little over a mile in length, the weight of an extended cable becomes so great as to exceed its weight-bearing capacity. I wanted to ask Nico if a cable had ever snapped but then decided I didn't really want to know the answer to that question.

We stepped into the secondary cage and proceeded all the way to shaft bottom at 50 level, about 12,000 feet below the surface and 6,000 feet below sea level. The heat built up very fast at these deeper levels. After dropping off a couple of managers at 50 level, the cage came up to 46 level, where we stepped out. That was when I noticed that the sides of the plastic jerry can that Duane had carried down to sample the service water had caved in from the greater air pressure. The air was noticeably heavy and my breathing had now become labored. At Western Deep Levels I had assumed that the increase in air temperature was due to the geothermal gradient, but I was wrong. Nico explained to us that it was the adiabatic heating. The ventilation air moves too rapidly down the shaft to exchange heat with its environment, but it adiabatically heats as a result of the weight of the column of air above it. At shaft bottom we were at more than two atmospheres of pressure. At depths greater than about 9,000 feet, the air temperature exceeds 30°C, and you are no longer able to cool yourself by sweating. To operate a mine at greater depths you need underground air refrigeration plants to maintain an air temperature of 28.5°C, where cooling by the evaporation of sweat is still efficient. These plants take the service water, which has already been chilled on the surface, chills it further, and then injects it into the ventilation conduit in the tunnels in front of the fans. Without cool air spilling from the ventilation conduits that snake through miles of tunnels, the air temperature would quickly climb as the overwhelming mass of rock radiates heat into the relatively tiny tunnels. At 46 level (a depth of 10,500 feet) the local geothermal gradient of 9.5°C per kilometer raises the rock temperature 30°C above the seasonally averaged surface temperature of 18°C or a virgin rock temperature of 48°C. This is of course heat from the interior of the earth. The deeper you go, the hotter it becomes. Even though East Driefontein is a relatively "cool" mine, the air temperature would rise to 48°C (118°F) without the ventilation. The chilled air enters the stope at one level where the miners are hard at work removing the ore, it descends down the stope and exits at the next lower level, 200 meters down, where it then begins heading back towards the fan drift. The fan drift is a two-mile-deep, straight vertical shaft with fans the size of a small apartment building sitting astride their entrances to the surface. The wind velocity in the fan drift exceeds 70 mph, and as it carries the hot, moist air to the surface, adiabatic heating kicks in and the moisture condenses. The air speed is so great, however, that the water droplets form clouds that remain suspended in the fan drift. The last place you want to be in a mine that suddenly loses power is at the bottom of the fan drift, where you will quickly experience a three-inch rain shower in as many seconds.

As we followed Nico out of the station with the wind at our backs, he explained that if the ventilation stopped it wouldn't be the heat that would kill us first but more likely the toxic gases—CH_4, CO, hydrogen sulfide, ammonia—from the blasting (I could vouch for that) and CO_2, of course. "At East Driefontein we have four massive fans located at shaft no. 2 that pull all the air through the tunnels," he explained. We had seen shaft 2 many times with its column of steam extending hundreds of meters into the air like the steam rising above the hot springs of Yellowstone National Park. Nico's team carried small, battery-powered air pumps with miniature LED detectors that monitored concentrations of the two most prevalent gases, CH_4 and CO. He pulled out his to show us a cigarette-pack-sized metal box with a tiny wire cage on top. They monitored at all advancing tunnels and stopes to make sure that the ventilation was sweeping these gases and, equally important, radon gas from the decay of uranium out to the surface. They also monitored the dust load, particularly because the ore was so radioactive, to ensure that there was no potential detriment to the miners. We were not so concerned about the radioactive dust as we were the microbial contamination carried by the ventilation air and dust. There were no HEPA filters on the mine's ventilation system. Nico went on: "The critical parameter is the explosive limit, an estimate provided by this monitor. The combination of the CH_4 with mine air O_2 is highly combustible." "You've got CH_4 down here?" I perked up, hoping for an affirmative. The CH_4 could be from methanogens, perhaps the autotrophs we sought. "Nothing significant," he replied. "Oh" was my only comment, swallowing my disappointment. He also told us that, as at many other mines, the miners at East Driefontein carried rescue packs, the portable catalyst-carrying masks that scrub air of CO_2 for a few hours for a trapped miner. "We also use refuge bays, which is where we are heading just now," he said over his shoulder as he motioned us to cross the tracks.

We had begun walking down one of the tunnels radiating away from the secondary shaft like the branches of a tree from its trunk. We passed through the gate and by the benches with the coats hanging on the wall above them. We picked up our pace down the side of a tunnel underneath the ventilation conduit passing from a murky twilight into the hot, inky blackness. As we walked, I used the opposite tunnel wall to focus my headlamp when BANG! The side of my hardhat slammed into ventilation conduit junction. It didn't hurt, but I was shocked and embarrassed. It was as if the mine was telling me to wake up and pay attention. Other than that, it was amazingly quiet. The only sounds came from the diesel trains on the levels above us, rumbling sounds that reverberated through the access tunnels and reminded me of a passing thunderstorm. It was surprisingly dry, like Mponeng was during my 1996 visit, and the air smelled of dry rock dust. A rock wall peeked from behind a screen of wire pinned to the rocks with rock bolts cemented in place. No shotcrete here. The geological structure was clearly exposed through the myopic vision of a miner's headlamp. The only visible water ran in the gutters on both sides of the track, and we were clearly going upstream. In that fateful meeting in his former Gold Fields office, Rob had constructed a detailed diagram of the water circulation system. "The service water, as it is called in mining parlance, is used for drilling, dust suppression, and air cooling. The miners also drink it, although they aren't supposed to do this. It comes from the regional water supply that was used to supply the farms in the area before the construction of these super deep mines. It's expensive, so all water that is pumped into the mine is captured and returned to the surface, where it is neutralized to reduce corrosion and disinfected, usually with hypochlorite, to one ppm of free chlorine, and bromine. The same is done with the condensate water from the fan drifts. All of this water is chilled to 17°C, returned to the shaft where it flows down the pipes, and is released out to the mine stope from the distribution valves at each level. In some mines we've even used this falling water as a source of hydropower. Because the tunnels radiate outward at a slight upward angle, all the water returns to the shaft bottom in a settling pond, where it is pumped up to an intermediate pumping chamber (IPC) at one kilometer depth. At the IPC, a borehole into the dolomitic aquifer augments the service water with just enough groundwater to make up for any water lost from evaporation. A second pump then takes the water up to the surface to start the cycle again." So the gray, silty water at my feet was returning from the tunneling and mining operations in front of us and heading toward giant settling ponds at the base of the shaft behind me. Well, at least I would never get lost down here, at least too badly lost.

Because East Driefontein no. 5 was so new, only a few tunnels and stopes existed at depth, but in older mines, such as Tau Tona, where tunnels radiated across a 300-square-kilometer area, far greater quantities of chilled water and air were required to cool the mining environment. This circulation system had intriguing implications for the microbial contaminants since the water that was used to cool the air was probably the principal transport vector for microbial contaminants entering the mine. Our highest priority was to be able to determine whether any microorganism recovered from these mines was indigenous to the geological formation, or part of the community brought in by the mine. This situation was decidedly different from my previous experiences with drilling. The service water was disinfected, which could mean that the mine had done us a favor and killed many of our microbial contaminants. The mine monitored the service water twice weekly for fecal coliforms, presumptive coliforms, and total viable aerobic heterotrophs. According to Dr. Williams, the surveys almost always returned values close to zero. Since the water came from the surface, we would have expected that aerobic organisms, perhaps those typically found in soils or shallow groundwater, would be present and that any obligate anaerobes would be absent. The refrigeration might also encourage the growth of psychrotolerant organisms, like the ones found in Earth's polar regions, and deter the growth of mesophiles and thermophiles, like the ones found in the rocks around us. The service water was fresh and had a neutral pH, ruling out acidophiles, alkaphiles, and halophiles. Of course the dust suspended in the service water running down the gutters beneath my feet could also contain indigenous thermopiles that were either facultative anaerobes or at least O_2 tolerant. Since there were not enough tracers in the world to inject into the mine's circulation plant, the best we could do was to use the Taylorsville Basin approach of IMTs and characterize the biodiversity present in the service water. Duane would try to enrich for any mesophiles and thermophiles in the service water, but we also had to look at the service water's 16S rRNA gene library. Any 16S rRNA gene signature found in the rocks or water dripping from boreholes that couldn't be located in the service water could confidently be assumed to represent an indigenous microorganism. BANG! "@##@!!" Another ventilation conduit, and this time my helmet was squashed over my eyes. That one hurt. I had to stop thinking and start focusing on what I was doing.

We were just coming up to the refuge bay after a half-hour-long running walk. Nico, Duane, and Joost shot ahead as I was staggering down the track pulling my head back out of my helmet. We opened the heavy metal door and went inside. The bay was a very Spartan affair, with concrete benches, water pipes, compressed air, telephone, and room enough to comfortably seat thirty or more miners. Nico showed us how we could seal the door against the outside air. "This way if a fire cannot be contained, the mine can shut down the ventilation air to starve the fire of O_2, while the miners are safe in the refuge bay." On the platform outside we began pulling the equipment out of our backpacks to collect a sample of this service water so we could begin to understand what types of microbial contaminants were carried into this environment. The meters on our field probes fogged immediately as their cool surfaces condensed the moisture from the air. We had trouble labeling our glass bottles with permanent markers for the same reason. The water was cool, only 17°C, and had a neutral pH, very low salinity, and high amounts of dissolved O_2. "It looks good enough to drink," Duane mentioned as he peered at it through a clear-glass serum vial. "That's why they disinfect it at the water plant. They're not supposed to drink it, but it's easier than packing down two liters of water," Nico replied. Our chemistry kit revealed a very high nitrate concentration. "That would be coming from the explosive residues," Nico said.

After a couple of hours, we had finished collecting all our samples, and we packed up our gear and continued our steady march upriver into the heart of the mine. Most of the miners we encountered walking back along the tunnels were dressed in nothing more than shorts, their gumboots, hard hat, headlamp, and gloves. We, by comparison, looked like a hazmat cleanup crew. In another few hundred meters we came to a wooden door that blocked the tunnel. I tried to pull it open and it didn't budge, although it didn't look that heavy. Nico reached past me with both arms and gave it a jerk, and all of a sudden I felt the wind at my back again. "This is a ventilation door. We use these to control the direction of air flow through the mine," he shouted above the whoosh and whistle of air. Holding the door ajar, we all passed through and into a much more sauna-like setting. He walked past us and the door slammed with a bang, the silence returning with the heat and the acrid smell of burnt rock. A few short strides and we came to a side opening in the tunnel. "This is the cubby hole," Nico pointed out. On the tunnel face and poking through the rusting wire screen about three feet from the floor was a decrepit pipe, and out of the pipe poured the tiniest trickle of water. More like a stream of drops. We estimated a flow rate of about one liter per hour. This was not exactly what we had been hoping for, and Duane was trying to hide his disappointment. But then as he started examining the pipe more closely, he noticed that the black oily ooze that appeared to be stuck inside it didn't just look like some leftover grease for drill rod lubrication. "I think this is could be a biofilm," he said as hidden disappointment converted to overt excitement. He used a small spatula to gingerly separate it from the pipe and collect the tiniest of samples, placing it in a 1.5-milliliter centrifuge tube. "We could check this out using one of the scopes at Wits," he said enthusiastically. He took the residual drop of water on the spatula and placed it on his tongue. "Man that is salty." "Really?" I remarked as I pulled off my glove and put my finger below the stream to get a drop. Indeed, it was a nice seawater-like salinity, not bitter. Nico positioned his methane detector over the hole and could just detect a whiff of methane coming out. We pulled out our meters to check the salinity, and it was off the scale. The pH was 5, more acidic than the service water, but not dangerous. I was relieved, because it definitely was not service water. Duane took photos while Joost recorded our observations in our field notebook.

As we took notes, Nico explained that this was a diamond drill hole. As the mine extends its tunnels and crosscuts by drilling and blasting to get to the ore horizon, they "cover drill" ahead about 100 meters from cubby holes. This cover drilling was designed to intersect any fractures containing CH_4 and hot water so they could be vented and drained prior to the drilling and blasting for the tunnel. This borehole intersected the contact with the Spotted Dick Dyke, a Ventersdorp-age igneous intrusion, and has continued to leak water for years. We all looked at Nico. "Seriously, spotted dick?" I asked. According to Nico the rock apparently looked like the British dessert, spotted dick, so the geologist named it Spotted Dick Dyke. I thought to myself that the geologist must have been a homesick expat Brit. Duane and I returned to puzzling over how we could anaerobically collect the water for microbiology. Clearly we would have to put a plug into the hole first. I suggested that we then attach the sand cartridges I had used two years earlier at Piceance Basin. Duane wasn't satisfied. He wanted a clean water sample from which to filter the DNA to compare with the service water. Nico was getting antsy. "We need to start back if we're to catch the 11 o'clock cage up." We took some photos while Joost jotted down some readings for the water, and then we packed up and started back. As we walked down the tunnel Duane and I discussed the cover drilling. If we could figure out a way of collecting the water from cover drilling holes before the miners had completely drained the fracture and advanced the tunnel, we could collect pristine water samples. We would need a rapid response team, however, and we would have to coordinate closely with the shift managers. BANG!!! "Dammit to hell!" I readjusted the lamp on my helmet and started focusing on walking again.

About halfway to the station for the subvertical, we heard a shrill whistle behind us on the track. Piercing the blackness was one light in the middle of the tunnel approaching swiftly. "Step back against the wall," said Nico as he motioned us back. As the locomotive pulled up beside us, Nico chattered to them in Fanagalo and then waved for us to follow as he walked to the rear of the train. At the end was a cage on wheels, and we slipped our backpacks off and crawled in. Nico had just hailed a cab. The train covered the distance quickly and dropped us off at the substation platform, where many white-suited managers had gathered with a considerably larger number of much-less-clad black miners. After the cacophony of chirps and tweets, a cage arrived. The ride up was as quiet and quick as it had been coming down. As we rose through the cold wind tunnel of the shaft collar, I realized that during that short trip my overalls had become soaked with sweat. We had just paid a visit to the realm of thermophiles and were none the worse off for it.

THE CROCODILES AND THE CASTLE: SEPTEMBER 1, 1998, TWO MILES EAST OF FOCHVILLE

We returned to the no. 2 shaft East change room where we had a quick shower to remove the mine grime, felt the pleasure of donning our dry civvies, and dropped off our sweat-stained gear, which was courteously laundered. From there we took our samples, fridge, and incubator to the lab and began unloading. After another knuckle-cracking handshake, Peter Pretorius informed us that our gas cylinders had finally arrived. I dashed over to the mine supply and returned an hour later beaming with satisfaction and a van full of N_2, CO_2, and H_2 gas cylinders and, the most precious prize of all, a mixed N_2/H_2 gas cylinder. With help from Peter's black assistant, "Jon-a-tin," we dragged the gas cylinders upstairs to the lab. "Where was Vee-tech when you needed him?" I mused to myself. Now we had gas for our glove bag and for gassing some of our media tubes. But after wrestling the cylinders upstairs, I discovered that our American pressure regulators didn't work with South African cylinders! Aaaargh! We dashed to Carletonville, where we found an excellent mine hardware supply house, snagged some regulators, and returned. Just as I was pulling up to the lab for the third time that day, one of Rob's go-to guys from Turgis, Duane, and Joost were negotiating a 500-pound crate with the anaerobic glove bag through the second-floor window of the lab using a diesel-powered front loader. I didn't say anything; I just watched and shuddered every time the crate lurched as if threatening to leap to its death. But the boys succeeded, and the glove bag was safely secured in our increasingly lab-like attic.

With the glove bag came the remainder of our gear from South African customs after Turgis had posted a sizeable deposit. As we unpacked, however, it became clear that the individual customs agents had felt compelled to impose their own unofficial tariffs, which included several of Duane's expensive caving backpacks, pocket calculators, and the like. Duane was apoplectic to say the least. We worked into the night unpacking the glove bag, hooking up the gas, reestablishing our dodgy email contact with the outside world, and unpacking the samples we had collected. Then, as Duane hooked up the bag's vacuum pump and switched on the power, a fuse blew in the circuit breaker box, plunging the lab into darkness. "It's not my fault; it's wired correctly," Duane exclaimed as Joost and I scowled at him. We found the breaker box downstairs; as we flicked the circuit breaker, the lights came back on. Realizing that we were fighting fatigue, I offered to treat the boys to dinner.

We pulled out on the N12, traveled a mile down the so-called highway until we came to a zebra-striped Dirbars biltong hut, where we turned left, taking a back road to Fochville.[9] We traveled another mile or two past small farms, and the large headgear of another gold mine loomed in the distance to our left. As we came around the bend in the road and dropped down into a small dry river drainage, we saw a medieval castle on our right. Not your typical Afrikaner farmhouse, but a real medieval castle. As we pulled in through the boom gate of the Crocodilian Estates and past the security guard, we saw a half-finished tower with a parapet, drawbridge, and moat to our right and what was clearly the restaurant to our left, with dozens of vehicles in a red-brick parking lot. We could hear the American rock 'n' roll music leaping out of the restaurant. As we approached the entrance, we crossed a bridge and could just make out a crocodile head peering shyly above the water of another moat to our left. The cash register on the right was manned by a strikingly good-looking woman who guided us past the stuffed crocs to our table. We were ravenous and ordered Greek salads and croc steaks with peri-peri and monkey gland sauces. While waiting for our food to arrive, we thought we'd just quickly step into the bar and check out all the hullabaloo. A rugby game was in progress, the white South Africans' sport, and a fanatically cheering fan base surrounded the TV, all speaking Afrikaans. I squeezed toward the bar trying to catch the bartender's attention and apologized for my lack of Afrikaans. He replied, "Just call me Piet. Can you pronounce that?" He was an extremely charismatic man in his mid-40s but with the most devilish gleam in his eyes. "Hey guys, come over here, what can I get for you? Where are you from? What are you doing here?" motioning to Duane and Joost. He started handing us some Windhoek beer as we all introduced ourselves. After the introductions, he saw his next victims walk into the bar and went to greet them. An hour and too many Windhoeks later we stumbled back to our tables, where the very obsequious black waiters dressed in fezzes and leopard-spotted waistcoats had been patiently waiting. They didn't seem the least upset by our absence, treating it as if it were routine. I tucked into my croc steak as I took in the décor, which had a medieval/renaissance flavor to it with lots of silky curtains and oil paintings. My head was still swimming from the beer when I examined more closely the painting of a mandolin-playing court jester hung across the table from us. The jester bore a remarkable resemblance to Piet, the bartender we had just met. The meal was sumptuous and upon paying our bill we found out that an entire motel existed behind the restaurant and the bartender was in fact the owner, Piet Botha. We had found the perfect place to house, feed, and entertain the first wave of scientists who would start arriving in another month as part of our plan for a full frontal assault on the carbon leader in the stopes of East Driefontein.

CHAPTER 6

HUNTING FOR WATER AND CARBON

Soon Hans had driven his pick two feet into the granite wall. He had been working for over an hour. I was writhing in impatience. My uncle wanted to use more extreme measures. I had trouble restraining him and he'd already grabbed his pick, when suddenly there was a hissing sound. A jet of water shot out of the wall and smacked against the rock face opposite. Nearly toppled by the impact Hans couldn't help giving a cry of pain. I realized why when I dipped my hands into this jet of liquid then uttered a loud exclamation in my turn. The spring was scalding hot. "This water's at the boiling point!" I yelled. "Well, it will cool down," my uncle replied. Steam was filling up the corridor; . . . soon we downed our first swallows. "Why this water has iron in it!" "Splendid for the stomach," my uncle remarked, "plus it has high mineral content. This is like going to the health resorts in Spa or Toplita."

—Jules Verne, Journey to the Center of the Earth[1]

MINERS VERSUS LOTS OF WATER: SEPTEMBER 6, 1998, KLOOF GOLD MINE

On September 6, 1998, nine kilometers southeast of East Driefontein and at a depth of 3.1 kilometers, miners at the Kloof no. 4 shaft were drilling and blasting a new tunnel toward the southwest that would take them to the hanging wall of the Ventersdorp reef, another carbon-rich, gold-bearing zone, when one of the tunnel drills intersected water. The unit manager, Faunie, requested that the diamond-drilling team contracted by the mine drill a ring of cover holes. They arrived during the next shift, set up drills, and drilled their first cover hole to six feet. The driller then pulled out and they set a casing with a high-pressure valve, which was standard procedure, before returning to the drilling. The cover drilling was going smoothly until the drill bit reached fifteen feet behind the tunnel face and water exploded out the two-inch-diameter hole. Beams of scalding hot water were shooting in every direction with a high-pitched squeal. The pressure of the water had jammed the drill rod in the hole, and the casing valve couldn't be closed. The water jets drove the drillers off their rig and forced them to retreat back down the tunnel. The tunnel quickly began to fill with clouds of steam as the intense heat of the water quickly overpowered the ventilation system. Faunie called in a Proto team to help the drillers move the drilling machine away from the hole, extract the drilling rods, and close the valve to stop the flow.[2] Even for the heat-acclimatized Proto teams, the battle to move the drilling rig took hours. The teams had to rotate in and out because the excessive heat would debilitate a person in five minutes. Water had begun to spill into the drainage dikes outside the tunnel, and the water level was already beginning to rise in the sump pumps back at the no. 4 shaft, two miles away. If the water overpowered the sump pumps at the bottom of that shaft, the electrical system would quickly be shorted, electricity throughout the lower levels would shut down, and all the miners would be trapped. The Proto teams had to work fast if they were going to avoid a major catastrophe. The water was above their knees and scalding their Vaseline-coated feet and legs.[3] Working frantically like sailors sealing a leak in their submarine, they budged the drilling rig away from the borehole. They tried hammering the drill rod with other drill rods to dislodge the rod from the borehole, but as they swung the heavy steel rods down they bounced futilely off the jet of water as though it was surrounded by invisible armor. The only solution was to use the valve itself to nudge the rod free and hope that it didn't impale anyone once it shot out from the hole. Even as Faunie directed the attack on the valve, he wasn't sure if the valve itself would stand up to such pressure, whatever it was, for the pressure seemed higher than any he had seen or heard before, especially at this depth. After twenty years underground, Faunie had seen and heard a lot. They risked that the entire tunnel face would give way, and then all would be lost. But they had no other option. With a yard-long drill rod wedged through the valve handle, four of them pulled and pushed the valve head. If the valve didn't hold, it would blow up in their faces and there would be no chance of sealing the borehole. They strained again, but the heat was sapping their strength. The valve moved a fraction of an inch. In all likelihood the mine would have to be evacuated; if so, they hoped that the pumping systems could handle the extra load until the fracture drained, if it ever did. Faunie ordered one team out as another team replaced them on the rod. With their feet anchored against the rock wall, they pulled with all they had and nudged the valve again. The trouble-causing rod flew from the borehole like a jet-propelled harpoon, ricocheting off the tunnel wall and right into Faunie's face. He was out cold when the other miners dragged him out of the tunnel bleeding badly from one eye. As he was transported to the cage, each team rotated into the tunnel to nudge the valve a bit more closed each time. Finally they had the choked jet of water down to a healthy spring. But water also seemed to be escaping from the cracks around the casing and other fractures in the wall at the tunnel's end. As Faunie was taken to the medical center, he was consumed with the nauseating nightmare of the entire tunnel face bursting open in a river of hot water and what had been a truly remarkable rescue turning into the worst disaster in South African mining history.

But the face did hold, and during the following seven months special nine-foot stainless steel casings with double stainless steel valves were installed for the cementation teams. Using a high-pressure packer, they pumped cement into the fracture for three months. Railroad carloads of cement were towed down to the site every day, twenty-four hours a day, until a hundred tons of cement had been pumped into the rock face and seventy tons of chemical grouting were injected into the fissure. More cover drilling confirmed that the rock immediately ahead of the tunnel was relatively dry; hot water flowed through the boreholes but not at a threatening rate. But at 150 feet behind the face, along the Glenharvie dyke, the cover-drilling team intersected another fracture, which had a flow rate of 12,000 liters per hour and a pressure of 250 bars. Such a high pressure meant that the fractures were receiving water from the Transvaal dolomite aquifer, 2,500 meters (8,200 feet) above them. Rather than repeat the disaster that had befallen West Driefontein in the '60s,[4] they stopped any plans for further tunnel advancement. On the lower levels they started a drilling campaign to determine whether the high-pressure, water-bearing fractures also occurred down there. They did. After months of probing, drilling, and tunneling, they identified a monster "water pillar" and gave it wide berth as they tunneled around it to the hanging wall for the Ventersdorp reef.

ASSEMBLING THE TEAM: OCTOBER 1998, PRINCETON UNIVERSITY

Ironically, as Kloof Gold Mine's epic battle with water was being fought in secrecy six miles southeast of them, Duane and Joost continued organizing the lab and exploring various other mine shafts on the Driefontein property in a desperate search for water. I had returned to Princeton at the beginning of the fall semester to discover that the National Geographic Society had funded my grant application from earlier that summer to cover the travel costs for a team of approximately a dozen microbiologists that I now had to assemble. Since each microbiologist usually specialized in a particular type of physiology, each had his or her own special media recipe.[5] I had figured I would have a "bring your own media" field party. I also needed the combined expertise of a team of microbiologists to make certain that we did not make a misstep in this first project. I invited Rick Colwell, who had worked on sulfide-oxidizing bacteria before being swept up into the SSP. He was also a fanatical soccer player and coached his boys in soccer. With all of the pickup soccer games with the local mine teams happening just outside the back door of our house, he would find himself in soccer nirvana. Besides, I also wanted to soothe any ruffled feathers from the *Scientific American* fiasco three years earlier.[6] I tried to coax Karsten Pedersen to join us, given his experience and his interest in radiolysis. He sent in his stead Svetlana Kotelnikova, whom I had met in Davos at the ISSM '96. She was originally from Russia and had more experience underground in fractured rock than any of us. She brought media (including radioactive CO_2!) designed to test for CO_2-fixing autotrophy. This was the Holy Grail for which she was looking. I invited Bill Ghiorse, who was the founding father of subsurface microbiology, carried a wealth of knowledge on field sampling, and was keen on finding thermophilic, neutrophilic, manganese and Fe-oxidizing, and CO_2-fixing bacteria. I invited Mary DeFlaun for two reasons. First, she was interested in isolating thermophiles that were capable of oxidizing the xenobiotic organics that were common in hot industrial effluent. The second reason was to see what she thought the prospects were for developing South African environmental remediation strategies using bacteria. I invited Gordon Southam, a microbiologist from the University of Northern Arizona, in Flagstaff, who had been a graduate from Terry Beveridge's lab. Like his advisor, Gordon was extremely talented when it came to transmission electron microscopy (TEM), a tool that we desperately needed in this project. The extremely high magnification afforded by the TEM enables one to examine the formation of minerals by bacteria, and Gordon's hope was to find evidence for the precipitation of gold by bacteria in the Witwatersrand Basin gold layers. If some of our isolates were capable of doing this, then the mining industry just might take notice. I wanted to bring Barbara Sherwood Lollar in, too, because she could provide the ability to measure the isotopic compositions of the CH_4, both the $\delta^{13}C$ and δ^2H,[7] if we could find any, and any other hydrocarbons and also to determine whether they might be degraded or oxidized by bacteria. Her stable-isotope approach would nicely complement Svetlana's radiocarbon tracer experiments. We were also struggling with how to obtain reliable gas samples, samples that we could use to measure the H_2 concentration, a crucial measurement for our radiolysis theory. Barbara could not join us, but in her place she sent Greg Slater, her graduate student. And, of course, I invited Tommy Phelps, Susan Pfiffner (now Tommy's wife), Tom Kieft, and Jim Fredrickson. I knew I could count on all of them for their anaerobic green thumbs and for developing a field plan for the tracers. All had agreed to come, and with my team in place, I started scheduling their trips to South Africa, notifying them of the vaccinations they would require, moving the money to Duane's bank account in South Africa, and preparing them for the ordeal they'd have to undergo to get their equipment through customs. Finally, I asked them to bring paraphernalia and swag from their respective institutions. It didn't matter what it was, but T-shirts were particularly popular among the miners and their managers. We had already started distributing Princeton University geoscience department T-shirts with big Smilodon faces printed on the front, to Nico and others who had helped us.

At first, everyone from the mine managers to the shaft bosses and shift managers had suspected that we were spies for the U.S. DOE looking for radioactive contamination in the ore processing plants. Earlier in '98, some scrap metal sold to a United Kingdom scrap yard by a South African dealer had set off alarms, and the IAEA had tracked the scrap metal to one of the Driefontein gold mines. Of course we did bring a Geiger counter with us because we were interested in knowing just how radioactive the carbon leader was. After all, we sought to test the radiolytically supported deep biosphere hypothesis. But despite reassurances from Andreas Knoetze, the Core Resources manager, that we were nothing more than geeky scientists "looking for bugs eating gold," we were treated with suspicion. Duane encountered a drunken British expatriate at the Driefontein Recreation Club who was convinced that we were American spies sent to scout locations for the disposal of our own high-level nuclear waste![8]

Fortunately for us, we had Anna-Louise Reysenbach on our team. She was a microbial ecologist who worked on thermophiles from hot springs and deep-sea black smoker vents and was originally from South Africa. We had met when she was at Rutgers University, and we shared a common passion for horseback riding. If she wasn't riding, she was diving in a submersible to the bottom of the ocean or hiking in Yellowstone looking for hyperthermophiles believed to represent our most ancient ancestors. She had since moved to Portland State University, where the environment was far more conducive to her outdoor activities. I invited her to join us on our expedition, but she declined, and instead offered to drop in to visit the mine and advocate for our cause. So on the way back from discovering a new type of deep-sea vent in the Indian Ocean, she stopped off in her home country, visited Duane and Joost at East Driefontein, and smuggled in some more sampling supplies for me. Because she had family members still involved in the South African mining industry, the various managers patiently listened to her explain why extremophiles are so interesting and even useful. After several interviews around East Driefontein, the managers all took her advice about being more supportive of the "suspicious looking Yanks."[9]

ROCK, WATER, AND THE JAWS OF DEATH: OCTOBER–NOVEMBER 1998, EAST DRIEFONTEIN GOLD MINE

Setting up our lab in the mine's grade plant, however, turned out to be a disaster. The lab was plagued by hourly accumulations of dust from the rock-crushing equipment below and the noise was beyond description. If this was not enough, the rains had come. Suddenly, the desert of burnt grass and ash-covered terrain that surrounded the mines was flooded by a sea of lush green Highveld grass that was being mowed by grazing cattle everywhere you looked. Each day would dawn with a clear blue sky, but by the afternoon the enormous white thunderheads began to assemble ominously overhead with darkening, disapproving frowns. At 4 p.m. they would unleash a torrent of rain with a furious fuselage of lightning that would flash across the sky every two to three seconds. It was not surprising, given the ferocity of the attack, that the grade plant suffered daily power outages and flooding. Even though we had lined the lab with plastic sheets to deter the dust from coming in, the rain just pooled on the sheets overhead and then drained onto the floor, poking holes in our meager defenses. In the end, we decided to use the lab at the grade plant only as a rock-processing lab to house the anaerobic glove bag, which was immune to contamination from the outside. Duane and Joost had decided that the living quarters with its much better environmental control would be used for the more sensitive microbiology work.

In one of Duane's frequent email reports from the field to the growing email list of collaborators, he described the day when they moved the equipment from the grade plant building to our house.

The foreman over there is one Peter Pretorius, who is working and doing heavy lifting despite having stitches in his abdomen (which he felt compelled to show me this morning) from hernia surgery last week. Peter graciously offered us the use of 6 of his workers for the project. These guys are heroes. They managed to transport the unwieldy, 150 kg laminar flow-through hood down several flights of metal steps in a groaning mass of muscular arms and fumbling feet. Being directly in the path of this bionic monster of my own making, I was glad to see it arrive into our trusty red Combi Van, which I was just told is officially stolen at this point, because the rental agency requires a monthly imprint of my credit card and I haven't seen them since I arrived in August. I must look into this. To make a long story short, when the hood arrived at our house/lab, it was 1 cm too wide to pass through the door. It seemed to take some time for my team of non-English-speaking miners to figure this out, however. Eventually, the door was removed hinges and all, as hinges are evidently not provided with removable pins in South Africa. Some 3 hours after unplugging it in the grade plant, the laminar flow hood was installed and functional in its new home, except that since several of the men had climbed over it at various points, the glass sides no longer fit. A trip to the grade plant lab to obtain the 20-ton jack from our rock splitter should make short work of this tomorrow.

In that regard, I tested the rock splitter last night. This is a piece of equipment fabricated locally based on an excellent drawing from Rick Colwell. Two massive steel jaws are forced together with all the persuasion a 20-ton hydraulic jack can muster. Anything in its path is sure to yield. I tested this on a large unwieldy chunk of waste quartzite. Much to my delight, the "Jaws of Death" were able to fracture the very hard rock without difficulty. The jaws advanced with every pump of the handle and then all at once like a rifle shot, the rock parted in a cloud of dust and with remarkable precision. I was thus able to obtain a square chunk of virgin rock from the interior. Rob Wilson feels that with practice, we will be able to fashion very fine stone tools and projectile points. So perhaps with this, our operation has reached a level of sophistication on par with the world's great prehistoric peoples?

Just now as I write, we have had a pretty significant seismic "bump."[10] These thunder claps from below are a never-ending source of fascination for Joost and I. We often engage in a play by play afterwards trying to determine if this one was a dud or not and from where it may have originated. So far, I've noted that one will miss them altogether when out running and that they are especially noteworthy when bare skin is attached to something grounded (i.e. when in the shower or on the toilet). They are especially disconcerting in the grade plant lab as the entire building is prone to sway a bit. Very strange. It is currently midnight and the roosters all around the shanty town are crowing. Your man in the field signing off.[11]

The new arrangements worked out very well for email, which functioned more reliably now that we had a stable phone line. The living room was crammed with equipment sitting on plastic sheeting. Sampling bottles could be sterilized thanks to the autoclave loaned to us by Jenn Alexander. She had also loaned us two incubators, one of which was rotary, for the price of replacing the latter's motor. We could now count on these incubators for our enrichments, given that we had much more reliable power than at the grade plant. Finally, we could assemble field kits over early morning coffee. The microscope lab was now reassembled in what had been my bedroom, and Duane fascinated the miners by showing them what was growing from his enrichments. Seeing the gram-stained little microbes jostling around on the microscope slide was a big hit.

We were prepared for handling rock samples, but we were coming up short in terms of sites where we could collect water samples. When the National Geographic Society expedition would arrive in December, a lot of microbiologists would be looking for subsurface water samples. Nico knew of another significant water source at 3.5 kilometers depth. It was located in an emergency evacuation tunnel that connected the no. 5 shaft to the no. 4 shaft. The tunnel was two and a half miles long and the water source was in the middle. The only problem was that the tunnel was not ventilated. To go there we would have to wear dry-ice jackets. As enthusiastic as we were to give it a try, the mine just could not see sending a bunch of overzealous scientists who had little heat tolerance down an unventilated tunnel for a sample of water despite all the indemnity documents we had signed.

It was Duane's inquiries at West Driefontein mine next door and a bizarre sequence of serendipitous events that led to Duane and Joost's eventual encounter with significant amounts of fissure water. Duane explained to me that the West Driefontein mine was definitely "Shelbyville."[12] Everything over there was the same, yet somehow different. One aspect of West Driefontein that was important to us was that it had intersected a lot of water. In fact, the West Driefontein no. 6 shaft once catastrophically flooded in 1968, when its developers had mined too close to the overlying dolomite aquifer. The story of how the mine was saved is a favorite in local lore and was the basis of the movie *Gold*. The subsequent dewatering of the associated "bank compartment" led to major subsidence events,[13] and massive, at times lethal, sinkholes appeared with some regularity. A few years earlier, one had opened up only 300 feet from the mine's head offices, and it was aggressively dealt with. A massive sinkhole, hundreds of yards across at the rim, had just opened up under one of West Driefontein's tailing dams and essentially swallowed not only a large portion of the mine tailings, but also a long segment of the retaining wall. One of the West Driefontein geologists had fallen into this sinkhole and dislocated his shoulder a few weeks earlier. This accident was to Duane and Joost's advantage, since the geologist was not allowed to "work" underground but was, therefore, free to escort Duane and Joost to various underground sites in the mine that had reported water.

Their first trip into the West Driefontein no. 2 shaft revealed a rather old, relatively shallow mine. The floors tended to be very oily, and the air often seemed foul with diesel fumes from the locomotives. The no. 2 shaft tunnels at 12 level, at a depth of about 4,300 feet, revealed abundant iron seeps. Some of these were associated with seep water that poured rather diffusely from the walls and ceilings over considerable distances, giving the effect of raining underground. It was not immediately apparent to them just how such 20°C water could be sampled with anything resembling anaerobic technique, because no weeping or running holes or pipes were available, and it was not immediately apparent what the source of this water might be, because there was considerable mining activity in the overlying levels and the pH was suspiciously acidic. Shaft no. 2 was also being prepared for decommissioning in two years, and the shaft pillars (the areas of rock nearest to the shafts) were currently being mined.

On their next outing to West Driefontein, Duane and Joost were taken to the no. 5 shaft, a deep secondary shaft, which could only be reached by a two-mile train ride from the no. 2 shaft. Twenty-six level, which was 6,000 feet deep, was drier than 12 level, although one crosscut had very abundant water raining from the roof. The water was so abundant that Duane filled a ten-liter plastic jug in a few minutes simply by placing it directly under a fast-flowing crack in the ceiling. Obviously, contamination from the air was a huge concern since each water droplet fell through so much air (perhaps nine feet), but it was a place to start. They filtered the water for lipids and DNA and used some of it to inoculate dilutions of the various media. Unfortunately, they had lost their thermometer during the long trek, but the temperature was like that of warm bathwater (mid-30 degrees C?). The pH was 3–4, and the dissolved O_2 came in at 5–6 ppm. At this level, there were also great accumulations of apparent elemental sulfur covering the walls in many places. The sulfur was always associated with small cracks and fissures in the walls and often dry and feathery in appearance. When the sulfur was associated with water, the pH was always in the range of 3 to 4. This pH is the classic symptom of acid mine drainage, except that here it was occurring more than a mile beneath the surface.

Then the West Driefontein geologist told Duane and Joost that two weeks prior to their visit, the miners had intersected water in a cover drilling operation at 38 level in the no. 6 shaft. The location of this cover drilling operation was in a very short tunnel that was advancing into "virgin" rock. There literally was no other mining activity at any level above or below this for kilometers in all outlying directions. The tunnel was a little stub maybe 150 feet inward from the tunnel from which it had been excavated. The first cover hole drilled here intersected water and was plugged. The second cover hole, however, hit high-pressure water and CH_4 at 230 feet beyond the face. They told Duane that this was one of those situations in which the drilling rods had been blasted out by the water pressure and someone had "almost lost an arm"! The water had been running ever since. Initially they measured the flow rate at 3,600 liters per hour, and on Friday when they visited the site, Duane had estimated that it was still pouring out at about 1,800 liters per hour. The depth was 6,500 feet, and the water was 45°C, had no detectable O_2, and smelled of rotten eggs. Joost measured 2 ppm H_2S with the portable water-testing kit, and he did not detect any nitrate or free chlorine (i.e., the water was not service water); the pH was around 9 (i.e., the water was not acid mine stope water). Although there was a known fault in the vicinity, the drill hole had not yet reached it, but the hole had intersected a small zone of fractured quartzite. The mine's plan, therefore, was to let the hole run until it drained out and then seal it with cement. They had anticipated that it would drain in two weeks, but they were surprised by how long it was taking to dewater this structure, so Duane made arrangements to get back there the following Monday with the full sampling kit. Once Duane, Joost, and the geologist had returned to the surface, Duane phoned me with the fantastic news. We both felt that the sulfide had to be produced by thermophilic SRBs, and the best way I could think of to confirm that would be to measure its $\delta^{34}S$ isotope composition. But neither Duane nor I had ever collected sulfur isotope samples. I rang up Lisa Pratt, who lately had been focusing her research on sulfur isotope analyses in her lab to ask her if 2 ppm was enough for isotope analyses, and whether she would be interested in analyzing the sample if Duane could get one. She replied yes to both questions and quickly shot an email to Duane with the instructions on how to construct a zinc trap to strip the H_2S from the water. When they returned that Monday, Duane and Joost had a geology assistant and a porter at their disposal and were told they could stay as long as they needed. They went down two shafts and then took a train two miles to a tertiary shaft, where they dropped down to 38 level and then hiked another 1.8 miles over very rough terrain. Unfortunately, when they arrived at the end of the developing tunnel, they had no ventilation, which almost proved their demise. Nevertheless they completed the sampling and managed to haul back ten liters of water, from which they were able to filter water for DNA, freeze water for chemistry, inoculate growth medium, and, most important, extract the sulfide samples for isotope analyses at Lisa Pratt's lab.[14]

The Western Driefontein no. 6 shaft was known to be seismically active, and during their trip underground they were "bumped." The first bump struck while they were in one of the stopes near the no. 6 shaft. They were told it was centered somewhere at Mponeng, which was about seven miles to the west. Duane emailed me later, "You could hear and feel it approaching for a spookily long period of time and then it moved on. This, of course, was above the ambient din in the mine, so it was pretty loud. We learned later that it was quite a shaker with a Richter value in the mid-3's. On the way out, we could see places where rock bursts had taken a section of roof or wall out, but always the heavy steel mesh retainers held the rock back. Quite amazingly, very little damage was done to the mine, although lots of debris had rained down from the roof." While they were waiting for the second cage on the way out of no. 6 West, a second big bump hit with a massive bang. This was also in the 3's and was centered right there in the no. 6 shaft pillar where they were. One of our sampler friends from East Driefontein mine, Raj Nair, was also underground in the stopes at the no. 5 shaft during the same event and told us that "the drillers all turned off their machines upon hearing the rumble, waited for a second or so of eerie silence, then the bump pushed through. The miners could actually FEEL the rock contract as the seismic wave passed. They then all looked at one another in silence for another 5 minutes, and then they all just turned on their machines and went back to work."

Duane returned to Princeton University for more supplies and "gifts" for the miners and to work on a water sampler. He had settled on using a soda-pop syrup canister because it was all stainless steel, could easily fit within the autoclave we had in South Africa, could hold twelve liters, was leak tight, and when full weighed only thirty-five pounds. George Rose, the department's machinist, manufactured a leak-tight cap and vent valve and purchased some bayonet connections. Duane could then fill the empty sterilized canister with filtered argon gas so that it was totally anaerobic. To fill it with borehole water, he would connect a sterile tube to the bayonet valve as if he were connecting it to a syrup dispenser, and as the water dribbled in, he would occasionally release the gas through the vent valve. With his anaerobic water sampler for the cover boreholes, he returned to South Africa hoping to be able to put it to the test.

TO HELL AND BACK, OOPS! DECEMBER 1998, EAST DRIEFONTEIN GOLD MINE

As members of the team arrived in December, we had to get them cleared through the medical examinations, chest X rays, and steppies. Joost was in charge of making sure each had mine gear. Two other people joined us as part of our National Geographic grant agreement. We hired Louise Gubb,[15] a photojournalist from Cape Town, to take photographs of the sampling expedition. Kevin Krajick,[16] a freelance writer, volunteered to join us in hopes of selling his article to *Discover* magazine. This immediately raised alarm bells with Rob Wilson. He had to make sure that permission for their underground trips was cleared all the way to the top CEO and that they must agree to the mine's right of refusal on anything they publish. But if han-

dled properly, Rob felt that Driefontein Consolidated might appreciate some good press. At the time we thought little of this, but we seriously underestimated the impact that Krajick's article would have on
the mines.

Our target was the carbon leader, which at East Driefontein formed
a single black band compressed between quartzite. The geologists
had told us that this carbon leader material was the richest ore in
terms of gold (and incidentally uranium) on the Earth. Some of these
samples were said to grade at 5,000 grams per ton (compared with
the more typical values of about 10–12 grams per ton found throughout the mines). The stopes were the locations where we would actually sample the carbon leader. We had discussed with Nico how to
introduce tracers (potassium bromide or rhodamine) into the drilling water used by the miners, but the water was under very high
pressure (two miles worth). We weren't as worried about the drilling
water, however, as we were about the ventilation air. The microbial
and fungal spore load carried by that air must have been huge, and
as soon as the newly blasted rock face was exposed, the spores would
be colonizing it. This source, combined with the water they used to
suppress the dust, was the most feasible route for contamination of
the carbon leader ore. After some negotiation with the shift managers, Nico arranged that our sampling team could go into the stopes
about three hours after blasting. This was the amount of time required to ventilate the blasting fumes before sending in the miners.
We would go into the slopes with the first wave of miners and take
our samples before the microbial colonists had a chance to grow.

We met at the house/lab that afternoon so that Joost could hand
out all the mine gear and equipment we had and go over everyone's
assigned task. Early the next morning we piled into the van and
headed for the sausage grinders at the no. 5 shaft. The mine had arranged a train to take us to the stope at 36 level. It was a quick trip
and we stopped in the tunnel that had the hole in the side. Helmet
lamps, gloves, and gobbles on, we, like Alice down the rabbit hole,
crawled through the man tunnel one by one. The carbon leader at
this mine dipped at a 30° angle and was no more than a tight coal-
like band that was at most a centimeter thick. To mine it effectively,

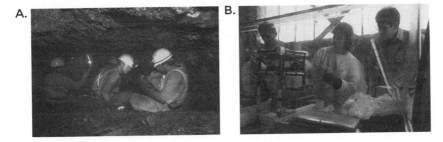

FIGURE 6.1. A. Duane Moser and author in the East Driefontein stopes between levels 36 and 38 collecting samples of carbon leader. B. In the attic overlooking the ore crusher, Svetlana Kotelnikova works with the "Jaws of Death" on carbon leader samples, with Jim Fredrickson inoculating some aerobic media as author looks on (courtesy of Louise Gubb).

the companies excavated only about a yard-thick slice of rock around the carbon seam. The mine blasted a little further into this thin layer on a daily rotation, outward in both directions from the central raise, which was nothing more than an approximately two-yard-high rock slide along which a steel cable dragged an iron scoop. The miners would move the blasted rock to the slide, and the scoop would drag the rock to an ore shoot that would drop the rocks 100 yards below to a lower level. During the daily cycle, they drilled, set the charges, blasted, ventilated, cleaned up by moving the rock from the walls to the slide, drove in supports for the roof, and started all over again. We had arrived just as the cleanup had begun. Inside, the stope was cluttered and extremely noisy with the scoop clanging up and down the slide, men hosing down the walls to remove the muck and dust from the blast, and materials being winched in and out on overhead cable assemblies, all in very tight quarters. Dust, noise, the acrid smell of burnt rock, steam rising from the wet surfaces, and darkness all conspired to cloud our senses (figure 6.1A). We fumbled our way down the slide, careful not to touch the winch cable running just over our heads, occasionally grabbing one of the hydraulic jacks used to support the ceiling and gingerly skirting the ore shoot. Falling down the ore shoot was certain death. I could not make out the carbon leader because of the dust, but I used the Geiger counter as my guide. As I moved it over a smooth section of rock wall that had just been cleared by the miners, it started clicking madly. Duane moved up behind, pulling a chisel out of its autoclave bag and a small ball-peen hammer. But before we could move, Raj Nair, who had been guiding us down the stope, came up and shouted into my ear. "Not here. There's no supports. You pull that piece and the whole wall could cave in on us." "That would be embarrassing," I shouted back, trying not to think about the fact that the supports were the only thing that separated my stack of bones from eight thousand metric tons of rock immediately above my head. Invisible to our eyes, Louise Gubb was rapidly clicking her camera, busily capturing all of this on film while dodging miners—not an easy task given that the only lighting was randomly wandering beams from all of our helmet lamps. Above us at the top of the stope, a deafening roar started up as drilling began. We shuffled another hundred yards downslope, where it had been cleared and the miners were already setting the hydraulic jacks used to support the roof. Soon the drillers would be arriving and we needed to speed up. I quickly surveyed the mine face and found our target. Duane handed me a spray can, and I painted two large orange squares around our carbon leader and adjacent quartzite. "Oh, tracer girl, where are you?" I shouted. Mary DeFlaun crab-walked down out of the darkness and over to us and pulled two squirt bottles from her backpack. She sprayed one containing the fluorescent microspheres over the surfaces marked out by the orange squares followed by a good dousing of rhodamine dye over the squares. We then tried to bring one of the high-pressure hoses over to spray the rock, emulating the mining process as much as possible. If microbial and fungal spores were landing on the rocks' surfaces and getting forced into the cracks as the miners sprayed the rocks, these tracers should reveal their pathways. But the hose wouldn't reach, so Raj organized the miners to carry water from the hose to the rock with their helmets and throw it onto the rock. In the midst of all this kerfuffle, a crawling crowd of visiting VIPs dressed in white coveralls and guided by one of the mine managers had shuffled down the stope behind us. They started pointing at us, and the mine manager in Afrikaans said something like "There's all the crazy scientists

I told you about looking for gold-eating cockroaches in our carbon leader." They all started laughing and then continued crawling down the stope. I thought to myself, "All we need is the Mad Hatter to show up next." Then Duane started to chisel out the carbon leader while I held a fresh autoclave bag underneath. Gingerly he wedged the chisel and levered a huge chunk of the leader, plop, right into the bag, which I quickly sealed and placed in our backpacks. Duane then went to work on the quartzite. Finally with about 20 kilograms of samples in hand we started backing out of the stope toward the raise slide. I warned Mary to be careful crawling around the ore shoot that we had negotiated coming into the slope. She replied, "What ore shoot?"

I squeezed past her, led them down to the bottom of the slide, and stepped out onto 48 level. Leaning back against the tunnel wall, we paused to swig some water. We had been in the slope for more than an hour, but it had seemed like five minutes. As I raised my water bottle to drink, sweat poured out of my gloves. I hadn't realized I had that much sweat in me. As we walked back to the cage, I could feel the sweat sloshing around in my gumboots. Upon our arrival at the cage, I sat down on one of the benches, pulled off my gumboots, and poured out a good liter of sweat from both. On the surface we headed for the grade plant lab, where we had prepared the anaerobic glove bag for our first rock samples. Duane had conceived a clever approach to avoid cross contamination inside the glove bag after discussing the problems we had encountered at Cerro Negro and Parachute. We pulled out the rock samples and he quickly spray-painted their surfaces a bright orange color. We then slid the rocks into the aluminum torpedo-tube air lock and pumped out and exchanged the air with the H_2/N_2 mixture. Duane and I then worked on them with the Jaws of Death, collecting the orange-flecked outer surfaces for tracer analyses and eventually recovering an internal nugget that was free of any orange specks. We spent the entire afternoon and early evening crushing up both sets of rocks and placing the coarse powder inside tens of serum vials. We could hear the rain pounding on the roof and it was getting late, but we wanted all the samples in their respective vials, sealed from any possible cross-contamination (figure 6.1B). We passed them out of the bag and the team headed back to the house/lab, where we placed them in the fridge. Tomorrow we would start the growth experiments, but now it was time to head to the Crocodilean Estate, or the "Croc" for short, for a well-deserved late dinner.

When I arrived at the room in which Nora and I were staying, I knocked on the door. When she opened the door I could tell by the look on her face that there was no way I was going to walk into the room in my mine coveralls. What had been clean, bleached-white coveralls were now tan colored with sweat and reeked of a pungent mixture of mine water, diesel fumes, and burnt rock. She demanded that I strip right then and there. "The underwear as well!" she said emphatically, because they too were stained brown. She then escorted me into the shower and handed me the soap: "And don't come out until you pass the sniff test." With me dressed in fresh clean civvies and smelling like a human, we then joined the crowd that had gathered in the main dining room of the Croc, which included Bill Ghiorse, who had just arrived, Svetlana, a greatly rejuvenated Kevin, Louise, Jim, Duane, Joost, Mary, and her husband, Steve. Piet had been keeping them entertained and lubricated waiting for our arrival, but now the drinks were on me as we tucked into crocodile filet with monkey gland sauce.

The next morning I gathered up my mine clothes from where I had left them scattered on the front porch to take them to the house lab for washing, but for the life of me I could not find my underwear. Then I looked across the lawn to see Piet's cocker spaniel waving what was left of my odoriferous, brownish underwear around in the air as though he had just discovered a new toy. "Guess I will skip washing those," I sighed and went to the Croc for breakfast.

Tommy and Susan arrived the next day, and Tommy suited up and joined Duane for a trip down to Duane's salty borehole on 46 level. Susan took one look at our lab and the scattered stacks of enrichments distributed around the former living room and in the incubators and shook her head. She was determined to clean up this mess and she needed help. She looked over at me and asked, "Have you ever prepared microbial media?" I told her I had not and frankly felt like the fifth wheel in the lab, not knowing how to help. She gave me a pitying look and handed me a pair of latex gloves, "Well, you better put on these." She grabbed a pair for herself, and we started to work. I had not had a chance to work with Susan before, but during the next week she put me through a microbiologist's training camp. She showed me how to prepare phosphate-buffered solution to use in desorbing bacteria from our crushed rock samples and then how to syringe the solutions into the different media without leaking air into them. This is a handy little trick for working with anaerobic media outside a glove bag. To avoid contamination during inoculation, she showed me the syringe flamethrower technique. Basically it is a syringe full of ethanol that you light with a cigarette lighter and then squirt on the sealed tops of Balch tubes to sterilize them before inoculating them with the rock extracts.[17] Then she showed me how to do serial dilutions. We constructed a filtered gas source for adding H_2 gas to the media. Because we were running out of media and Balch tubes, we took some of Duane's old inoculated tubes that appeared to be dead, autoclaved them, and used them as well. We inverted the Balch tubes to make sure that air would not leak into them and stuck them in our 60°C incubator. One week at Pfiffner's boot camp for microbiologists and I had learned more field microbiology techniques than in the previous five years.

Meanwhile Nico had set us up with a second sampling attempt. He had arranged one of the drillers to obtain some cores of carbon leader for us from the stope. Greg Slater had arrived from Canada and made some polyvinyl chloride (PVC) core samplers into which we could place the cores and fill with argon gas. Argon being denser than air, we could slide the cores into the tube, and as long as we kept the tube vertical it would provide an anaerobic environment.

Tommy, Greg, Susan, and I went underground the next morning to the stope to begin coring. Tommy and Susan stopped on the way to sample a biofilm, while Greg and I went ahead. We ducked into the tunnel entrance of the stope and slithered down to where we thought we would meet the drillers, who had not yet arrived. We scrunched up on one of the more level spots and turned out our headlamps. We could feel the ventilation air cooling the sweat on the back of our necks while we sat there in the dark. It was surprisingly quiet because no miners were in the stopes yet. It was pitch black; so dark I couldn't see my hand in front of my face. You could hear some low-frequency beats coming through the walls, perhaps due to a loco moving around, but I didn't know. Way down below us I could see headlamps bobbing around, flickering on and off like a swarm of fire flies, but the ventilation carried their sound away from us. I turned on my headlamp and looked over my shoulder. Greg was sound asleep on the rocky, sloping floor. I shook his pants leg and he started up. "The drillers are here," I said. We slithered down to find Tommy and Susan with them. They had set up a bucket of rhodamine dye into which they stuck a pump that sucked up the dye-laden water and pumped it into the shoe holding the core barrel to the rock. It was a pneumatic drill powered by compressed air, and it screeched as the core bit into the rock. As they cored, the cores were pushed out of the hole to the back end of the barrel by the pressure of the water, where we quickly grabbed them with sterile gloves and placed them into Whirl-Pak bags. We then removed the lid to the PVC tube, slid the bags inside, and quickly replaced the cap. The cores came out in discs initially. The intense pressure gradients that existed at the mine wall, from atmospheric pressure to at least 800 times that in the space of a yard, had fractured the rock. Pink water was everywhere in the stope, including on our clothes, but at the end we had probably obtained the most defensible samples in terms of being able to estimate the microbial contamination. Then suddenly the pink water pouring out of the end of the drill pipe cleared. Warm water started pumping out along with another core disc. We had intersected fracture water less than a meter from the face of the stope. It wasn't high pressure but it certainly was not our drilling water. Susan and I quickly swooped in to collect the water for lipids and geochemistry. With rock and water in hand we could see how badly contaminated the rock was before the blasting cycle had begun. Over the course of the next hour, we managed to core one meter into the freshly blasted wall before the explosives team arrived to set the charges for the day's blasting. We slid down the stope to the bottom and popped out at 48 level. As we started back to the cage carrying our precious PVC tubes, Susan stopped to collect a water seep and biofilm from the tunnel wall. One of the samplers stayed with her as Tommy, Greg, and I went on down the tunnel toward the cage. By the time we had arrived we noticed that a crowd of miners had begun to gather. For some reason the cage was delayed. We stood at the back leaning up against the railroad cars to wait for Susan. An hour later the crowd had grown into several hundred tired, frustrated, and increasingly agitated miners, all wanting to know when the cage would arrive. By now the managers had formed a barrier with interlocked arms to keep the miners from pushing into the cages ahead of them. More miners were showing up to cage, and finally the mechanical canary tweets indicated that the cage would soon reach this level. Still, no Susan. Now we were getting seriously worried. We could not leave her down here, but trying to find her in this maze of tunnels would be impossible. Nor would we dare push our way up to the front to join the managers. Finally we saw Susan come in from a side tunnel unexpectedly, and she nonchalantly headed to the surging crowd of vocal and increasingly hostile miners. Her curly brown hair was sticking out beneath her hardhat and she was smiling her "Mona Lisa" smile while she chatted with her sampler, completely oblivious to the danger we thought she might be in. We rushed forward to try to get to her and stop her, but just before the three of us could reach her the miners parted in front of her. Like Moses parting the Red Sea, they all politely stepped aside as Susan and her sampler penetrated deeper into the phalanx of miners. The three of us were gobsmacked, but we quickly jumped in behind her wake so we wouldn't be left behind, and we soon joined the managers stepping into the cage. Susan looked around and then saw us behind her. "Well, where have you all been?" she asked as we clambered into the cage.

While we were burrowing away in the stopes at East Driefontein, Cos Kumahlo, the senior ventilation engineer at Kloof Gold Mine, contacted Duane. Cos told him that they had a water problem and had heard from Nico that our team was interested in collecting water samples. Could Duane come over and collect water from a very unusual deep-water incursion? Duane grabbed his stainless steel containers, hopped in the Combi and dashed up the N12 to

Kloof while we returned underground for another round of carbon leader drilling. The next day he told Jim and me about the flooded tunnel with bulging walls and 60°C water at 41 level of the Kloof mine and what he thought was the best water sample yet. He had even managed for the first time to obtain good water samples for isotopic composition of the noble gases neon, argon, krypton, and xenon that we could use to determine the age of the water. In fact, as we would soon discover, Duane had made the most significant discovery of the expedition.[18]

In July 1999, *Discover* magazine published Kevin's article about our East Driefontein expedition. Appropriately, on the front cover it was called "Journey to the Center of the Earth." Unfortunately for us, the title of the article inside the cover was "To Hell and Back." Kevin had done a sterling job of accurately portraying the sampling effort at East Driefontein and some of the history of the SSP, but the editors at *Discover* just had to make it more exciting for the lay reader by embellishing the narrative with numerous references to a hellish environment in the mines. The title and the frequent references to hell were the mistakes that Rob Wilson had hoped to avoid. Gold Fields senior management was not pleased, as we would soon find out. Oops!

SUBSURFACE LIFE BECOMES A GLOBAL BIOSPHERE: JUNE 1998

Just before we started beavering away in the South African gold mines, an article on microbiology was published in the *Proceedings of the National Academy of Sciences* (*PNAS*) that represented a tipping point for the subsurface biosphere. This paper provided for the first time an observation-based estimate of the total global biomass of prokaryotes to make the point that eukaryotes, including ourselves, were not only outnumbered but also outweighed by prokaryotes.[19] Reading the paper reminded me of one of D. C. White's more colorful characterizations of prokaryote abundance: "If your body was run as a democracy where each cell had the right to vote and they voted by party, the bacteria would easily win every election." Included in their calculations was a jaw-dropping estimate of a subsurface biosphere mass of $4-5 \times 10^{17}$ grams of carbon, rivaling that of the photosphere mass, in which trees represent most of the biomass. The oceanic biosphere was small potatoes by comparison.

The authors used the results of the DOE SSP studies at Savannah River and Karsten Pedersen's work in Sweden, but the most globally significant data set was that derived from marine sediments, specifically that from John Parkes's group at the University of Bristol. John and his colleague Brian Cragg began by studying the depth profiles of total cell counts and activity in cores of ocean-floor sediments off the coast of Peru in 1990 by the Joint Oceanographic Institutions for Deep Earth Sampling (JOIDES) drilling ship, *Resolution*, during leg 112.[20] They found that the number of bacteria at depths of from 3 to 1,650 feet (10 to 500 meters) below the ocean floor varied between 10^4 and 10^8 bacteria per gram of marine sediment.[21] In comparison, agricultural soils typically contain more than 10^9 bacteria per gram. Given that the ocean floor coveres 70% of the earth's surface, their work began to constrain the size of the subsurface biosphere globally.[22]

Assuming the range of bacterial populations reported above, the estimated global mass of subsurface bacteria corresponded to 10^{17}–10^{18} grams of carbon. Since the estimated surface biosphere mass represents approximately 4.6×10^{17} grams of carbon, the subsurface biosphere could represent a substantial portion of the global biosphere. The most important aspect of the *PNAS* paper was not so much what it was about, but who the authors were. William Whitman, David Coleman, and William Wiebe were marine and soil microbiologists, but they had never worked on subsurface samples. They had no vested interest in the subsurface biosphere. Their acceptance and use of the subsurface data accumulated over the preceding twelve years marked a turning point in the perception by the broader scientific community. Studying the subsurface biosphere had now become a legitimate enterprise and a permanent fixture of microbial ecology. No longer would reports of living bacteria from the deep be relegated to obscure scientific journals or popular science news media.

But what of the oceanic crust beneath the seafloor sediments? For the oldest oceanic crust, where the temperature increases an average of 15°C for every kilometer of depth, the maximum depth of life could reach seven kilometers. For continental crust, where the temperature increases an average of 25°C for every kilometer, the maximum depth of life could average four kilometers. In 1992, Martin Fisk, a geologist from Oregon State University, joined an Ocean Drilling Program (ODP) cruise to the Juan de Fuca ridge. On board was a Norwegian petrologist, Harald Furnes, from the University of Bergen. Harald had a graduate student who had discovered, in basaltic glass from volcanic breccia (hyaloclastite) on the island of Surtsey off the coast of Iceland, strange worm-shaped dissolution features that she was convinced must have been created by microorganisms (figure 6.2).[23] The only problem was that there were no obvious signs of extant bacteria inside the wormy vesicles. Harald was unconvinced. But after looking at the basaltic glass of the seafloor pillow lavas being cored by the *Resolution* he found similar delicate, flower- and tube-like dissolution features in the glass, only they were much more dramatic and extensive. He started grabbing anyone willing to take a look underneath the microscope at these strange squiggles in the glass, convincing himself that his student was right because no one else seemed to have a clue (or even cared) as to what else they could be. Soon the rest of the research team on board began to avoid Harald, with the exception of Martin. Something about the budding stalks preserved in glass struck a cord in his very being. It brought to life the normally sterile topic of petrology. These trace fossils may have been readily dismissed as abiotic in origin, if it weren't for the presence of DNA at the tips of these features.[24]

Later in 1998, after the *PNAS* article had been published, the U.S. Science Advisory Committee met to discuss the future of the ODP and the new International Ocean Drilling Program. They invited Tommy Phelps to attend the meeting, and he became the first microbiologist and DOE national lab employee to be a member. During this meeting he presented the tracer techniques that he and others in the SSP had developed in the late '80s, showing the committee just how easy they would be to implement the protocol on the JOIDES drill ship, *Resolution*. He emphasized how important the proto-

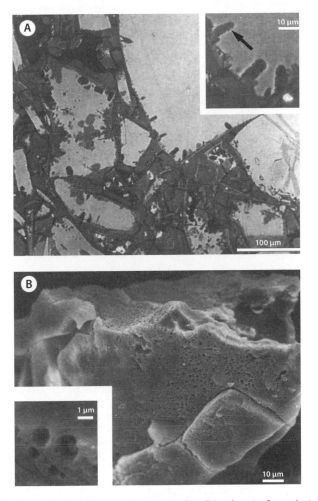

FIGURE 6.2. Dissolution features in oceanic basaltic glass (palagonite) attributed to microbial oxidation/dissolution. A. Back-scattered electron microscope image, in which brightness is correlated with atomic weight. Inset shows micro-tunnels into palagonite. B. Secondary electron image of palagonite with dense array of dissolution pits (from Thorseth et al. 1992).

cols were to resolving whether the bacteria that John Parkes and Brian Cragg had been counting were indigenous to the sediment and not drilling contaminants. Soon after this meeting, the number of scientists investigating the subsurface biosphere would triple in size.

CHAPTER 7

THE SUBTERRANAUTS

"It's just a forest of mushrooms," he said. It may be imagined how big these plants that love heat and moisture had grown. The vegetation of this subterranean region was not confined to mushrooms. "You are right, Uncle. Providence seems to have wanted to preserve in this huge hothouse the entire flora of the Secondary Period of the world."

—Jules Verne, Journey to the Center of the Earth[1]

LIFE IN EXTREME ENVIRONMENTS: JANUARY 1999, BEATRIX GOLD MINE, NEAR WELKOM, SOUTH AFRICA

In January of 1999, just after Duane and Joost had cleared out the grade-plant lab, Duane received another phone call, this time from the Beatrix gold mine, which was located more than a hundred miles south, in Free State. They had a water intersection at 0.7 kilometers (2,300 feet) that had lots of methane. He decided he would collect one last sample for the project and drove down to Beatrix, which was just a little north of Bloemfontein. In his email to me he described a fantastical tunnel along his way to the borehole, which the miners called Gorillas in the Mist. There he had come across "mushrooms the size of hubcaps! Just like in Jules Verne." Although my primary focus had been on the ultradeep sites in search of thermophiles, the expedition had revealed a surprising range of biological diversity in the mines and an abundance of water that diminished with increasing depth.

After Duane had returned from South Africa and we had paid all the bills, it was clear that the cost of the expedition had greatly exceeded our NSF budget and Duane would soon run out of salary.

NSF and NASA had just announced another call for LExEn proposals while we were in South Africa, but the due date was only one month away. This time they were looking for long-term sites for interdisciplinary research in extreme life. Duane and I decided to pursue his suggestion of having small teams on site that could quickly mobilize upon receiving a phone call from various mines to sample a fissure-water intersection. Such teams would require sufficient funds for a field laboratory that would remain active for several years, allowing scientists and students to visit, collect samples, and record observations. We needed to work in teams of three to four people rotating down to South Africa on an approximately four- to eight-week basis. The mine manager from East Driefontein Consolidated sent an encouraging letter, and Rob Wilson suggested several other contacts from the other mines who he felt would be supportive of our teams. Jenn Alexander was also game for another round, having become interested in methanotrophs in the mines. Duane and I therefore assembled a team from our side of the Atlantic composed of eight universities and two DOE labs. Most of the team members had already been involved in the South Africa project on a voluntary basis; this time we could get them some funding. Over the next four weeks we hustled to get data from our samples that we could put into the proposal. Susan quickly ran the PLFA assays on the rock and water samples to give us some idea of the biomass concentration and bacterial composition. We had some positive thermophilic enrichments on the water samples that we had collected as well, but no 16S rRNA data. I had also come across a second article by Beda Hoffman that provided the equations I needed to calculate the rates of radiolysis in the carbon leader.[2] Jim Fredrickson had just measured the gamma radiation tolerance of the *Thermus* SA that Tom Kieft had isolated, and it proved to be quite radiation resistant. I could now calculate how much H_2 and O_2 would be produced by the time 50% of the *Thermus* cells would be killed by the in situ radiation. I then emailed Tommy to help me with the estimate of whether this was enough H_2 for a *Thermus* population to double. Tommy assumed that only 3% of the energy derived from the H_2 would go into the biomass, but even this very conservative growth yield would be more than enough to replace the cells killed from the radiation. We had finally completed the calculation we first started at Susan's and his home in the winter of '93, six years earlier. The energy required just to repair the DNA damaged by radiation would be even less. He concluded in his email, "For excellent maintenance of a drunkard bug having to maintain 3 copies of DNA at a time (wife+2 mistresses) I would want ½ the food for growth, but ORNL management would expect 1000 to live on less. Let me know the errors of my ways. TJP." I wasn't entirely sure how to interpret those last two sentences. Was he referring to our *Thermus* or to polyploidal *Deinococcus radiodurans* or to himself or to all of the above?[3] But our back-of-the-envelope calculation seemed to support the hypothesis that the energy generated by radiolysis would more than offset the radiation damage suffered by subsurface microbial communities.

The proposal was submitted to NSF on March 1, and it was a long shot. Not everyone in the group felt we should have applied given the preliminary nature of our first results, but Duane would soon be on the streets of Princeton with a tin cup, if we hadn't. As Frank Wobber would have said, "You have a snowball's chance in hell." While we awaited the outcome, we carried on with analyses of the samples collected in the previous three months. Jim's new postdoctoral visitor, Ken Takai, swiftly set to work on extracting DNA from the filters that Duane had collected of the fissure and service water. Duane focused on the racks of positive enrichments in our lab but was having difficulty sustaining them on transfer to fresh media. It seemed that as soon as we transferred the culture out of the amended fissure water and into fresh media, they would die. In August, much to our amazement, we were notified that it was back to South Africa that NSF and NASA wanted to send us. The unusual requirement of the award was that all the money went to me at Princeton University, and from there it flowed out to the other universities and DOE labs. So I spent the rest of August setting up subcontracts between Princeton University and the other institutions.

In September 1999, we began planning on how to transform our dreams into reality and create a field laboratory where scientists and graduate students alike could work with South African students so they could develop some microbial skills. While in South Africa, Susan had talked about the idea of training black South African students, referred to euphemistically as the "previously disadvantaged" students. We had been drinking a delicious vintage of Pinotage at the Croc at a dinner party in celebration of Piet and Pietra's twentieth wedding anniversary. Piet held it at the Castle, his future conference center that was still under construction. Over the sounds—many raucous cheers and stories were being given about Piet, Pietra, and the Croc—we had discussed Susan's idea. She proposed bringing minority students from the United States to work and socialize with the South African students with support from NSF. Susan was a natural-born teacher and she clearly enjoyed showing the mine samplers who had helped her underground some basic microbial sampling skills. Suddenly cheers went out from the seventy or so guests as another toast was given to the Bothas, and Tommy, Susan, and I raised our glasses as well. South Africans were simply amazing; once you got to know them their hospitality seemed boundless.

Now that we had money to return to South Africa for five years, we needed to find the South African students. I had contacted program managers at the National Research Foundation (NRF) of South Africa (South Africa's NSF) before I left South Africa and now reached out to them for suggestions. They had programs to support students' travel abroad but not for travel within their own country to a workshop. So I prepared an application to the NSF International Program for funds to host a conference that would kick off the project and to which we could invite as many South African scientists and engineers as we could muster who might have interests in collaborating with our effort and from there develop a student training program. After all, it was their water—or the mines' water, depending upon your point of view. They should be interested in what type of microorganisms it contained and how they could be used to clean up the toxic legacy left behind from deep metal mining. We would include a trip underground to East Driefontein, because they had seemed pleased with our previous trip. The earliest we could submit the proposal, however, was February 2000, and the earliest we could have the workshop was November 2000 in order to mesh with the semester break and exam schedules of the South African universities. Jenn Alexander was, unfortunately, retiring from the University of Witwatersrand and could not host the conference there. Rob tried other faculty but grew increasingly frustrated because he couldn't understand why no faculty from his alma mater were interested. He contacted Rudy Boer, who had taken a faculty position with the Institute for Ground Water Studies, at the University of the Free State, in Bloemfontein. He and Frank Hodgson, a long-time consultant to the mining industry on groundwater issues, agreed to host the conference. With our South African academic hosts solidified at last, we submitted the proposal.

During the spring, more 16S rRNA data was pouring out of Ken Takai's hands as he worked feverishly on the filters of the fissure water and focused on amplifying the Archaea. It was not straightforward because of the low total biomass and the apparent paucity of Archaea, but he had succeeded in obtaining a 16S rRNA sequence of something close to a *Pyrococcus abysii* from the Kloof hot-water sample collected by Duane. *Pyrococcus abysii* was an anaerobic hyperthermophile that had been isolated from a deep-sea hydrothermal vent collected at a pressure comparable to that of the Kloof borehole. Finding the DNA signature of a submarine-vent hyperthermophile in fracture water 10,000 feet beneath the surface at Kloof was compelling evidence that indigenous microorganisms inhabited the fractures. The fracture-water sample from Beatrix yielded a new clade of Euryarchaeota, which Ken called SAGMEG, for South African Gold Mine Euryarchaeota Group. The new archaeal sequences were so novel that Ken had to completely reconstruct the archaeal genealogical tree. I decided to start placing this data—along with tables on the geochemistry of each site, geological maps, cross sections, and the field photographs and microscopic images of each positive culture—on a website for easy access to all the different groups. Constructing data depositories on the Internet was new to me, but I was assured that our department servers could be trusted as long as I provided everyone with a username and password. I struggled to come up with something I could easily remember but that no one could possibly guess and finally it hit me: username = *dirbars*, password = *biltong*. Unless they frequently drove down the N12 past East Driefontein, no one could possibly guess those. Besides, all the time we were there, we saw no one actually buying or selling biltong at the zebra-striped Dirbars Biltong stand on the side of the road. I collected the data in a report to the National Geographic Society and started sending the data reports to the mine manager at East Driefontein Consolidated.

I needed a nom de guerre for email communications to all of our various research groups, like Rick Colwell's Les Chutres signature for our Parachute campaign. I tried Biomining Buanas, but it seemed too racist, sexist, exploitative, and colonialist all at the same time. I thought of Anna Louise plying the depths of the oceans in a submersible. Surely she would be called submariners, I guess. So I decided our nom de guerre should be Subterranauts. As spring progressed, Duane headed out to PNNL to work up the 16S rRNA eubacterial clone libraries for his favorite borehole in East Driefontein. Ken Takai, in his spare time, had isolated a novel thermophilic *Clostridium* that had the highest pH tolerance of any thermophile on record. This and other results I communicated to East Driefontein Consolidated from the Subterranauts. I had not received any replies, however, to my reports for over a year and a strong sense of déjà vu descended upon me. It was soon thereafter that Rob Wilson, who was again ready and willing to play the role of our liaison to the mines, informed me that Gold Fields wanted no part of us. The frequent references to "hell" in the *Discover* article the previous year had apparently incensed the management, who felt that the article had impugned the Gold Fields reputation for mine safety. They, who like myself, had not seen a draft of the article, considered it a damning indictment of their safety record. They wanted no part of American scientists, photographers, and writers in their mines again. In August, NSF's International Program awarded us the grant for the workshop, but Rudy's health had tragically deteriorated during the spring with the onset of Parkinson's, and he was unable to act as host.

I was looking at the September calendar for 2000. I now had money, considerably more this time, from two NSF grants, one a year old, but I had no mine, let alone mines, in which to work and no South African university at which to host the workshop. I needed to fix this now or I would be returning the awards to NSF. Rudy suggested that Rob contact Derek Litthauer, the chairman of the Department of Microbiology, Biochemistry, and Food Biotechnology at the University of the Free State. Maybe they could host the workshop.

After talking to Rob, Derek Litthauer discussed the project with his former graduate student, Esta van Heerden. Esta was originally from Durban in the Natal Province but had been lured to the University of the Free State as a Ph.D. candidate by an exceptionally generous bursary. She was just finishing the final stages of her dissertation on a yeast protein that had lipase (fat-eating) properties. She was being groomed for a faculty position in her department, where she would become their second female faculty member, a rarity in South African academic circles. When she was visiting the University of Nice and performing analyses of yeast proteins, she had interacted with a research group examining thermophilic acidophiles and was amazed at how little was known about these organisms. No one in South Africa was doing this type of research. It had left an indelible impression upon her psyche, but when she returned to the University of the Free State, she pondered how to get started in that field. This is what we would call a serendipitous moment in time, but she had to hurry, as the only time for the conference would be in December, which was less than three months away. Esta contacted Rob, who contacted me, and soon we were making arrangements for the workshop. She would be sure to invite other microbiologists from other universities, including the University of Witwatersrand, University of the North (a historically black South African university), University of Natal, University of Cape Town, University of Western Cape, etc. to attend.

NSF was also eager for us to recruit minority, in particular black American, students into performing research on this project in an attempt to increase minority interest in the field of earth sciences. The Geosciences Division of NSF funded part of the workshop. The program manager, Rich Lane, had had enormous success with outreach to African students in a previously funded grant from the Carnegie Institute of Washington on seismic tomography of the South African continental crust. He proposed that we bring high-profile, black American scientists into the workshop. The problem was that there weren't that many black Americans in earth sciences. You could probably count them on the fingers of both hands. But Rich was very, very persuasive, and he found two of the best—Wes Ward and Brendlyn Faisson. He convinced them to attend a workshop in the middle of South Africa with two months notice.

Wes Ward was a noted planetary scientist who worked on Mars photogeology at the USGS Astrogeology Center in Flagstaff, Arizona. He was very charismatic with an almost spiritual countenance, but with a laugh that could flood a lecture hall. He was teased about his laugh: "You sound just like Gene," they would say, referring to Gene Shoemaker, the center's founder and a geologist made famous by his early work in the Apollo program, meteorite impacts, and hunting of Earth-crossing asteroids. "That's because Gene's my father," he would reply. He was right, of course. At one time or another, Gene and his wife, Caroline, were the adoptive parents of many geologists and Caltech students passing through Flagstaff, including me. Which I guess meant that Wes and I were sort of stepbrothers.

Brendlyn Faisson was an accomplished geochemist at ORNL and knew Tommy and Susan quite well. She specialized in environmental geochemistry, including the extraction of petroleum from coal by microbial processes. But she also had a long-abiding interest in the exploration for life in our solar system, having taken classes from Harold Morowitz, the lead scientist on the Viking Lander life-detection experiments.

Tommy and Susan were definitely up for going to South Africa again. Mary DeFlaun, who was working on an "Africa room" for her new house, was delighted to come over, if only for a day, to talk about the microbial isolates obtained from the first samples at East Driefontein. Gordon Southam agreed to fly in with Wes. Barbara Sherwood Lollar couldn't make it, but she would send down a new graduate student whom she had recruited to work on the project, Julie Ward. The one-and-a-half-day workshop was filling out nicely.

BIOPIRACY IN THE TWENTY-FIRST CENTURY: DECEMBER 4, 2000, UNIVERSITY OF THE FREE STATE, BLOEMFONTEIN, SOUTH AFRICA

Three months later, on December 4, Duane, Susan, Tommy, Gordon, Mary, Wes, Brendlyn, Julie, and I flew into Bloemfontein, "the fountain of flowers." Bloemfontein was very much like any college town in Midwestern USA in the middle of the semester. KFC, Spur, Steers, McDonalds, and other fast-food restaurants decorated Nelson Mandela Boulevard, the town's main street that passed by the campus. The grounds were a mixture of old Afrikaans style and modern architecture comprising mostly white buildings on beautifully adorned grounds with fragrant bushes in full summer bloom and bespeckled with students both black and white. I think we were all a bit taken aback by the department. It was a one-story, white stucco building with outdated wet labs and a few ancient PCs and emanating a third world flavor. As it turned out, however, the Department of Microbiology, Biochemistry, and Food Biotechnology held the yeast culture collection that was used in brewing South Africa's favorite beverage. So it did have lots of cold rooms, incubators, and a fully equipped lipid laboratory. We stayed just outside of Bloemfontein at the Bains Game Lodge, a suburban-style game park that housed primarily zebra and ostriches. Its celebrity ostrich strolled around the main lodge/bar area displaying a featherless left arm to all the guests. Apparently during one of the braais, this over friendly ostrich wandered a bit too close to the grill and caught fire. A few stalwart Afrikaners managed to tackle it while it was running around with its wing in flames, wrestled it to the ground, and smothered the flames, but its feathers had not grown back.

Rob, Esta, and Derek, good to their word, had assembled an impressive number of South African academic and industry participants in the university's central auditorium. During the morning of the first day Duane, Gordon, Susan, and Julie superbly summarized our results from the previous year and how our results could prove useful to the mining industry. Duane focused his talk mainly on the 16S rRNA sequence identities of the uncultivable microbes from the fissure water, results that Ken Takai had just submitted for publication. The data validated not only the concept that the fissure water intersections carried indigenous microorganisms, not mining contaminants, but also that some of these microorganisms, like the *Pyrococcus abysii* from Kloof, resembled those of deep-sea vents, just as Anna Louise had predicted two years earlier. The amazing thing was that the Witwatersrand Basin had not seen marine ocean water in two and a half billion years. How did *Pyrococcus abysii*, an obligate hyperthermophile, travel from a deep-sea vent in the middle of the ocean to freshwater fracture two miles underground in central South Africa?[24]

The session broke for lunch and then in the afternoon turned to biotechnological applications with a representative of BHP Billiton, a metal mining firm, speaking about their significant advances in bioleaching technology. Brendlyn provided an excellent presentation on coal liquefaction using microbes. It was during her talk that the representatives from SASOL, South Africa's utility in charge of coal gasification via an industrial scale Fischer-Tropsch process, sat up with ears perked. Don Cowan, an outstanding microbiologist who had recently moved from Great Britain to the University of the Western Cape, followed Brendlyn's presentation with a more theoretical topic. He promoted the idea of simply removing the genes from the environmental DNA in pieces and transforming them with plasmids into *E. coli* to build expression libraries. As he explained, if researchers are interested in a specific activity, say lipase, then they challenge this library with an assay for that activity. This approach enables them to get around the inability to grow the microorganism in the lab, at least in principle. The last talk of the morning was by Mary. She presented the results from her characterization of a *Geobacillus* she had isolated from a hot water seep in East Driefontein that demonstrated a thermophilic lipase activity. She also summarized the physiology of the *Thermus scotoductus* isolated by Tom Kieft and *Alkaliphilus trans-*

vaalensis isolated by Ken Takai.[5] Esta was immediately intrigued by the metabolic plasticity of the *Thermus* and wondered if it could be adapted to reduce other more environmentally toxic metals in the environment. As the audience began moving toward the lobby and refreshments, one of the South African participants quickly ran up to Mary exclaiming, "I've been looking for thermally stable lipase for my entire career and you found it in the first isolate you made from a mine? That's incredible." Mary was beaming with pride, "I know. Well, of course. That's what we do," she said shrugging as though she did this all the time.

After the break the workshop focused on the biodiversity controversy, and temperatures discernibly rose in the room. Three months earlier, just when I thought I had fixed all my problems, I had received an email from one of the invited South African participants, a microbiologist from one of the universities in the Natal Province. They had just received their invitation from Esta. They accused me of violating the Convention on Biological Diversity (CBD) and, in fact, of breaking South African law during the research we had already conducted at East Driefontein Consolidated. I was clueless about the CBD.[6] But as I quickly discovered upon searching the Internet, the collection of natural samples beyond national borders was guided by the CBD, and we were being accused of "biopiracy." Our goal with this workshop, however, had been to encourage and train South African academic researchers to be the discoverers and patent bearers. So I felt we were compliant in principle. The problem, however, was the ownership of the bacteria. Were they the property of the mine, or of the government of South Africa? The NSF was explicit that any international research grant must be in compliance with the CBD, even if the U.S. Senate had not yet ratified the treaty. The very thing that I had sought to nurture was now being used as a weapon against us. "At first we were DOE spies, and now we're biopirates, AAAARRRR!" I muttered to myself as I called Rob Wilson to work on damage control. "Ach, look, I'll contact a lawyer who has written our contracts with the mines. He knows this IPR [intellectual property rights] stuff forwards and backwards." Rob's lawyer, who had agreed to come to the workshop and to give a talk on IPR and CBD, had arrived. So we had our ace in the hole in case the afternoon discussion came to harsh language.

The first presentation accused me of instigating a program of bioexploitation and proposed instead to create a homegrown, South African institute of biotechnology that would not only serve the role that we were proposing to perform in our LExEn project, but also perform gene expression arrays to screen mine bacteria for useful properties and provide an archive for interesting microorganisms. It was a South African vision for South African bioexploitation that did not require any assistance from our group.[7] Fortunately in the presentation that followed, the lawyer whom Rob had brought in for the meeting was quick to point out that the South African legislature had not in fact voted on any laws even concerning ownership of plant- or animal-sourced biochemicals or genes, let alone those from subsurface bacteria. He also emphasized that ownership of any intellectual property resulting from the characterization and application of microbes from the mines had to be negotiated between the mine company and the academic institution. Some of the people in the audience were murmuring among themselves, clearly not satisfied, but I was breathing a lot easier. Rob had once again saved our bacon. Fortunately this exchange was followed by a presentation by two geophysicists, one from Wits University, Sue Webb, and the other from University of Cape Town, Marian Tredoux, on how they, as partners in the NSF-funded seismic crustal tomographic project, had received funding from the NRF of South Africa to support student bursaries, which they used to recruit and retain black South African students in geophysics as part of a "high flyers" program. They provided examples of the success stories and the failures in an almost how-to-do-it formulation that stimulated a lot of discussion over dinner later that evening.

Wes's talk on Martian surface geological features was the last talk of the day, and he was clearly the hit of the workshop. For the South African students it was a rare treat to listen to a planetary scientist who was an expert on the forces that shaped the surface of Mars. Af-

terward they crowded around him, white and black alike, like group-
ies around a rock star. It was gratifying to see how much interest the
South African youth, having grown up under sanctions, expressed in
the exploration of space, and they hung on Wes's every word as if he
were an astronaut.

That evening while the others were out dining, Tommy, Susan,
Mary, and I sequestered ourselves at the Bains Game Lodge and pow-
wowed, while feeding treats to the zebra roaming around in the
compound. Not surprisingly, the workshop had revealed the internal
politics of competing South African universities. One group was
composed of the University of the Free State and the University of the
North, the historically black university. The other group comprised
universities from the Cape and Natal provinces. The problem for us
was that we knew Esta and Derek only by their actions, which to date
were honorable and gracious, but they had limited experience in microbiology. The other group had technically better-equipped institutes and microbiologists with established credentials. The University
of Witwatersrand's microbiology department—which was the logical
partner given its proximity to the deepest mines, its resources and
student base, its preeminence, and our previous collaboration—never
showed up at the workshop. The long-term success of the project
hung in the balance, and we had no clue as to how the South African
funding agencies and legislated CBD rights would swing, but we had
to make the decision then and there. If there was just one thing that
we had all learned from working in Frank Wobber's program, it was
to make sure you nurture the young and needy. Tommy strongly endorsed Esta, and we all agreed and hoped that the god of microbial
ecology would bless the decision. To get Esta started we would try
using the Research Experiences for Undergraduates (REU) program
at NSF. Susan had devised a brilliant plan of organizing six-week-long
workshops to be held at Free State in which about a dozen minority
students from the United States would be brought over and teamed
up with black South African students to analyze samples from the
mines while learning about environmental microbiology, geochemistry, and hydrology. It would jump-start Esta's program by recruiting

students for thesis projects and would provide for an interesting cultural interchange between the two groups of students.

The next morning I closed the workshop by summarizing our ideas to fund a student education program, to comply with the biodiversity treaty through archive sample distribution, to distribute our current thermophilic isolates to any South African scientist upon request, and to promise not to pursue any biotechnological applications, leaving those to our South African collaborators. Later that day the group guided by Sue Webb headed north to a platinum mine owned by AngloPlat in the Bushveld Complex, the Northam Mine. Because Gold Fields had shut its doors to us, Rob had convinced his connections at Northam Mine to let our group travel underground there as part of the workshop. He had previ-ously worked on cooling systems for this mine, and because of its ex-tremely high geothermal gradient, it looked like a promising target for hyperthermophiles.

While the rest of the conference attendees drove north to Northam, Rob and I drove to Joburg to talk to the management of Gold Fields. But first, we needed to shop for housing around Carletonville, some-place where we could set up a field lab. The past couple of years had not been good to this community. The low price of gold had led to cutbacks and unemployment, and AIDS had taken a visible toll, par-ticularly on the black South African communities servicing the mines. Carletonville's hustle and bustle of '98 was replaced with a laconic listlessness and an undercurrent of crime that threatened the security of our field lab even though high-quality housing was available for a pittance. Ashen-faced and gaunt black South Africans were a com-mon and unsettling sight on the roadside. So we looked at locations removed from the local cities. As we drove from one mine to another along the N12, we passed big billboards advertising the importance of using condoms during intercourse. I couldn't believe my eyes. In a mere six years this society had gone from one where you couldn't purchase a *Playboy* magazine, to one where billboards were preach-ing safe sex. We turned right onto a road that led into the smaller, and hopefully safer, village of Glenharvie. We pulled up to a three-

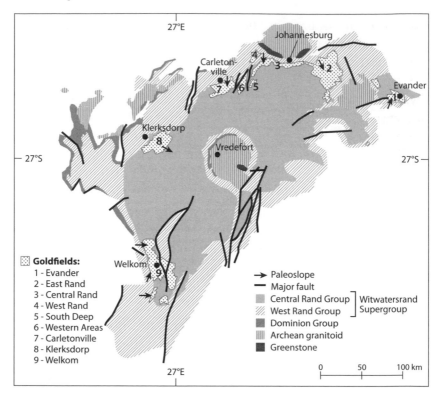

FIGURE 7.1. Map of the Witwatersrand Basin showing limits of surface outcrops and locations of mines and the Vredefort impact structure (from Frimmel 2005).

bedroom, one-bath house with garage and carport leased from Kloof mine on its outskirts. Across the street from it was an open field of scrub brush and grass that climbed up to a ridge of Pretoria Group metavolcanics.[8] It had a well-trimmed backyard with high wall and a secure-looking gate and fence in the front. The icing on the cake was that Glenharvie was only a heartbeat away from the Kloof no. 4 shaft, which was the source of the hot fissure water that had yielded our *Pyrococcus abysii* clone. It was also centrally located so that within two to four hours we could be at any mine in the Witwatersrand Basin (figure 7.1) or in the Bushveld Igneous Complex. The three bedrooms and sitting room could easily house the field team, and the garage could readily be converted into our field lab. "This is it," I said to Rob. We got into his car and drove back to Joburg.

The Kloof gold mine was run by Gold Fields, who considered me persona non grata. So the next day Rob and I met with the head of Gold Fields and essentially begged him to let us back on the property. This he finally agreed to do, but only after a formal memorandum of understanding (MOU) was negotiated between the company and Princeton University and the other eight participating institutes mentioned in our LExEn proposal. That would be utter torture for me, but we had no choice and agreed to the terms.

In February 2001, Duane, Julie, Lihung Lin,[9] and other assorted graduate students and postdocs returned to South Africa, and with the help of a local contractor, converted the carport in our new field house into a wet lab with walls, cabinets for chemicals, and a sink with a geezer for hot water.[10] Of course the sink drained into the backyard, so we had to be somewhat circumspect as to exactly what we were pouring down the drain. They assembled our anaerobic glove bag, which we had shipped back to South Africa, and placed it along with incubators, a freezer, and autoclave in the garage. Tommy had constructed a gas-mixing station and O_2 stripper from hardware-store parts and had shipped them to South Africa. The small one-room, one-broken-bath servant's quarters behind the garage had long been vacated, a relic of the apartheid era. This became the ideal storage room for all the sampling vials, pipettes, filters, bottles, and various other sampling devices. We used the dining room for computer communications, as a library, and as a microscope station. We also recruited an undergraduate, Mark Davidson, from the University of Wits microbiology department.[11] I also made sure that we hired a house cleaner and gardener to keep the house from deteriorating into a frat hall. To reduce costs and legal hassles, Rob even leased us his wife's Saab.

The first team established email communications and, by hosting a few backyard braais, soon became friends with the various mine geologists from Beatrix, Kloof, and a gold mine located far to the east of Joburg, known as Evander. Evander Gold Mines was run by Harmony Gold Mining Company, a competitor of Gold Fields. During the initial forays that spring, these mining geologists led our teams to old weeping boreholes and containment dams. We used an array of plastic fittings picked up from the hardware store in Carletonville to seal the boreholes, which came in a surprising range of sizes, and directed the water flow through sterile tubing into the stainless steel syrup canisters, since this seemed to have worked at Kloof earlier. We took a different approach with the media that each team brought, especially the anaerobic media. To minimize contamination and air exposure, we took the Balch tubes, preloaded with media, down with us to the boreholes and filled them directly from the tube through a small syringe needle and valve.

The field house was soon buzzing with activity as water samples were collected, filtered, and preserved and media tubes were inoculated and incubated. But then, just as we had hoped, phone calls started from the mines, reporting fresh intersections. In these cases, however, the size of the borehole was never known beforehand, and the flow rates were simply too high, and in some cases too hot, for our flimsy plastic fittings. Fortunately we were able to purchase some telescoping cementation compression packers from a local mining cementation firm, Cementation Products.[12] These metal pipes had an ingenious design that allowed a person to simply insert and twist the packer into any borehole, and they provided a high-pressure seal, typically up to 15 megapascals. We took several of these back to the machine shop at Princeton, where George Rose reconstructed them out of stainless steel, some with plastic tube liners, and then constructed a manifold out of a high-temperature, autoclavable plastic that could be screwed into the packer. The complete package enabled us to collect multiple samples simultaneously while never exposing the sample to air and avoiding contamination of the water. We could also regulate the flow rate with a huge ball valve at the end, turning the manifold into a low-flow-rate sampler so we could finally obtain reliable measurements of the dissolved O_2 and uncontaminated gas samples. This technique reduced the time at the rock face, which was crucial because these fresh intersections were often miles away from the shaft and time at the rock face was often limited to less than an hour before we'd have to turn around and head back for the cage. The packers and the new manifold—christened the Octopus because it had eight ports, six of which typically had Tygon tubes hanging from them while being used—worked like a charm even though they added considerable weight to the backpacks. Samples began to pile up in the refrigerators and freezers of the field house, so every time a team member flew out of South Africa, they took back coolers full of samples. We then shipped these samples to the other collaborating institutes for analyses. Surprisingly there were few mishaps in the lab other than the occasional finger jab. The worst incident was when Mark managed to melt his jeans one day while preparing a sodium hydroxide media supplement, leaving him in the embarrassing position of driving back to Joburg in his boxers.

One of the new team members was Dr. Johanna Lippmann. Jim Fredrickson had brought her work to my attention with a recent paper of hers. At the University of Potsdam, she had performed noble gas isotope analyses on rock cores collected at depth and was able to determine from these analyses the age of the pore waters. She was currently working as a postdoc in Martin Stute's lab at Lamont Doherty Ocean Observatory in New York, so I contacted her to see if she and Martin might be interested in this project. She was and had provided us with a few copper tubes to collect noble gas samples for dating the fissure water during our previous expedition. Now we actually had some funding from NSF, and after we'd worked out the finances, she was on the plane to South Africa with a box full of copper tubes.

A series of expeditions focused on Beatrix Gold Mine, where Ken Takai had discovered the novel euryarchaeal group, SAGMA 1.[13] Samples were collected over a period of weeks, especially from the newly sunken shaft no. 3, which was one mile deep. A horrendous CH_4 explosion occurred at Beatrix Gold Mine on May 8, 2001, however, that killed twelve miners and shut the mine down for several months. Fortunately none of the students had been underground there at the time, but it was a wake-up call to me that not all the mines were as safe as East Driefontein had appeared to be. From that point forward I switched my body clock to South African time, waking up at 4 a.m. to call the field lab and check to see if everyone had come up from underground in one piece.

All the activity by foreigners staying at the Glenharvie house had not gone unnoticed by the locals. Often we were mistaken for doctors, and the gardener would occasionally bring by a friend who would beg for some medication to help his family. For simple wounds we could offer some alcohol and aspirin, but they were more often seeking antivirals for HIV, something the South African government at that time was reluctant to provide. Then there were the break-ins, usually to steal stereos, cameras, and computers, which stopped as soon as we hired a company to install a security system. South Africans had come to rely on these private security firms, because the only places you could reliably find policemen were at the numerous speed traps, where they were "radaring" motorists.

Meanwhile I was steadily losing my hair running between Princeton's and Goldfield Ltd.'s lawyers finalizing an MOU that would grant us access to the latter's mines as long as we passed all our papers through them prior to publication and, most important, did not pursue any commercial applications based on the data or microorganisms retrieved from these mines. After eight weeks of faxes and MOUs flying back and forth between the States and South Africa, we finally had our agreement with Gold Fields in place. We could now reenter Kloof.

Fieldwork had progressed smoothly with teams of two and three rotating in and out of the Glenharvie house/field lab through the austral winter until September 11. The al-Qaeda attack on the World Trade Center occurred just as Lihung was flying from South Africa to JFK. Fortunately his flight was redirected to Rome, Italy. Although I was not too concerned about Lihung having to stay in Rome a couple of days until international flights resumed, I was worried about possible al-Qaeda attacks in South Africa because of their track record, and in particular, I was worried about two students who were alone at the Glenharvie house.[14] I asked Rob to check on them, but Rob assured me that everyone in South Africa was in shock about the 9/11 attacks as well and that there were no signs of al-Qaeda sympathizers at work in South Africa. To the contrary, the feeling of empathy toward Americans was running higher than he had ever seen it historically.

THE SACRED BALANCE: OCTOBER 1, 2001, KLOOF GOLD MINE, GLENHARVIE, SOUTH AFRICA

Two weeks later, after Lihung was safely back in Princeton and international flight schedules had returned to normal, Duane and I headed to South Africa to join Amy. I saw the smoldering hole in the Manhattan skyline that used to be the iconic World Trade Center as I crossed the Verrazano-Narrows Bridge on the way to JFK. Like many, my heart was still tender from the loss of so many souls on that day, and I was very wary on the flight over to Joburg. Whenever any male passenger looking remotely Middle Eastern stood up and left his seat, everyone tensed. But there was a job to be done and our target was Northam platinum mine, which had reported some fluid intersections, and we wanted to follow up the November sampling of the previous year with a complete sampling suite. Upon returning to the field house after gathering supplies for the journey north, we noticed a van parked in our driveway. Parking in the street, we walked down to the house and no sooner set foot on the doorstep than the door opened and a Japanese-looking man with a Canadian accent greeted us. It turned out to be David Suzuki, an acclaimed Canadian geneticist recognized around the world for his many television shows on science and the environment. I remembered him from the CBC's *The Nature of Things*, which I had watched avidly when I was a postdoc in Toronto in the early '80s. He and his television production team had arrived in South Africa wanting to go underground with us to one of our sites. He wanted to feature it as an episode on the miniseries he was producing called *The Sacred Balance*. Duane offered to take them to the borehole site where Ken Takai and he had first found *Pyrococcus abysii*, down on 41 level of the Kloof no. 4 shaft. Given the stature of David Suzuki and the strong promotion of environmental preservation and management in *The Sacred Balance*, Rob was able to convince Gold Fields that this would be a total public relations triumph for them. Kloof mine pulled out all the stops and spent two days working with the film crew. They assigned Arnand van Heerden, the assistant chief geologist at the no. 4 shaft, to guide us underground and make certain everything went off without a hitch. The night before the shoot, we all met at the Croc, where Piet and Pietra exceeded their normal ebullience in entertaining our special guests that evening. Over croc steaks, peri-peri sauce, and Pinotage we made it clear to David's team that we also needed to actually obtain samples from this site as well. The next morning we arrived at the no. 4 shaft parking lot in full battle gear, backpacks full of sterilized bottles and carrying stainless steel canisters preparing to enter the sausage grinders when we were motioned to a side door that they unlocked for us and David's film crew to pass through. They then shot several takes of us getting into and out of the cages crammed with miners. I simply could not believe it. The mine was actually holding up the cages for David's team to do the filming. When we finally arrived on 41 level, we did not go straight to the site; instead, Arnand had arranged for us to travel through several tunnels to the stopes, down to a tunnel they were extending. We moved down the tunnel like a multicellular amoeba, with the lighting crew and the cameramen racing ahead of us to set up the shot ahead while we steadily marched along, stepping on the railroad ties, David in the lead, followed by me, Duane, and Amy. As we approached the camera crew, we carefully ignored them while gazing wonderingly at the dark, flinty rock surfaces pretending to look for biofilms. "So this is what playing an extra in a movie is like?" I whispered to Duane as we marched along. After we passed the camera crew, they broke down the camera, grabbed the lights, and raced ahead of us again to the next shot. We eventually came to an inclined tunnel to take us down another 100 meters to 45 level, the deepest in the mine. We could not get there by the cage. Instead we took what looked like a converted rope tow bar from a ski lift that carried seats with stirrups. We stepped up to the launch pad of this ski lift holding our backpacks in front of us, and then as the chair came up, swung around the head of the ski lift, and turned to go down, one by one we grabbed the bar with our right hand, quickly swing our left leg over the saddle, jumped on, pulled our feet up into the stirrups, hugged the bar around our backpack, and went down the inclined tunnel, bobbing, swinging, and swaying in the saddle. We were warned before mounting this contraption that whatever we did, we should not put our fingers at the top of the bar or we risked getting them sliced off by the wheel that connects the chair to the steel cable. Suddenly, we heard a scream above us, and the rope tow came to a stop leaving David, Amy, Duane, and me swinging in the tunnel. Unfortunately, one of the cameramen had grabbed the cable and the tip of his finger had been quickly sliced open. They were frantically staunching the flow of blood from his finger. He was placed on a bicycle cart[15] and taken swiftly back to the cage and up to the surface to the medical center. Despite one missing cameraman, David's team kept moving toward our target. When we arrived at the tunnel and started moving downward toward the end, the heat quickly became unbearable. It was a sauna. We dropped our gear near the borehole. Arnand checked the temperature and for CH_4 and then waved the camera crew in. We partially unpacked before beating a retreat back to the cooler portion of the tunnel. Although the mine had turned on the ventilation, it was clear it had not been on for long enough. Fortunately back near the head of the tunnel was a tap for chilled service water that we used to douse our heads. The camera crew worked in shifts to set up the lights and camera around a valve that Andrea had cracked open. Water was now pouring out, bringing more heat into the tunnel. As the camera crew ran back and forth to set up, another one of them tried to figure out where on me to put a microphone and, unfortunately for me, decided to put it beneath my hard hat. David, Duane, Amy, and I then moved in toward the water. Duane, Amy, and I performed our routine for water sampling. We attached the Octopus to the valve, and then using the field probe, Amy and I measured the temperature, pH, etc. and collected biosamples and samples for geochemistry and gas analyses. All the while Duane was filling his stainless syrup canisters. Andrea, who seemed immune to the heat, was watching everyone, taking people who were overheating to cool off at the service water tap. All the while, David was asking me questions to which I would reply again and again and one more time until the director was satisfied. As other people were taking turns cooling off, the microphone in my hard hat meant that my brain was beginning to boil. But David stuck it out with me, and when we finished the shot they took me out. I ripped the helmet off my head before they had a chance to remove the microphone and plunged head first into the shower. David was amazing, however, because he showed little signs of wilting under the heat despite the fact he was twenty years my senior. We rode back on the bicycle cart together, with my head still swimming. At the surface we found out that the medical center had successfully restored the cameraman's fingertip. Duane, Amy, and I headed back to the field lab to process the samples and break down the equipment and clean, sterilize, and prepare it for tomorrow's trip to the Northam platinum mine. We then drove into Joburg to Rob Wilson's house. Rob and his family hosted a dinner party that evening, which was attended by his colleagues from Turgis Technology, David, and his production team. David's crew seemed quite pleased with their day's work and showed us some of the footage of lots of very sweaty and red faces, fumbling fingers, and equally fumbling lines. I could see that I had better stick to my day job and leave this television stuff to a pro like David. The next day David headed back to Canada while his entourage headed to Sun City, supposedly to collect context footage. The three of us piled into our Saab and headed for Northam platinum mine. Several hours later we passed Sun City on our left as we continued on the road to Thabizimbi.

Upon returning to the Glenharvie house three days later, Duane, Amy, and I set about wrestling with a new problem. Before leaving for the United States in September, Lihung and Amy had been shown a new tunnel located at 8,860 feet depth at Driefontein no. 9 shaft. During his sampling run, Duane found that the tunnel contained a series of flowing boreholes, including one large-diameter exploratory well that descended vertically downward to 11,000 feet. He said brackish water was flowing out of it like an underground artesian hot spring, and frequently the water's surface roiled from gas eruptions. The microbiologists who specialized in sampling terrestrial hot springs often debated whether the bacteria they were collecting had ascended from the subsurface plumbing that fed the hot springs. But here we had an actual underground hot spring. The top of the borehole contained a thick black biofilm where the water poured out, and downstream in the water-filled channel along the tunnel floor, whitish filaments of sulfide-oxidizing bacteria waved with the constantly fluctuating current. Although this hot spring was located deep beneath the surface, we could not assume that the bacteria residing at the surface of the borehole were from the subsurface fractures feeding this conduit. Nevertheless, it was our "Dream Borehole" —our deepest, hottest one yet with a seemingly endless supply of water where we could test the concept of using artesian hot springs to sample the subsurface biosphere.

But how could we reach deep down more than 2,000 feet into this borehole to collect the least contaminated samples near those fractures? Duane went down to the hardware stores and fished around until he came up with the material for a bailer. The design was simple, portable, and robust enough to work. It was constructed out of a six-foot-long rigid PVC pipe and section of thick-walled, transparent vacuum hose that would not collapse under the pressure of 2,000 feet of water. The bottom end had a nipple that let water flow into the tube, and the top end had a nipple to let water flow out. The tubes were connected with hose clamps. Inside were two of the biggest, heaviest steel ball bearings that Duane could find, each sitting in a slightly narrower tube squeezed by a hose clamp. As the bailer descended into the hole, the water flowed into it, pushing up the ball bearings and filling the tube. When the bailer was pulled up, however, the ball bearings would slide down the tube and sit in their tube nest, sealing the entire PVC bailer from the outside water. The dissolved gas should ascend to the bottom of the top steel ball bearing. Duane had cut a small hole in the top tube and inserted a small butyl rubber stopper where a syringe could be inserted to remove the gas without risk of air contamination. As I returned to Princeton University, Tom Gihring arrived to help Duane and Amy with the sampling of the Dream Borehole. Tom was a microbiologist who had just completed his B.S. degree at the University of Wisconsin, where he studied hot springs in Yellowstone National Park and was currently working with Jim Fredrickson's group at PNNL.

Getting down to our Dream Borehole, however, was challenging. The no. 9 shaft was formerly the property of the West Driefontein Gold Mine when it was competing with the East Driefontein Gold Mine. It was little more than a mile from the no. 5 shaft. When the West and East Driefontein gold mines consolidated, construction of the no. 9 shaft had ceased. Duane, Amy, and Tom took the cage down to 24 level. Because shaft 9 was not in production, miners were few and far between, which made the cage ride pleasant for a change. But to get down to 38 level, where the Dream Borehole was located, they had to ride in a Kibble. The Kibble was an enormous iron bucket twenty feet deep and ten feet wide that hung on chains in the secondary shaft and was used to haul rock out of the deep levels. It was easily big enough to carry them, their escort, and their equipment, which not only included the bailer and the usual wide assortment of sampling gear, but also a steel cable winch with 3,000 feet of cable for the bailer and the biggest packer that Cementation Products had. This packer was 3¾ inches in diameter and designed to seal the big HQ exploration boreholes. We had modified it with the usual autoclave-proof plastic fittings to obtain clean samples. In terms of speed, the Kibble compared to the cage was like a slow ferry compared to a hydrofoil. But it didn't take long to make it down 2,600 feet to 38 level, which was virtually abandoned. A short hike down the tunnel took them to a cubby that was much larger than the usual ones. The ceiling was high enough for the long drill rods that the mines used when drilling down a kilometer or more. Part of the head frame that had been used for the drilling was still there, rusting away. Gurgling away near the center of the cubby was the Dream Borehole. They set up camp for the next few hours with each person taking turns to man the winch. Duane carried out a few test runs, filling the bailer with rhodamine dye and lowering it 150 feet down into the borehole and then pulling it out. The rhodamine was still there, which meant that the steel balls were sitting tight enough not to let it leak out. Then they sent the sampler down 820 feet. He jerked the steel cable 100 times to dislodge the steel balls and let out the water and dye that was inside and to allow the borehole water in. They then pulled the bailer up to find clear water and a slug of gas at the top, which they quickly

FIGURE 7.2. Tom Gihring and Amy Welty collecting dissolved gas samples at 1.7 miles depth in Driefontein no. 9 shaft (now sealed off with cement) in October 2001. We nicknamed this borehole the Dream Borehole because it penetrated to 2.1 miles depth, giving us a 2,000-foot vertical profile of the deep biosphere (courtesy of D. Moser and Tom Gihring).

removed with a gastight syringe and transferred to evacuated serum vials. They jiggled the water from the bottom directly into sterile, anaerobic serum vials. Duane's junkyard bailer worked like a charm. They began profiling the borehole water as though sampling a lake – lowering, jerking, and fishing out the bailer, then transferring the gas and water into serum vials (figure 7.2). They sent the bailer down into the darkness, deeper and deeper, tracking the depth on the steel cable until they finally hit an obstruction at 2,125 feet that kept them from going any deeper. It must have been a borehole collapse. At these depths and under such great pressure the boreholes seldom remained stable for long. When they pulled the bailer up, it was filled with a black fluid. The water was choked with iron sulfides. Somewhere around 1,500 feet depth the borehole water changed dramatically in color. What they had been sampling at the surface of this slow-flowing artesian well was clearly not the same as what they pulled up from below that depth. What they would discover in that black, murky water would surprise us all.

WELCOME SCIENTISTS: THANKSGIVING HOLIDAY 2001, HOMESTAKE GOLD MINE, LEAD, SOUTH DAKOTA

One month later and ten months into the field project, over the Thanksgiving 2001 holiday, we all met at the Golden Hills Inn in a very snowy Lead, South Dakota. I had also invited Jim, Rick, and Terry Hazen, who had moved from SRP to Lawrence Berkeley National Laboratory (LBNL), to attend as well. Still we were a small contingent compared with the several hundred physicists who had gathered there along with John Bahcall.[16] Driving up from Rapid City into the Black Hills I was impressed by just how rustic and true to its frontier heritage the local culture was. There were quite a few hotels located in Lead and the neighboring town of Dead Wood, a popular gambling resort, but there was no mistaking which the Golden Hills Inn was. On its marquee out front posted twenty feet in the air in big red letters were the words "Welcome Scientists." The first time I had ever seen that phrase on a hotel marquee.

Our own meeting began the day before that of the physicists, with each research group presenting its findings. Duane recounted the success of his bailer and the importance of the Dream Borehole. Johanna presented her isotopic data that suggested the saline fracture water was incredibly old, older than anything she had ever seen. I was totally blown away by her data. She was reporting $^{40}Ar/^{36}Ar$ ratios of tens of thousands for the dribbling, salty borehole in East Driefontein. Those ratios were as high as what I had measured from many rock samples in my own lab, and they were higher than any previous analyses of groundwater, indicating that the water could be hundreds of millions of years old. The noble gas data were incredibly valuable because they demonstrated that the fracture water had not been exposed to mine air, otherwise the helium would have escaped. The 16S rDNA analyses of some of the shallower fracture-water samples yielded a surprising diversity of bacteria and a limited number of Archaea, but the deepest water samples yielded nothing at all. We already knew, on the basis of some fission-track apatite data,[17] that the rock units had not cooled down to their current temperatures until 40 to 90 million years ago. Was it possible that these deepest fractures had not yet been colonized by fracture fluid carrying bacteria from the surface? Could this be our holy grail of Darwin's warm little pond in the subsurface where the processes of genesis would be revealed?[18]

As if to answer that question, Julie Ward from Barbara Sherwood Lollar's lab at the University of Toronto presented the $\delta^{13}C$ and δ^2H analyses of methane, ethane, and propane gases based upon the initial suite of samples she had collected. At the shallower mines of Beatrix and Evander from the same boreholes at which the 16S rDNA had detected Euryarchaeota, the division of Archaea that would include methanogens, she had found microbially generated CH_4 and very low H_2 concentrations. These results were consistent with our radiolytically supported autotrophic community, although they did not prove it conclusively. But in the deepest mines around Carletonville, the same ones with the incredibly old noble-gas ages, she had identified abiogenic hydrocarbons. They were not the result of thermal alteration of Archean marine sedimentary rock in the Witwatersrand Basin, nor were they from the overlying CH_4-rich coal beds, as all the mine geologists believed. The helium isotope data also indicated that this wasn't CH_4 coming from the mantle, apparently precluding Tommy Gold's hypothesis. This observation was profound given that the Vredefort Crater was 190 miles in diameter, which was thirty-six times larger than the Siljan Ring impact, and was only thirty miles away from these mines. If any meteorite impact could have fractured the continental crust and released mantle CH_4, it would have been the one that produced the Vredefort Crater. Instead, it appeared that the CH_4 and higher hydrocarbons were originating in situ by some type of Fischer-Tropsch reaction.[19] The Fischer-

Tropsch reaction required a lot of H_2 and CO. Julie reported that the fractures with the abiogenic hydrocarbons had enormous H_2 concentrations, up to 10%, just as they had found in the deep mines of Canada. Barbara speculated that this H_2 must be derived from ultramafic serpentinization reaction. The only problem was that the Witwatersrand Basin had no large sequences of ultramafic volcanics, or komatiite, like the Abitibi Greenstone Belt of Canada, the terrane where she had worked.

I suggested that perhaps radiolysis was the source of the H_2 and, if so, the Fischer-Tropsch reactions could be ongoing, even today in the fractures. If Fischer-Tropsch reactions were continuing, then perhaps even Wächerhäuser "origin of life" reactions were also occurring, and we were finding evidence for it in our deepest fractures.[20] As Carl Sagan pointed out in *The Demon-Haunted World*, the "absence of evidence is not evidence of absence."[21] We could not accept our apparent absence of DNA from our deepest fractures as evidence of total abiogenesis, because the filtration of two to twelve liters of water was simply not yielding enough biomass for 16S rDNA characterization even when using nested PCR.[22] Besides, Greg was keen on analyzing the $\delta^{13}C$ isotopic composition of the PLFA using approaches he was learning from John Hayes at the Marine Biological Laboratory at Woods Hole. But he estimated that he would also need far more biomass than that contained in twelve liters of water. Tom Kieft and I came up with the idea of placing a filtering manifold directly on the boreholes themselves and letting them flow for several days in order to filter thousands, maybe even tens of thousands, of liters. The question was whether we could build a filtration system that could handle the pressure and temperature and that wouldn't blow up if the filter became clogged.

Our most challenging problem was that we were collecting only planktonic microorganisms, and the dogma was that most of the microorganisms were attached to the mineral surfaces along the fracture plane or in the host rock. We had already used the Todd Stevens "bug trap" idea: Duane had attached cartridges with crushed rock to flowing boreholes in order to enrich for the microorganisms that preferred mineral surfaces. It appeared to work. Gordon had stunning scanning electron microscopy (SEM) images of microbial colonies weaving polysaccharide nets across the mineral surfaces. But Gordon wanted to image fracture surfaces directly to see the extent of biofilm deposition in these deep fractures. Our experience at Piceance Basin, however, told us that drilling fluids always contaminated these fractures. If the drillers were really good, however, and we knew approximately when a water-bearing fracture would be intercepted, we could get them to turn off the drilling water. Then when they cored into the fracture, the rock core should shoot out the barrel with the force of the fissure water and we could quickly stuff it into a Whirl-Pak. In fact, mines provide the only opportunity to perform this experiment because they usually drill under pressure in reverse circulation mode,[23] not overpressure, positive circulation mode (see figure 2.6), as with surface rigs.[24]

If we then added tracers to this drilling campaign we could look for the microorganisms in the rock pores. To focus the search we could use the sulfur-35 (^{35}S) autoradiography that Krumholz and Suflita used so successfully at Cerro Negro. But that was for sandstone collected at 650 feet depth with 10% porosity. Now we were talking about a quartzite at more than ten times the depth and with ten times less porosity. It seemed farfetched to expect to find living microorganisms trapped in such impermeable rock. But if they were there and they were being fed by radiolysis, then Johanna could date their entrapment age by analyzing the noble gases dissolved in the water of the rock pores. Now we were talking about a situation very comparable to that of the halophilic Archaea we had isolated from fluid inclusions in salt and to that of Russell Vreeland's 250-million-year-old *Bacillus* that he had isolated from WIPP salt.[25]

Despite these issues, we had in the first year already collected a formidable geochemical data set on the same water from which we had characterized the microbial diversity over a range of depths using the 16S rRNA. This was something we did not have at either the Thorn Hill or the Parachute drill sites. We were now in a position to evaluate the bioenergetics of the environment by calculating the energy available for different microbial metabolisms and comparing it to the dominant metabolisms inferred from comparisons of our environmental 16S rRNA data to that of well-characterized isolates.[26]

The following morning we all drove up to the Homestake mine to visit the deepest level, 8,000 feet below the surface. This was a record for North America, but only a moderate depth by South African standards. We had brought a lot of our sampling gear. We dressed casually, with the mine providing steel-toed boots, hard hats, headlamps, and an orange breathing apparatus for stripping CO from the air in case of fires. The safety manager had told us in the orientation meeting that morning that methane gas was never a problem at Homestake. He was a bit puzzled by the disappointed looks on our faces upon hearing this news. We then "tagged in," taking a small metal coin with a number on it so the mine could keep track of how many and who went down and who came back out. I was surprised that 1960s-era technology was being used at what was still an operating gold mine, when the South Africans had switched over to cards and optical scanners decades earlier. We then climbed into a small, rickety cage to ride down the Ross Shaft to the 4,850-foot level of the mine. As we slowly descended and rattled around on wooden guides, I couldn't help but sing to myself the "Dungeon Song" from one of my favorite childhood movies, *The 5,000 Fingers of Dr. T.*[27] I also wondered how long it had been since this cage had been inspected.[28] When we reached the 4,850-foot level, the air was comfortably warmer than it had been on the surface, perfect T-shirt weather. We stepped out of the cage into a gray, dirt-covered tunnel, turned a corner, and walked about a hundred yards to step into an even smaller secondary cage, which the Homestake miners called no. 6 Winze. As grimy and dilapidated as the 4,850-foot level looked, this was the level where Raymond Davis, Jr., of the University of Pennsylvania had constructed the first neutrino detector in the early '60s. He had reasoned that to detect the elusive neutrino he had to place his detector deep beneath the Earth's surface to shield it from interfering cosmic rays. Down one of those tunnels was the Davis cavity, where he had performed the experiment. We proceeded to creep down to the basement dungeon of Homestake mine, the 8,000-foot level. The 8,000-foot level was nothing more than a single tunnel that ran for three-quarters of a mile. As we walked along that tunnel, we encountered prolific biofilms blooming out of boreholes from which water flowed at steady rates. We found all the classic signatures of microbial activity. Lots of orange biofilms characteristic of Fe-oxidizing *Gallionella* mats. This may reflect the fact that the Homestake mine was located in metamorphosed banded iron formation. White mineral crusts decorated the gray walls, indicative of calcite deposits, which meant that the water was carbonate rich. Tommy stuck a probe in the water streaming from one borehole. The temperature was 54°C, which surprised us. It was much hotter than the water temperature at similar depths in South Africa, indicating a geothermal gradient that was almost double that of South African mines. Near the end of the tunnel we came to a T-junction with a tunnel that headed away from the main tunnel and had water flowing out of it, but dirt and rusty ventilation hardware had been piled to the ceiling, preventing us from entering. Close by, however, were prolific whitish-yellow biofilms characteristic of sulfide-oxidizing bacteria that were spilling out with the water from the boreholes; however, the pH was neutral and the salinity was very low, indicating that this was likely young water recharging from the surface. Tommy and Susan inoculated some media for methanogens, while I scooped up some of the yellowish white biofilm and collected water for geochemical analyses. Johanna set to work collecting water in copper tubes for noble gas analyses from one of the boreholes (figure 7.3).[29] The prospects of finding interesting organisms and performing experiments here looked very good. A little paint and some floral arrangements and the place would be good as new. As long as we renovated that creaky cage, we could have an underground research lab that would rival Karsten's Äspö URL. But little did we know then that this was going to be the last trip for any of us to the 8,000-foot level of Homestake mine.

We returned to the surface with our samples, not realizing just how precious they would soon be, and back to the Golden Hills. The next day the physicists arrived en masse. Some of us stayed to listen to talks on dark matter, weakly interacting massive particles, neutrinos, double beta decays—none of which I understood. Most of the

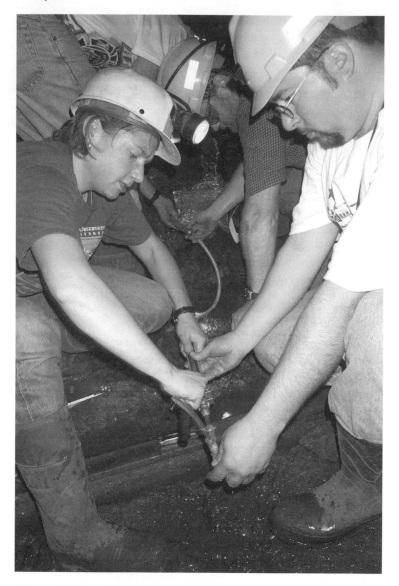

FIGURE 7.3. From front to back: Johanna Lippman, Duane Moser, Tommy Phelps, and Jim Fredrickson collecting noble gas samples from flowing borehole water on the 8,000-foot level of Homestake gold mine, November 2001.

physicists were reporting on results obtained from other underground laboratories located in Italy, Japan, and Canada. After Davis's discovery, these countries had jumped into the neutrino detection business by building bigger detectors located at even greater depths. Both NSF and the U.S. DOE funded the astrophysicists involved in this research to travel to other country's URLs to perform research. During the early 1980s George Kolstad, the geoscience program director of DOE's Basic Energy Sciences, advocated for a deep underground laboratory for the United States, which they named Earthlab. But the initiative went nowhere, and by 1987 the working group behind it had disbanded. Interest in an underground lab began to emerge in 1993, when Congress pulled the funding for the superconducting supercollider for the high-energy physicists. Now the physicists were convinced that with Homestake gold mine they could establish the deepest underground lab yet, enabling them to take the lead in the searches for dark matter and double beta decay. They could send beams of neutrinos from the accelerator located at Argonne National Laboratory, through the Earth to detectors located in Homestake in order to measure some property of neutrinos. All of this was somehow related to the standard model of the universe. At least this is what Wick Haxton, a physicist from the University of Washington, tried to explain to me between the sessions. Not only were the scientific planets coming into alignment, so was the politics. Tom Daschle, the Democratic senator from South Dakota, was the Senate majority leader. The cost of the underground lab would be about $1 billion, equivalent to one NASA flagship space mission, and that kind of money could come directly from Congress with Daschle's support. Wick felt that with the additional interest from the earth science community and especially the scientists pursuing deep life, the funding of the National Underground Science Laboratory was a sure thing. With all this flattery coming from physicists, I couldn't help but get a bit swelled-headed. But as Wick walked away, Tommy punctured my overinflated ego by whispering to me, "There is no way in hell those physicists will ever let us use their lab."[30] During the next day as we discussed our plans with the physicists, it was clear that what I had hoped could be a synergistic collaboration was feeling more and more like a shotgun wedding. But the advantages of having an underground laboratory were so compelling that we all decided to persist. With 125 miles of underground tunnels from 300 to 8,000 feet deep, Homestake mine offered unparalleled opportunities for in situ experiments. But we had little time to discuss this because Tommy, Susan, and Tom were heading immediately to South Africa to collect the last round of samples and to secure the field lab before the mines shut down for the long Christmas holidays.

Once the mines had reopened in late January, I sent Stephanie Devlin, a research assistant in my lab, to South Africa to reopen the field lab and join Adam Bonin, a Ph.D. graduate student of David Boone's. Sharing David's enthusiasm for isolates, he noted that many of the Balch tubes that had been left in the incubators since the summer were finally showing signs of growth and yielding distinctive morphologies beneath the lab's microscope. They then got calls from the Masimong and Merriespruit gold mines, which were located near Welkom. Both of them had water at two kilometers depth, much deeper than at Beatrix gold mine, which was close by. Over the next few weeks Stephanie and Adam journeyed down there. Bloemfontein was an hour away, so Esta sent one of her students, Saun Knoessen, to assist in the collection. It was after one of these long drives that Adam and Stephanie were having a well-deserved dinner at the Croc. It was a slow Monday evening, but late, and Piet and Pietra had already gone to bed. Adam and Stephanie pretty much had the restaurant to themselves. While they were dining, a car silently slid past the boom gate (the guard was not to be seen) and down the dirt road past the Croc bungalows to the house at the back of the estate. Someone stepped out of the car, walked up to the door, and knocked. Piet crawled out of bed thinking it must be one of his staff bringing the cashbox. The gunman fired point blank into his chest as soon as he opened the door. Piet was thrown to the floor by the blast while the gunman took a shot at Pietra. The gunman then rushed out, jumped into the car, and sped off into the night. Everyone in the restaurant was oblivious to what had just happened, so quickly and quietly had the attack occurred. Adam and Stephanie were just finishing their dinner when the police and ambulance arrived. They started questioning the guests but sent Adam and Stephanie quickly away. Pietra had fortunately survived the shooting, barely, but Piet had not. Our jolly jester was dead.

CHAPTER 8

A LOT OF BREAKS AND
ONE LUCKY STRIKE

But to my mind this mass of liquid must have leaked little by little into the bowels of the earth, the water obviously coming from the oceans above and making its way down some crevice. However, I'll concede that this crevice must be currently plugged up, because this whole cavern—or, more accurately, this immense tank—would have filled completely in a fairly short time. Maybe, after it had to do battle with underground fires, part of that water converted into steam. Ergo the clouds hanging overhead and the release of electricity that creates storms inside the earth's bedrock. This interpretation of the phenomena we'd witnessed struck me as satisfactory, because no matter how stupendous nature's marvels may be, they can always be explained by the laws of physics.

—*Jules Verne, Journey to the Center of the Earth*[1]

THE UNDERGROUND MICROBIAL ART GALLERY:
MAY 2002, EVANDER GOLD MINE, SOUTH AFRICA

Stephanie and Adam had taken a flower arrangement to Piet's funeral along with the collective condolences of the Subterranauts before they returned to the United States. Rob had assured me that Piet's murder was likely a contract hit and not racially motivated. Although the Truth and Reconciliation Commission started by Mandela had made great strides in reducing the racial tension in South Africa, the manner of Piet's death still unnerved me. Enough so that I decided that we should pause the field campaign for a while until we learned more. Mark and Rob would keep an eye on the Glenharvie house for us and let me know if there were any other problems, shootings or otherwise. I needed to find a new home for our field lab, one that was in a safer environment. The next time I was in South Africa, I would approach Derek and Esta about the possibility of moving the field lab to the safety of their department, permanently.

Meanwhile we were dealing with the problems we had identified at the Lead, South Dakota, meeting. A smaller group of us had settled on the new filtration design, and I was awaiting the parts before returning. A new Ph.D. graduate student of mine, Bianca Mislowack, began working with Tommy and Tom on designing an in situ cultivation device for boreholes in preparation for the austral winter field campaign. Finally I began talking to Arnand van Heerden at Kloof Gold Mine and Colin Ralston at Evander Gold Mine about obtaining cores for microbial studies over the next year.

Colin Ralston, the chief geologist of Evander Gold Mine, had become enthralled with our microbial hunting expeditions, or safaris, as he called them. He had spent his life in the gold mining business working his way to the top of his trade at Harmony Gold Mines, but he was also an avid naturalist and took it upon himself to guide our teams to the various locations underground, especially in his home shaft, no. 2. The Evander Basin was a separate basin on the east side of the much larger Witwatersrand Basin but contained the same gold-bearing geological units. Getting there from the Glenharvie house/lab involved a 150-mile drive and required a very early morning departure in order to take the 7 a.m. cage down.[2] Fortunately the security at the Evander gold mines was not as intense as at the ultradeep mines in the Carletonville region. The mines were also older and the ore deposits shallower, only going down to a depth of 6,500 feet. Evander Basin was crisscrossed by east-west faults as well. The combination of shallow depth and numerous faults meant that Evander mines had loads of fissure water. There were nine shafts distributed across the basin, two of which were flooded, and Colin had managed to get our teams underground at all the operating shafts: 9, 5, 2, and 8. We really loved shaft 8 the most because it was the deepest and because the miners there were sending a long exploratory tunnel into virgin rock on the northern side of the basin, where no other mines existed. Colin was a very persuasive advocate to the management of Harmony Gold Mines for our underground safaris. If anyone could arrange for a drill rig to collect cores for microbiology, Colin could. He had even managed to get our team into the Middlebult coal mine, or colliery, which mined the Karoo coal and was located just south of the gold mines at a depth of only 2,300 feet. Because the Evander mines were shallower, and thus cooler than the ultradeep mines near Carletonville, their tunnels harbored a wide variety of insects, including albino spiders and extremely large cockroaches. The mining version of urban legends typically featured a cockroach carting away a miner's lunch pail. Lihung had brought back biofilms from one of the Evander sites that even appeared to contain some type of worm.

In the beginning of May 2002 I returned to South Africa to reopen the Glenharvie house and had brought with me the new filtration gear. The first place to deploy the device would be the same borehole I had visited with David Suzuki in October 2001. Johan von Eden of Kloof mine, who had overseen this particularly troublesome hot spot ever since it was first encountered, informed me that the management had decided to permanently entomb this leaky tunnel in cement. If I wanted any more samples from there, I had to get down there now. The samples collected during the Suzuki shoot yielded precious little DNA, so hopefully the "massive filter," as we called it, would do the trick. There were actually two different types of massive filters, one for DNA and one for PLFA. I picked up the Saab keys from Rob, opened the field house, and reestablished email communications.

I picked up Lisa Pratt and Eric Boice from the airport in Joburg the next day. Over the next two days we went through our medical exams and steppies and getting our indemnity forms approved. The medical exams over the past four years had shown that the hearing in my right ear and the peripheral vision in my left eye were getting progressively worse. My performance on the steppies was also getting progressively worse. Gold Fields knew more about my medical history than my family doctor, and at this point I really didn't qualify to be underground at all, but as long as I signed the indemnity form, it was my problem. Lisa and Eric sailed through their exams with flying colors. Kloof set up special steppies for Lisa given that she was female, and she surprised them with just how tough she was. Our underground certificates in hand, we piled into the Saab and headed for the boom gate and the road back to Glenharvie.[3]

The next morning I visited the Croc to see Pietra. I found her sitting by herself at a corner table and asked if I could join her. This was my first time to speak to her in person since Piet's death, my first time to express my condolences in person and to thank them for making about forty foreign scientists so welcome. She shared with me what happened that night. The second bullet had actually grazed her head; she had been literally a hair's breadth from death. Fortunately their daughter had hid in the closet when the attack occurred. Now she was in the process of selling the Croc and the Castle and moving back to her farm. She extended her condolences about the 9/11 attacks. It seemed that the teasing, angst, and vitriol leveled at the know-it-all Yanks by South Africans had softened considerably and been replaced by sincere sympathy.

Later that morning I received a phone call from Johan, who had cleared us to go to the 41 level borehole, but the ventilation had just been started and it was still too hot to enter. He told me the dimensions of the pipe we would be sampling, so I had time to go to the Westernaria hardware stores to purchase the metal fittings I would need to hook up the massive filters. The rest of the time we spent cleaning, sterilizing, and prepping the filtration device. Johan had said it would be several days before the tunnel would be safe to enter. We contacted Colin to see if we could stop by and talk to the mine manager about obtaining some core samples, collecting some fissure water samples, and scouting a location for Bianca's in situ cultivator. We arrived at the Highveld Inn, where Lisa gave a talk to Colin and several of his geologist friends and colleagues about the work she had been doing in South Africa. Before we left for dinner, I called Johan at Kloof to check on the status of our visit. He informed me that we would have to go down the day after tomorrow. It was going to be a dash to get back to Glenharvie in time for the early cage the next day.

One of the attendees at Lisa's talk was a young mining geologist named Peter Roberts, who joined us the following morning and with Colin escorted us down to the 5,300-foot-deep 21 level at the

A.

B.

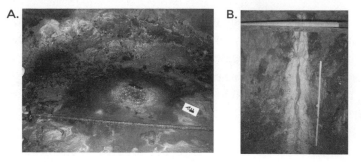

FIGURE 8.1. Biofilm photos of South African mines A. Acid seep with pH of 3 from Kimberley Shale (courtesy of the author). B. Alkaline seep with pH of 10.

no. 2 shaft. From the cage we hiked two miles east before reaching the tunnel that had so excited Duane during his visit the past November. As we started down the tunnel, Colin explained that we would be crossing one of the major faults cutting through Evander Basin and that could explain why so much water was flowing through fissures into the tunnel. He pointed out that some of this water could be coming in from mining operations overhead, but that did not matter to us because it was the biofilms that streamed from the flowing boreholes on the walls that amazed us all. The boreholes penetrated the sulfide-rich Kimberley Shale, and acidic water flowed out with a pH of 3.[4] Although the boreholes were big, several inches in diameter, they were choked with biofilm. One after another we used our headlamps to peer up into the boreholes. They changed color from brown at the borehole entrance to a bright orange some ten inches in and then a steel-blue gray still deeper. The borehole biofilms draped the walls like drawn curtains running down to pools on the tunnel floor (figure 8.1A). The biofilms were zoned with a black interior transitioning to a brown margin decorated by brilliant yellow crystals on the outer rim. Gray filamentous mats occurred at the center of the pool, grading outward to a granular bright reddish-orange annulus that suddenly turned into a swirl of petal-shaped yellowish, pink, and beige filamentous mats. Further down the tunnel we stepped into the Witwatersrand quartzites, where the pH of the water shot up to 9 to 10 and calcareous stalactites decorated the ceilings from overhead seeps. Here the biofilms hanging from the borehole had a whitish filamentous center grading out to a pinkish, fleshy tongue. Sometimes the tongue was so big, it was the size of a thigh muscle. The pink would grade outward into a translucent orangish-red film that ended against a triple-layer sandwich of bio-film that had a yellow layer against the rock and a middle, brownish layer, both covered with a white frosting of carbonate. We passed a dozen of these multicolored mats before finally arriving at the end of the tunnel (figure 8.1B). Each was like a Jackson Pollack painting—splashes of multiple hues framed by dark, flinty rock. Fortunately the CH_4 concentrations in the tunnel were negligible, and Lisa took numerous flash photos, but none captured their beauty as seen with a miner's headlamp. We collected water samples and harvested an experiment left behind by Duane from the terminal borehole and then packed up to return to the surface. As the seven of us, led by Colin, stepped cautiously to avoid the pools on the way back through the tunnel, I couldn't help but feel like Professor Arronax being led by Captain Nemo through the undersea forests off Crespo Island in Jules Verne's *20,000 Leagues Under the Sea*. When we finally reached the tunnel's entrance, I turned to Colin and told him, "You should cordon this off, turn it into a microbial museum and sell tickets. Each of these biofilms could be a thesis for a student. We should give each a name and a short scientific description." Colin was all for that. Who would have ever guessed that, hiding in such dank darkness, life could create such astonishing color and beauty. On the surface, we discussed coring plans for the next several months over some tea and sandwiches in Colin's office, and then we jumped into the Saab for the three-hour drive back to Glenharvie.

LOTS AND LOTS OF FILTERING AND CORING: MAY–DECEMBER 2002, KLOOF GOLD MINE, SOUTH AFRICA

The next morning at Kloof no. 4 shaft, Lisa, Eric, and I hiked the two miles to the borehole, led by Faunie. Stumbling down the dark tunnel, we could hear the tea-kettle whistle of the borehole long before we saw it. The temperature at the borehole felt as hot as it had been back in October, but everything else about the site had changed. The pipes and valves that had been present were gone. The steel ladder the cameraman had used for lighting had corroded so much that it was little more than a pile of rusty pipes. The steel pipes with huge leaky valves poking out of the metal-flanged rock wall made the alcove look like the torpedo room of a submarine. Faunie told us that they had replaced everything. Clearly Johan had not received that memo. Now we were supposed to collect our water from a high-pressure steel valve with a three-inch-diameter fitting. We carried nothing that we could attach to that. So we took some measurements, packed our equipment, and returned to the surface. Back in the field lab, none of our plumbing parts were big enough to match the huge valve on 41 level. After having lunch at the Westernaria Chicken and Nice Things, I found what I needed at the auto parts store: a meter-long radiator hose with clamps. After multiple alcohol and bleach rinses, I stowed it in the backpack. The next morning we returned to 41 level, this time with the shift boss, Chris. Chris had a limp from a previous accident, a boyish face, and charming personality, but when talking to the miners, and he talked to every miner he encountered, he switched into authority mode without a slip in the clutch. Chris shouted in Fanagalo at some miners, who promptly scurried off and returned with a couple of bicycle carts. We were then driven like royalty to our borehole.

The borehole was still there waiting for us, hissing in the insufferable heat. It was hotter than it had been yesterday, if that were possible. I had designed the filtration system with two filters held in shiny stainless steel canisters. Each could handle four liters a minute and could easily handle 120°C. But the clamp fittings I had could not handle the 200-bar pressure that this borehole could dish out. I mounted ball valves upstream of the filters to adjust the flow. Downstream of the filters I connected flow meters to record the total amount of water filtered. Past these I added check valves. If the flow for some reason stopped, I needed the check valves in place to keep the filters from getting contaminated by air and the pool of water that was always present at this site. Finally I added pressure regulators upstream to vent water, just in case the filters clogged. These were to keep the filtration system from blowing up and contaminating the filters. Assembly was a piece of cake in the lab, but under the steamy conditions at this borehole, the tubing I had used grew soft and the hose clamps kept loosening. It was proving maddeningly slow. Lisa grabbed me and pulled me away to a cooler spot while Eric and she went in to finish the connections. Finally, Faunie took his meter-long wrench and placed it on the valve. I grabbed the wrench and nudged open the valve. The hot water started spurting out of the tubes all over the place and from the outlets from the flow meters. Eric and Lisa were busy with the screwdrivers, sealing the leaks while I measured the flow rates. I nudged the valve open more and brought the flow rate up to eight liters per minute, the maximum recommended for these filters. The filtration system started vibrating, but it held. We piled rocks around it to make sure it would not fall over. I wanted to start collecting water samples, but Lisa vetoed that because of our heat exposure. "Let's collect the samples when we come to retrieve the filter tomorrow," she said. We packed up our tools and supplies to head back. All of us were soaked, but the bicycle cart got us back quickly enough to catch one of the last cages up. During the ride back I downed a liter and a half of cold water that I had brought down frozen. It was going to be absolutely crammed, judging from the crowd of miners waiting at the gate, but Flip, a positively massive shift boss, grabbed Lisa and stuck her in the back of the cage and shielded her with his two leg-like arms as the others squeezed in. I had seen Flip at work before packing cages. He would reach up with his two arms to the top of the cage, wedge his feet across the door, and compress the twenty-odd miners into the back half of the cage to make room for another twenty. So I was happy to see him on our side this time.

Back at the Glenharvie house we licked our wounds from the past two days. I placed ice packs against my head and took aspirin. Lisa was taping up some nasty-looking blisters on her feet. Eric was wrapping up his knee, which was causing him some problems. The managers wouldn't approve an early morning trip for us, but we could catch late cage down and come up in the early afternoon before the blasting time. At 4 liters per minute per filter, each filter should have about 6,000 liters through it by the time we disconnected them. If the cell concentrations were 1,000 cells per milliliter, then we should have about 5–10 billion cells for DNA sequencing. That should easily be enough for 16S rRNA analyses and for $\delta^{13}C$ analyses of the PLFA. Faunie took us to the site again, but unfortunately without the bicycle cart this time. When we arrived, the filtration system for the PLFA filter had been blown off, but the DNA filter was still intact. We checked both flow meters and it looked as though we had captured about 5,000 liters through the DNA filter. Faunie wrenched the valve closed. We disconnected the filter canisters and quickly screwed plugs into inlet/outlet holes to seal them against air contamination. Then we completed the water and gas sampling, gathered up the pieces of the filtering system, packed everything into our backpacks, and started back. Although the filtration parts were heavy, I was grateful I was not carrying back two 12-liter steel syrup canisters and that the borehole had done our work for us. After three contiguous days underground, we had begun to get a bit giddy, perhaps due to the elevated CO_2 levels or the heat or both. Let's just say that I was happy that our treks to the broiling room were over.

All of a sudden we heard a "boom" ahead and slightly above us. Actually, you could feel the mine detonations more than hear them. With the dull thud of an explosion, the compressional wave in the mine traveled down the tunnel and popped our eardrums. We felt more "booms" at regular, timed intervals. The mine had started blasting. Faunie, however paid no attention, and seemed quite nonchalant. As we continued our hike, the explosions grew closer and we began to feel the vibration in the rock beneath our feet. Then silence. As quickly as they had started they had ceased. We came to a blast barrier that barred our path near a phone and some empty red boxes, in which explosives are usually kept. Faunie called in to make sure that they were not planning to blast ahead and received the "usual" assurances. He sidestepped the barrier and carried on. After a second hesitation, we followed. Within about ten yards we walked right into a cloud of dust and very acrid smoke. Ammonia gas, undoubtedly from explosions. At this point we were becoming very hesitant about proceeding and retreated to "safety" behind the blast barrier. Faunie walked into the acrid mist, his headlamp quickly disappearing. After a few minutes his headlamp reappeared, and then his entire form stepped out of the cloud. "It's only a little ways, come on," and he turned back into the cloud. All of us took a lung full of fresh air and dove into the cloud. On the other side the air was as sweet as spring breeze. We proceeded another half hour before we reached the cage. When we reached the top of the secondary cage, Faunie called the banks man, who was in charge of the cages. Getting off the phone, Faunie told us it would be awhile. We all lay down on the cement to rehydrate and wait for the cage. When we finally arrived back at the Glenharvie house/lab, I pulled the wet filters out of their housings, slid them into sterile ziplock bags, and placed them in the freezer. I needed to modify the filtration system somewhat to make it easier to deploy, but overall it had worked astonishingly well. We had solved our first problem identified at the Lead meeting.

By August, Colin had arranged for us to obtain our first rock cores for microbiology and gas analyses. They would be coring from a tunnel on 24 level of the no. 2 shaft, their deepest level. We would be coring into one of their gold-bearing reefs. This was the first time I attempted this since three and half years earlier in the stopes of East Driefontein. This coring would be particularly challenging because we had to take high-purity N_2 cylinders, a vacuum pump, a pumping station, and an electrical generator underground! Johanna had contributed her stainless steel, leak-proof canisters for collecting cores. She gave us instructions on how to use these canisters. Mark, T. J. Pray, an undergraduate working in my lab, and I dropped this equipment off a week ahead of the drilling to give Colin time to transport it down to the drill site. We kept our fingers crossed that all would be intact when we arrived underground the following week. Twenty-four level was lower than could be reached by a single cage ride down, so from 21 level we went down a short tunnel from the no. 2 cage and stepped into an iron sled that was obviously designed to haul rocks up along the incline. It was large enough to comfortably sit six. The five of us with three backpacks and two additional samplers were comfortably snug as we slid down the 30° incline some

1,200 feet. When we arrived on 24 level, at 6,500-foot depth, the drillers were already at work collecting cores. A high-pressure hose of service water was used to cool the diamond bit, which was cutting away about 20 meters below our feet. Mark tried to inject the rhodamine dye solution into the hose, but the pressure was simply too high, and most of it squirted over Mark; nonetheless we managed to get some into the drill water. As the cores came out, we laid them out on sterile autoclave bags and quickly split them up into three groups. One group went into sterile Whirl-Pak bags immediately; these were placed inside sterile ammo cans that we gassed with the N_2. These were then placed into a cooler with frozen Blue Ice. The second group was placed into sterile Whirl-Pak bags immediately and then laid in a smaller cooler with dry ice. The third group was placed into the stainless steel canisters. We turned on the generator, which provided power to the vacuum pump, connected the canisters to the pump, and evacuated them three times, after which we replenished the canisters with N_2. After that we evacuated and sealed each canister. It worked like a charm, and within less than an hour we had obtained very nice cores, far cleaner and more intact than the fractured cores we had gotten from the stopes of East Driefontein four years earlier. The cores would remain under vacuum inside their stainless steel canisters, and over a period of weeks the pore gases would leak into the canister. From these samples Johanna would be able to determine the stable and radiogenic noble-gas isotopic composition of the pore gas and thus constrain the age of the pore water. We would also analyze the cores placed on dry ice for DNA after carefully processing and identifying internal core fragments that were not contaminated by the rhodamine dye. The cores in the ammo can were to be used for ^{35}S autoradiography. Altogether these data would tell us the origin of the CH_4 and H_2 gases in the fissure water and how old the gases were, so we were pretty excited and eager to obtain more of these types of samples over the next couple of months. We had licked the second major problem identified at the Lead meeting.[5]

After another trip underground at the Evander Mine, Mark, T. J., and I returned to Princeton, where Mark was enrolled in graduate school. Lihung Lin and Johanna Lippmann, who were to be on site for September, replaced us at the mine. No two better-trained people could have been ready to handle what happened next. As has been the rule since the inception of my research into subsurface life, destiny turned on a dime. After being there for a week, they received a call from David Kershaw, the chief geologist at Mponeng. They had just intersected a hot-water- and gas-filled fracture on 104 level, 9,000 feet down in the Ventersdorp metavolcanics. He wanted to know if they would like to come and collect samples. The next morning they were underground. Lihung clamped on the massive filters while Johanna collected her noble gas samples. There was considerable pressure through the valve and the temperature was 60°C! That was much hotter than the ambient rock temperature at that depth and clearly indicated that the water was coming from two and a half to three miles depth. The drill rig was still sitting at the site. They also collected some of the service water to check for drilling contamination of the borehole water. For two weeks Mponeng let them collect samples as the water temperature and the pressure dropped. Finally the mine tunneled through the fracture to reveal a huge quartz-filled vein. It was the luckiest strike for the Subterranauts to capture fissure water at such great depth and at a first intersection when there would be the least possible chance of contamination. But it wouldn't be until Lihung ran the clone library 16S rRNA gene sequences that we would realize how lucky we had been.

In the meantime, not only had Susan's application to NSF for support for undergraduate workshops in South Africa been successful, she had recruited a remarkable number of students from the United States for a weeklong workshop right before Christmas. Esta had also recruited a large contingent of undergraduates, many of whom were "previously disadvantaged" (a euphemism for black South African) from the University of the North. Esta had located an ideal conference venue, the Farm Inn, which was located in the rural outskirts east of Pretoria. Altogether we had about twenty-five students attending along with nine instructors and my wife. But where underground could we take such a massive crowd safely and also have enough places in one location to sample so that everyone would have a chance to be involved in the collection and measurements? Colin also wanted to take part in the meeting and offered a tunnel on 21 level of the no. 2 shaft, his "microbial museum," which was remarkably rich in biofilms. We all thought this would be an ideal spot for the field trip.

A DAY IN THE LIFE OF A WIFE OF A SUBTERRANAUT: DECEMBER 16, 2002, FARM INN, PRETORIA, SOUTH AFRICA

Esta sent out the invitations to the South African academic community announcing that we were having a one-day workshop on December 16, 2002, as part of a one-week educational/training workshop. We held presentations from our research groups in the morning on the microbiology, geochemistry, groundwater ages, and the water, gas, and isotope chemistry of the various mines. This took place in an outdoor meeting room under a traditional thatched South African roof. Over lunch we had some student posters, along with clips from the three TV programs that had been shot the previous year at the Evander, Kloof, and Mponeng mines. Tom Kieft, Mary DeFlaun, Gordon Southam, and Tommy Phelps had also shown up to lend Susan a hand.

At 4 a.m. the next morning it was pitch black. The students and instructors piled into three vans that Esta had rented to transport them to Evander Gold Mine. Nora and I were going to follow in our rental car, but at the last minute I had to run inside to return the key to our room. I was back out in less than a few minutes but the vans had already left. I jumped in and Nora headed after them, but it was clear that Colin must have been breaking all the speed limits because after a couple of turns we couldn't see them at all. I quickly pulled out our very worn South African road atlas that we always carried into the field with us and using the overhead light started to plot our route to Evander. We were driving down small two-lane roads past farms in what I thought was the right direction. The problem was that there were no signs on the roads, and we had to navigate more by dead reckoning, which was difficult in the dark. We crossed what

FIGURE 8.2. First student workshop underground sampling trip in December 2002 at Evander no. 2 shaft (courtesy of Evander Gold Mine).

I thought was the N17 and went down another small asphalt road that suddenly turned into a dirt road. By that time it was daylight but heavily overcast. Unusual for this time of year I thought. That is why it had been so dark. Just then the gas alarm went off in the car. We were out in the middle of the veldt on a dirt road about to run out of gas with no town in site. For the next half hour we kept driving down the dirt road with no houses around, looking anxiously at the red light on the gas gauge. All of a sudden we crested a hill and were looking down at the R580 and Secunda with a gas station dead ahead. When we arrived, I was able to get on the pay phone and call Colin's office. They had just arrived themselves, and he sent Peter Roberts to lead us to the no. 2 shaft, which was just down the road.

We arrived in time to suit up in clean white coveralls and gumboots. We were all provided pink hard hats, blue headlamps, and orange rubber gloves. We had never been so well dressed before going underground and had to take a group photo while our coveralls were still white (figure 8.2). We then all filed into the no. 2 cage in groups of a dozen. Once we arrived at the 21 level tunnel head, we split up into three teams and attacked three different boreholes. This was Derek Litthauer's first time sampling a flowing borehole, and I was giving him a good lesson on how to stay wet. But the water was pleasantly warm at 30°C and the temperature in the tunnel was just 29°C. Tommy and Tom were manhandling another borehole with a compression packer. Susan was running up and down the tunnel, visiting each of the groups and barking orders like a drill sergeant to different students to collect this sample and that sample. She also supervised the third group, which was focusing on collecting samples from the largest biofilm in the tunnel. She was everywhere at once. At the end we made sure that all the samples had been packed into the six assorted backpacks for the journey to the surface. Susan checked that each of the students had collected the biosamples they would use in their enrichment experiments once they arrived at Bloemfontein. When we surfaced, the day had turned beautifully sunny, but our white overalls had turned brown from the sweat and the dirt. The manager's shower room had been cleared for the females to use. Nora, Mary, and Susan joined the dozen female students, mostly black South Africans, there. We men filed into the regular miners' group showers. It was a rush to get through the showers in time to load into the vans and head to a local restaurant where the Evander Gold Mine hosted a luncheon. It went on for quite awhile, and the meaty Afrikaner hors d'oeuvres left Nora and me scrambling for dry toast and veggie tidbits. After thanking our Harmony hosts we began our six-hour car and van convoy to Bloemfontein. By the time we arrived on the N1, our sunny day had turned into an unusually menacing monsoon front. By the time we passed Kroonstad, we were being pelted with hail the size of softballs, and all the cars up and down the highway pulled underneath the few overhead passes and waited for the storm to abate. We had pulled over on the shoulder when Nora asked me if tornadoes ever happened in South Africa. I looked over to see her sticking her head out the window and looking up. "Then what is that?" she asked pointing up. I looked up to see the funnel cloud passing over us, spattering hail on our car and Nora. As soon as the freak mini-tornado headed into the veldt, the storm passed and we got back on the highway. Unfortunately, Colin, who had elected to drive his brand new diesel Mercedes Benz to Bloemfontein, was left with a smashed windshield and severely dented hood and roof. By the time we reached Winburg, the sun was staring directly into our faces. So intense was the sun, that even with her sunglasses on, Nora had to roll down windows of our car, stick her head outside, and look down to see in which lane she was driving. The sun set just as we pulled into Bloemfontein, sixteen hours after we had begun the journey from the Farm Inn. Esta had prepared a sumptuous banquet at her department of more Afrikaner hors d'oeuvres but was restricted from serving alcohol on campus to minors, so we had no Windhoeks or even a Pinotage to wash them down. For Tom, Mary, Nora, and me this was the last straw. We left and drove to the Loch Logan Waterfront for some delectable seafood and copious bottles of South Africa's best vintage. The next couple of days were spent with the students, showing them various microbiology lab techniques and discussing what they had seen underground. After that, the American students were packed onto the plane and went home. Tommy, Susan, and Tom left, along with Nora. I returned to the Glenharvie house for the last time. The week before, Tommy and Tom had placed most of the furniture on the front lawn for people to take, and to my delight it was all gone. Esta and Derek came from Bloemfontein with some students in a truck, which we then packed with all the lab equipment, emptying the garage and the little servant's quarters in the back. As they headed back to Bloemfontein, I locked up the Glenharvie house/lab. It had served us well for two years during which we had collected water samples and rock cores from more than 180 underground sites ranging from 2,200 to 11,000 feet deep. That collection more than tripled all the previous microbial samples ever collected from such great depths in the continental crust.

THE NEXT GENERATION OF SUBTERRANAUTS: JUNE 20, 2003, THE UNIVERSITY OF THE FREE STATE, BLOEMFONTEIN, SOUTH AFRICA

Six months later, Tommy, Susan, and I returned to the University of the Free State with more undergraduates from various universities in the States.[6] It was June, wintertime in South Africa, and we were running a six-week-long REU on subsurface microbiology. We had arrived early to get the field lab at Bloemfontein up and going. Tommy got the anaerobic glove bag working in one of Esta's labs on the first floor, while Susan and I were in the basement, where all the field gear, media chemicals, packers, filters, and the freezer for samples were located. Right next door was a closet containing all the mining coveralls, boots, hardhats, and gloves. It was an ideal and extremely secure field lab. The campus was safe for the students, and the accommodations, within walking distance of the campus, were equally safe. Upstairs were autoclaves, walk-in freezers, walk-in refrigerators, and walk-in incubators. Esta's group had grown, and new equipment was appearing in her ever-increasing lab space. She was able to obtain funding from the NRF to study the proteins expressed by some of the isolates we had given to her. This led her to pursue development of bioreactors based upon these enzymes and to continue screening mine samples for useful "extremozymes." In fact, that is how she referred to her research, exploring for extremozymes.

The next day we left the University of the Free State at 4:30 a.m. in three vans, three teams of students and faculty on an hour-and-a-half-long commute to the closest gold mines. An hour down the road, the van with Derek and Esta split off to the right, heading toward the no. 2 shaft of Beatrix Gold Mine, while the other two, one driven by me and the other by Susan, kept driving north toward the community of Virginia. In the dark we weaved our way around the hills into the mining town that was already abustle and drove up to the outer gates of Merriespruit Gold Mine. The geologist waiting to brief us cleared us through the boom gates, and one by one, we squeezed our gear and ourselves through the turnstile sausage grinder that separated the offices from the parking lot. After the obligatory safety lecture over mini-coffees, we pulled on our gumboots, strapped on our rescue packs, donned our helmets, checked our headlamps, and carried our backpacks toward the shaft to catch the 7 a.m. cage. Our group of fifteen huddled together outside at the base of the headgear while dawn was just breaking over the basalt-capped mesas of the lower Karoo Formation. Shivering in our mine coveralls as the frosty breeze bit us before it headed down the shaft, we welcomed the series of cricket chirps that preceded the arrival of the cage. After the cage tenders opened the doors to let an early shift trundle out, we sardined into the five-foot by eight-foot steel box, wedging our gear into any space available. They locked us in and we waited patiently as the wind rushed down the shaft beneath us, the students giggling nervously. The banks man dropped us, and we quickly accelerated on our way down the 6,500-foot straight plunge, more than twice the height of Angel Falls in Venezuela. The chill air was warming up quickly under the increasing pressure that was forcing us to pop our eardrums. The ride in total darkness was quiet except for when the cage shook against the guides and the occasional, brief, ear-splitting screech as we passed a high-pressure air vent valve.

A few minutes later we stepped out into the brightly lit, noisy station at 49 level. We switched on our headlamps, pulled on our packs, counted students, and proceeded to pick our way over the loco tracks and past train cars crammed with machine parts, drills, and all the tools one needs to make big holes in hard rocks. We followed our guide past the shotcrete-coated walls reinforced with steel rebar that anchored a grid work of steel mesh and passed underground store rooms, power stations, and machine shops, finally disappearing down one of the many dark tunnels radiating from the shaft. The air at our backs was cool, a mere 27°C compared with the rock temperature of 45°C as we stepped along the concrete ties. The bare-rock tunnel was dry, and as we moved our heads around, the cobble-filled, crosscutting channels of dipping sedimentary strata of a 2.9-billion-year-old fluvial deposit would appear in the beams of our headlamps. BANG! My head slamming into the ventilation duct reminded me yet again that I needed to focus on where I was stepping and not on the geology.

Earlier that morning in her office, surrounded by teetering piles of maps, rocks, and reports, Janet Larkin, one of the few female mining geologists in South Africa at that time, had pointed on the mine map to the crosscut located about a mile from the shaft where we would be going to sample the water pillar. This "water pillar" was a fault zone that in places contained a two-inch open fault filled with water.

Where the mine had encountered it in the past at much shallower depths, they had tried unsuccessfully to pump it dry before cementing it. She was certain that the water was coming from the Sand River that flowed above the mine.[7] At greater depths, however, the mining company had given it wide berth because of its potential ability to flood the mine. Unfortunately, the water pillar lay between the mine's deepest level and high-grade ore on an adjacent, recently purchased mine property. When the order came from Joburg to tunnel through the pillar at levels 49 and 50, the mine's engineers and geologists had plotted their actions carefully. As expected, when the tunneling crew crossed the rock buffer to the fault zone, water and CH_4 gas roared out of the drill holes at 10,000 liters an hour and further tunneling became impossible. It was then that Janet first contacted Adam Bonin to come down to sample. She had heard about our project through the mine geologist's grapevine. That was over a year ago, and the water and gas were still emanating from the boreholes with little indication of slowing down.

As we continued to hike, the moisture content of the air was palpably higher, and small trickles of water were flowing in the ditches on each side of the tracks. As I panned my headlamp to the ceiling I stopped. "What's up?" asked Susan, as she collided into me from behind. "I believe we're getting close to the water pillar," I replied. Above our heads and stretching for a hundred yards ahead were stalactites glinting, dancing, and winking in our lamplights. It was a beautiful gallery of delicate, sparkling, milky white soda straws, some up to a yard in length in the drilling cubbyholes on the tunnel sides. The stalactites were best developed on the fractures cutting across the rock ceiling. As Susan and I waited for the students to catch up with us, we stood gazing at our underground celestial firmament.

As we pushed past the "star" chamber, we came around a corner in the tunnel where a ventilation door blocked our path. Janet's assistant began cursing in Afrikaans, which when loosely translated meant "Who was the @##@!! fool that closed the @!#! ventilation doors to the !@#!$!! tunnel, now it's going to be roasting like !@#@!%! on the other side." To keep the mine's lower levels cooler, the shift boss had closed the vent doors to the water-heated tunnels of the water pillar.

Janet's assistant was right; a wave of heat struck us as he pried open the vent door. In the darkness ahead we could hear what sounded like a waterfall. The tunnel floor along the loco tracks got soggier and soggier, and pretty soon streams of water were flowing past us on both sides of the tracks on their way to the sump pump at the base of the shaft. As we followed the rails around a corner, they disappeared beneath a pool of murky water. We scanned with our headlamps for the other end of the pool, but the fog that hung over the warm pool blocked our sight. I stopped and pulled out our pH probe to check the acidity of the water, and it was perfectly neutral. Not knowing the depth I tentatively waded in, feeling the rail ties with each step. I could see stringers of white bacterial filaments and frothy bubbles floating on top of the muddy water, particularly around spots where the water seemed to roil from some hidden spring beneath the surface. With my next step I sunk down to mid-thigh. I had missed a rail tie and sunk deep into mud like quicksand. Susan pulled me back up, and we retraced our steps to the edge of the pool. Unfortunately we had not packed kayaks. I spotted a narrow rock ledge at the tunnel wall and stepped over to it. Hanging onto the wire mesh with my back to the pool, I side-stepped my way along the edge of the tunnel above the pool. I motioned to Susan and the students to follow. All of us carried on shuffling along our ledge, crossing beneath one raining fracture after another. Beneath us I could see massive rafts of white, orange, and black biofilms floating across our lake.

After about a hundred yards we came to the end of the pool and stepped back onto the side of the tracks, but the mud was so deep that we started sinking up to our knees. Fortunately a two-foot-diameter, unused metal ventilation conduit had been abandoned on the tracks, so we duckwalked on the top of that for another half hour. We began to smell the rotten-egg odor of H_2S when Janet's assistant pulled out his methanometer to take readings, waving us on once he verified it was safe. The roar of water became deafening upon approaching the boreholes, three of them all in a row. Gelatinous ooze, a biofilm formed by sulfide-oxidizing bacteria and referred to as snotites, hung to a length of about a foot below their metal casings.

The group of students under Susan's supervision stopped and began pulling sampling gear out of their packs.

Our team proceeded past the first two boreholes to the final borehole, about a hundred yards further at the end of the tunnel. The metal casing at chest level belched gas and hot water as we perched our backpacks on any dry spots we could find. Shouting to be heard, I directed the students to pull out their respective kits and follow the protocol we had rehearsed in lab. Each student was assigned a specific task: one to measure the water chemistry; one to collect microbial samples in sterilized, anaerobic serum vials; one to collect water for lipid analyses in big, twelve-liter, stainless steel canisters; and so on. I unwrapped the sterilized telescoping packer from its autoclave bag and began inserting it into the borehole. Immediately the water was shooting everywhere, but when I slowly turned the handle at the end, the packer swelled and sealed the hole, and the water was forced into a narrow jet that shot out of the end of the packer. I then screwed the Octopus onto the end of the packer. As the other students attached six sampling tubes and filters to the manifold, we closed the end valve to force water through the legs of the Octopus and began collecting water samples for aqueous, gas, and isotope geochemistry and microbial analyses. A jet of water blasted me in the face as the pipe that connected the Octopus to the packer split open under the pressure. A student came to my rescue, and the two of us took a fracture-water shower as we held the Octopus in place while the students finished taking measurements and collecting their samples. As the water dripped off my helmet, I made a mental note at this point to change the Octopus to an all-steel design. It would be heavier, but it could handle the pressure and temperature. Sampling done, we pulled out the packer, filled our backpacks with our samples, double-checked to make sure we left nothing behind, and began the long hike back to the cage, picking up Susan's team on the way. Their Octopus had remained intact, and they had connected to the borehole several experimental cartridges containing minerals to be left on for a week before retrieval.

Janet had assured us that no one would steal the precious experiment before our return, so Susan's team pulled on their backpacks stuffed with water and biofilm samples, and we all started the long trek back. An hour later we emerged from the tunnel into the bright lights around the shaft in time to squeeze into the twelve o'clock cage going up. Suddenly the cage dropped about thirty feet. The banks man was messing with us. We "eeked" and "aahed" as we yo-yoed in the dark at the end of mile-and-a-half-long steel cables that stretched like rubber bands. Then slowly we began the ride up to sunlight. Despite the thrill of their first trip underground, everyone was relieved to step out into bright daylight and ready for showers, dry clothes, and food, which Merriespruit Gold Mine generously provided in abundance. The geologists and engineers joined us for lunch and were curious what we hoped to learn from our samples. Could we tell them the origin of the CH_4 and H_2S gas and how old the water was? And there was the ubiquitous question, from where do the microorganisms originate?

"So where does all our bleeding CH_4 come from?" asked Deannie, the chief engineer at Merriespruit and editor of a local publication, *Mine News*. I washed down the rest of my veggie-patty-fried-thing with the remains of my third bottle of water and replied, "The 16S rDNA data from the filter samples and from a growing, living microorganism from the fracture water collected last year both indicated the presence of methanogens, or CH_4-producing Archaea. The $\delta^{13}C$ and δ^2H stable isotope composition of the CH_4 are consistent with the CH_4 being produced by these methanogens, and the geochemistry suggests that the CO_2 in the water is being converted to CH_4. Of course we knew all of this from Barbara Sherwood Lollar's work. This seems to be a common occurrence for the gold mines in this region at depths less than two kilometers." "Right, so you're telling me that the cause of some of the worst mine fires in South Africa's history are these microorganisms?" He looked at me doubtfully. "Well, yeah, but on a good note, your drilling water is full of methanotrophs," I replied. Perplexity swam across Deannie's face. "Methanotrophs are bacteria that eat the CH_4 and convert it back to CO_2 if you feed them O_2, which is also abundant in your drilling water, so if you circulate drilling water through your fracture zones you should be able to reduce the high CH_4 concentrations, in principle," I explained. "Hmmm, someone could make a bundle of rand with that idea," he answered as he positively warmed up to the idea that studying these little bugs in the fractures had some merit, and we had another convert.

Bellies full and rehydrated, we returned to our vans and left Merriespruit and the community of Virginia. In full daylight we could see that the hills had driven through in the early morning darkness were in fact slime dams. These sunbaked white, barren, earthen dams rose several hundred meters above Virginia and stretched across several kilometers. In fact, one of the dams had broken during a severe rainstorm, flooding the town of Virginia with 600 million liters of toxic sludge and killing seventeen people.[8] On satellite images the slime dams, or tailings ponds, appear to be pestilent boils erupting from the pores of mine shafts all over South Africa, each carrying a load of toxic metals from the ore-processing mill. Back at the University of Free State, we filtered the canisters for DNA and lipid analyses and inoculated degassed anaerobic media with water syringed from our biological samples. The team from Beatrix Gold Mine—which included Derek and Esta; one of my Prince-ton undergraduates, Jessica Garvin; a local undergraduate, Melindi Teboho; and two of Esta's graduate students—arrived five hours later. They looked like zombies from the tales of the crypt. They had endured a one-mile hike guided by a masochistic, macho-man mine geologist while they carried a six-foot ladder in addition to the regular field gear in the face of a withering, 30+°C, 100% humidity, and wind in their faces all the way into the borehole. This drill hole was located in the ceiling, hence the need for the ladder, and when the water was first intersected about six months earlier, it had flowed at a rate of one million liters per hour! Now it was a manageable 40,000 liters per hour, but the ventilation was not so good. By the time they had finished collecting all the samples from the borehole, their sweat was filling their gumboots. Apparently on the epic trek back, pieces of equipment, like the ladder, and articles of clothing were dropped one by one and left behind. Eventually, all the samples, including the two twelve-liter steel canisters, ended up on Melindi, who marched indefatigably onward while the rest straggled behind. For the rest of the workshop, Melindi was known as The Machine.

For the next several weeks the students extracted, amplified, and sequenced DNA, looking at the 16S rRNA gene that was used to determine "who" was living in their water samples. In most cases, however, the sequences were too dissimilar to come to any firm conclusion other than assigning to a particular family or, if they were lucky, a genus. The students also learned how to grow microorganisms using the massive walk-in incubators at the University of the Free State (they were also good for warming up after a brisk winter's walk outdoors).

While I was helping Susan on some enrichment cultures down in the field lab, she would periodically check her email. "Oh, my god! We got the astrobiology center," she screamed and started shaking her hands in the air as if she were scaring away mosquitoes. "You must be joking. Holy Toledo!" I said as I peered over her shoulder to read the announcement and my jaw dropped. I couldn't believe it. The angels surely must have been on our side. But now we had to move our field operations to the Arctic as soon as possible and start searching for life beneath the ice.

CORING THE BLACK DYKE: JULY 30, 2003, NO. 7 SHAFT, KLOOF GOLD MINE, SOUTH AFRICA

Before we left, however, we had to finish one last project. I had been working with Arnand van Heerden at Kloof Mine. Arnand had been the geologist who helped us at the no. 4 shaft with the David Suzuki film and he and his wife, Renee, had become good friends when our field teams were living at the Glenharvie house. Over the past two years he had guided us to some of the best (and most arduous) fracture-water intersections we sampled. The data recovered from Kloof Mine were fairly representative for the three-plus-kilometer-deep fracture-water intersections of that region.[9] At these depths, where the water temperatures were about 60°C, sulfate-reducing, thermophilic bacteria appeared to dominate the fracture's microbial population, but there were no methanogens. Noble gas and cosmogenic isotopic analyses yielded subsurface residence ages ranging from 30 to about 100 million years.[10] According to radiolytic theory, O_2 and H_2O_2 were produced, yet we never detected either in the fracture water. One possible explanation was that these oxidants converted the sulfide in the rock matrix to sulfate that diffused into the fracture zone, where it was consumed by the SRBs. But to test this hypothesis we needed core from rock adjacent to fluid-filled fracture. Another question that such a core could answer was, where were all the microorganisms living? Were they concentrated in the fracture water, on the fracture surface, or within the tiny pores of the rock matrix? But no sane mine geologist intentionally plans to core into a high-pressure fracture-water zone, because of its potential danger. Despite Arnand's best efforts, he had failed to convince his boss at the no. 4 shaft to undertake the drilling operation.

That May, Arnand had been promoted to chief geologist at the no. 7 shaft, the mine overlooking the Crocodilian Estates. He began to deftly manage his drill rigs so that one could be used to collect a core for us. Arnand told me that the chances were 95% of getting core from a high-pressure fracture zone near the Black Dyke in the no. 7 shaft near the end of July. We knew the risks involved in trying to predict when one would get a core in a working mine, but we had been there before. Our oldest water samples with the highest H_2 concentrations had been collected nearby two years earlier from the same fracture zone, and if the cores from the same fracture zone contained microbial life, then they would have been at least 100 million years old. The locations of the fractures were known and the distance to the fracture small, only twenty yards. Being the chief geologist, Arnand had control of the coring rig and could set the drilling pace. The miners had already penetrated the fracture zone, cemented, and tunneled through it two years earlier, but Arnand said he could back up the rig into a cubbyhole away from the tunnel and drill at an angle to intersect the fault far enough away from the grouting to guarantee no contamination. I asked my graduate students, Bianca and Mark, to book their tickets and pack their supplies and come down to the University of the Free State.

The date to start coring was set for July 27. Bianca and Mark arrived on the twentieth to prepare the equipment and to deliver equipment to Arnand so it could be taken underground to the drill site on July 24, a Friday; drilling would begin on Monday. I had to return to the States, so they would be on their own. Upon arriving in Johannesburg airport, Mark and Bianca found out that a miners' strike was scheduled for Sunday night, July 26. So much for that plan. Undeterred, however, Arnand, Mark, and Bianca persevered in the hope that the strike would be called off. Mark and Bianca arrived at the University of the Free State, where, with the help of Esta, Susan, and the students, they prepared the field equipment. Mark already had coring experience from the previous year at Evander Gold Mine. This time he would bathe the core in rhodamine dye and tiny latex microspheres after it was removed from the core barrel to identify zones where contaminating microorganisms may have penetrated the rock matrix. They would also archive core immediately in a dry ice–filled cooler for DNA analyses. As at Evander, some of the cores would be transferred into ultrahigh vacuum cylinders for performing isotopic analyses on the gas in the pores, and some would be transferred into anaerobic ammo cans and kept refrigerated. The latter cores would be used in experiments designed to detect living microorganisms in the rock matrix. All these different samples meant taking an ice chest, generator, vacuum pump, high-pressure gas cylinders, and vacuum canisters, as well as all the other field gear used in fracture-water sampling underground. Arnand had arranged for a special loco to transport this gear to the drill site. Miraculously, the diesel generator, vacuum pump, and vacuum canisters worked just fine after sitting in the lab at the University of the Free State for nine months. After a little effort, even the modified ammo can used to store the live core anaerobically worked fine. The rhodamine dye and fluorescent microspheres were all there. Bianca and Mark packed their humongous tanklike rental car to the brim. They had room left for only one telescoping packer. The remaining uncertainty was the size of the borehole casing. Which packer would they take? They picked the 46-millimeter-diameter one, and looking like the Beverly Hillbillies, they headed to Kloof Gold Mine.

Mark and Bianca met with Arnand and the drill team on Friday even though the strike was still scheduled for Sunday. Everything seemed a go. The miners confirmed that they could core the last meter into the fracture without using the drilling water, thereby eliminating contamination of the fracture surfaces. Obtaining a core that preserved a fracture surface was extremely rare, but obtaining one that wasn't contaminated with drilling fluid, so the fracture surface could be examined for microbial colonization, had simply never been done before. It was impossible to get such a core by drilling from the surface. Our experience with the Gel Core at Piceance Basin ten years earlier had taught us this lesson. This was going to be a truly one-of-a-kind opportunity. But then a safety officer seemed to question whether the generator was truly flameproof. Frantic phone calls and emails to me in United States ensued. Is there a certificate of flameproofness? Alas, we found out that our valiant generator, which had been used in coring operations at other mines a year earlier, wasn't flameproof at all. What to do? Arnand guaranteed the ventilation would keep the CH_4 concentration down and that it was probably not a concern. So down into the no. 7 shaft the equipment went. On Monday, the twenty-seventh, Mark and Bianca had driven into Johannesburg to get dry ice when they learned that the strike was averted at the eleventh hour! They promptly headed back to Kloof Gold Mine.

Capitalizing on their good luck, they descended the no. 7 shaft the next morning with Arnand and hiked for twenty-five minutes, until they arrived at the Black Dyke drill site, which was thankfully as cool and well ventilated as Arnand had promised. The equipment was safe and sound and coring was already proceeding. They set up the equipment around the corner from the drilling operation along the side of the loco track. Surprise! The drillers intersected gas and water at only four meters into the tunnel wall, but the water pressure wasn't too high. The drillers decided to keep coring through this first fissure to try to reach the actual fissure associated with the Black Dyke, which they were certain was at twenty yards. Meanwhile Bianca and Mark started collecting core as it passed out through the core barrel, dividing it into DNA, microbial, and pore-gas samples. Mark checked the methanometer readings before getting ready to turn on that generator. "Methanometer readings look okay, Bianca; YOU can turn it on," said Mark. The pump glugged for a few minutes then started humming, and the needle on the pressure gauge began to drop down. "That's one good pore-gas sample and we're still here!!" remarked Bianca.

Whoosh! At three yards the drillers hit a high-pressure water zone. A core of the fracture zone went flying out of the core barrel and down the tunnel and landed in the mud. "Well, I guess we will not be looking at that fracture surface," Mark said. As the fracture water cascaded out of the two-yard-long casing at 24,000 liters per hour and a temperature of 55°C, steam began filling the tunnel. The drillers checked their methanometers to make sure the ventilation was keeping up with the gas flowing out of the hole. Unfortunately, the drillers had to stop. At that flow rate they couldn't continue drilling, but they reoriented the drill rig to core a satellite hole targeting the fracture at four meters. The drillers put a valve on the end of the gushing borehole so they could shut down the water if need be. With the valve on, Bianca and Mark started collecting water and gas samples from the first hole. They now focused on the water chemistry. But the packer, needless to say, was too small. "I told you so," Bianca commented. They tried the backup, a huge-diameter plug, but it fractured under the pressure. So no gas samples for isotopic analyses and no massive filter cartridges for DNA.

With the drill rig in its new location, they were ready to go down the next day to finish coring. Overnight, however, a loco managed to clobber the drill rig, which had been sticking out a little too close to the tracks. Details are vague on this incident. But Arnand "Never Accept Defeat" van Heerden pulled a drill rig from another site, putting his job on the line, to help Bianca and Mark get those cores. It took another day to arrange with the loco driver to move it into place, set up, and start the drilling. Mark and Bianca returned again and acquired the precious vacuum canister cores at several points between two and four yards. When the drillers started approaching the four-yard fracture, they shut off the drilling water, cored a little further, and "katoosh!"—out into Mark's and Bianca's autoclave bag came the core preserving the fracture surface. They quickly chiseled off the fracture surface and preserved it with formaldehyde in a small vial. With the cores on dry ice, they returned to surface, where Bianca taped up the cooler and drove to the airport to fly out that evening, exhausted but triumphant.

Mark, meanwhile, had to return all the gear, which would be coming to the surface the next day, to the field lab at the University of the Free State. There Mark discussed the situation with Esta. With no DNA samples, noble gas water samples, or dissolved gas samples, we couldn't compare the core results to the fissure water, which was the main objective of coring the fracture. Esta agreed to come to our rescue, and she, Derek Litthauer, and Melindi "The Machine" modified the packer and the Octopus to handle the larger borehole diameter and high flow rates of the valved borehole. They planned with Arnand to go down that Friday, July 31, to finish the water sampling while Arnand held off the cementation team, which was anxious to seal the fracture. But the plan was put on hold because the driller's union decided to go on strike. Fortunately, without the drillers, the cementation teams couldn't cement either. We had managed to dodge another bullet. When the strike ended the following week, Esta's team went to work. They trekked down to the drill site, where fracture water had been pouring out for over a week, and began shoving the packer into the extra large borehole. But under the high water pressure, the packer wouldn't stay in the borehole. They pulled some of the chains from the ventilation duct, wrapped them around the pipe and packer, and tried to tie them in place, but the chains could not be tightened enough to seal the borehole. Finally, Melindi took off his mine belt, lashed one end to the end of the packer and the other around the borehole, and cinched the packer into the borehole. Water was still shooting out everywhere, but the packer stayed in place (figure 8.3).

FIGURE 8.3. Derek Lithauer, Meider Forster, Esta van Heerden, and Melindi (The Machine) Teboho, who used his belt to hold the filter onto the borehole while the fissure water from the Black Dyke fault passed through it. Shot at 3.2-kilometers depth in no. 7 shaft Kloof Mine (courtesy of Arnand van Heerden).

Two one-liter stainless steel–housed filter cartridges were hurriedly attached to the Octopus, and the team began filtering the water at very high rates on each filter. While the filtering was taking place, they connected a third hose to the Octopus and placed the other end of the hose in a bucket of water. A stream of gas bubbles appeared from the end of the hose. They submerged and inverted a small plastic beaker in the bucket and then maneuvered the open end over the gas stream. The beaker began to fill with gas. The bottom of the beaker, which was now facing up out of the bucket, was fitted with a valved needle. One after another they filled evacuated serum vials by pushing the stoppered tops onto the needle and opening the valve. Finally, they pulled the hose out of the bucket and connected it to a foot-long copper tube with clamps at both ends. Water shot through the tube, and using ratchet wrenches, they closed the ends of the copper tube, trapping the water containing the noble gases that would be used to determine the water's age. Four hours later, with the precious gas samples and the DNA filters in hand, Esta's team headed for the cage. "It was a piece of cake," Derek emailed me later.

CHAPTER 9

LIFE BENEATH THE ICE

In Sneffels Yoculis craterem kem
Delibat umbra Scartaris Julii intra
Calendas descende, audas viator,
Et terestre centrum attinges . . .
Kod feci. Arne Saknussemm

Go down into the crater of Snaefells Jökull, which Sartaris's shadow caresses just before the clends of July, O daring traveler, and you'll make it to the center of the Earth. I've done so. Arne Saknussemm.

—Jules Verne, Journey to the Center of the Earth[1]

THE MARTIAN CRYOSPHERE AND ARCTIC MINES WITH SLIME: SEPTEMBER 2003, PRINCETON UNIVERSITY

Six months earlier I knew that the chances of renewing our South African project without the LExEn program, which had ended in 2002, were slim to none. Tom Kieft had mentioned to me that NASA would be calling for proposals for more institutions to join their NASA Astrobiology Institute (NAI), with the deadline in the first week of March 2003. As soon as I had returned from the workshop in South Africa, I had begun crafting a proposal that would examine subsurface sites that were more analogous to Mars, that is, essentially to explore the deep, cold biosphere that exists in the polar regions of our planet.

When I was a graduate student at Princeton University, Sheldon Judson, the department chairman at the time, and his graduate student, Lisa Rossbacher, had first proposed that beneath the surface of

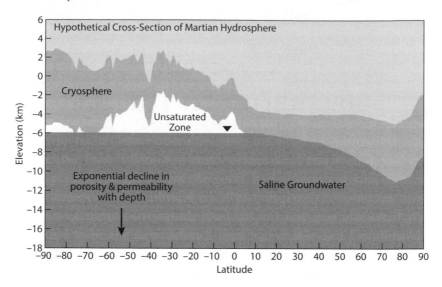

FIGURE 9.1. The Martian Cryosphere/Hydrosphere. Theoretical pole-to-pole cross-section of Mars today along 157°W longitude that illustrates the relationship between topography, latitude (mean annual temperature), ground ice, and subsurface brine (from Clifford and Parker 2001). At low latitudes, where the mean annual temperature is the highest, the thickness of the cryosphere is the lowest, but still three to four kilometers thick. Depending on the abundance of brine, the low-latitude cryosphere may be underlain by an unsaturated zone. Deep caves would be located here.

Mars existed a contiguous global cryosphere of frozen rock that extended to great depths (figure 9.1).[2] They thought that many of the features on the surface of Mars were related to the partial melting and refreezing of this cryosphere. Could a microbial ecosystem exist beneath a solid layer of rock and ice? Radiolysis suggested that it could, but we needed to sample beneath a cryosphere here on Earth to answer that question. The added advantage of working beneath permafrost was that it eliminated the possibility that the fracture water beneath the permafrost would be contaminated by small amounts of modern surface water. Permafrost was impenetrable, or so I thought. The only places on our planet where thick continuous permafrost existed were in Antarctica and in the high Arctic. Antarctica was not an option because it had no mines or boreholes penetrating its permafrost. Neither Greenland nor Iceland had continuous permafrost or mines. This left the prospect of finding some mines on the north slope of Alaska, in the Canadian high Arctic, or in Siberia that tunneled through permafrost.

Barbara Sherwood Lollar had suggested we contact Miramar Mining's Con gold mine and Dominion Diamond's Ekati diamond mine, both of which were near Yellowknife in the Northwest Territories, but neither of these mines had continuous permafrost. I decided to call Steve Clifford at the Lunar Planetary Science Institute in Houston, Texas. I had met Steve at the NASA Ames meeting in 1996, where we discussed drilling on Mars. I had read his papers describing his theory on the hydrogeology of Mars that he had published when he was at Cornell University. He had recently completed some interesting theoretical work on methane clathrates[3] and then switched from theory to actively developing ground-penetrating radar (GPR) methods for rovers to delineate water deposits in the Martian subsurface.[4] I thought he would know where in the Arctic one could access deep permafrost, places where he could test his GPR techniques. I asked him to join our NAI proposal. He agreed and suggested that we approach the Nanisivik mine on Baffin Island. It was a lead-zinc mine located 540 miles north of the Arctic Circle! Although the mine had officially closed the previous year, his contacts there indicated that it would remain open for a couple of years for storage. He also suggested the Polaris lead-zinc mine located on Little Cornwallis Island, 800 miles north of the Arctic Circle, but Polaris had shut down completely the year before. Our timing was unfortunate, because metal prices had dropped over the past few years, forcing the closures of these Arctic mines. Our prospects were looking bleak. Then on February 20, two weeks before the proposal was due, I received an email from Steve about a gold mine named Lupin operated by Echo Bay Mines in Edmonton, Alberta. A consultant friend of his had mentioned it to him. Lupin already had some researchers underground working on the chemistry of the briny fluids found there beneath a 1,300-foot-thick layer of continuous permafrost. Lupin sounded perfect for our NAI proposal.[5] The big item on our to-do list was to drill from the permafrost into the underlying fractured aquifer, collecting microbial core samples along the way, so we could evaluate the difficulties and potential outcomes that would be encountered by a similar exercise on the surface of Mars.

While we waited to learn our fate, we began to quantify the rate of subsurface radiolysis in the deep South African subsurface to see whether it was sufficient to support the biomass that we had detected in so many samples. As Lihung performed the irradiation experiments to determine how long it would take produce the H_2 concentrations we were detecting in the fracture water, I began digging into the literature to find the kinetic parameters we needed to check the theory against Lihung's data.

The idea that radiolysis might contribute to the long-term sustainability of deep subsurface bacteria was first presented in a Russian paper by Vladimir Issatchenko in 1940.[6] He was trying to account for the pink water encountered in Soviet oil fields near Baku at a depth of one mile. The pink color was due to the large concentrations of purple sulfur bacteria, more specifically, species of *Chromatium*. *Chromatium* are anoxygenic photosynthesizers that can oxidize H_2S in the dark. The problem is that in order to oxidize the H_2S, they need O_2. Issatchenko cited a paper by Vernadsky, who suggested that O_2 could be generated in the subsurface biosphere by the "splitting of water by X radiations of radium and mesothorium."[7] But subsequent Russian work had dismissed the pink water as caused by contamination of oil fields during flooding, and the radiolysis idea was forgotten.[8]

It was the papers by Danish and Yugoslavian radiation chemists from the late 1980s and early '90s that provided us with the relevant rates for the thirty-odd reactions involved in saline water radiolysis.[9] Surprisingly, these data indicated that H_2O_2 was the dominant oxidant produced. H_2O_2 is very reactive and readily oxidizes sulfide minerals to produce sulfate. Eric Boice found a Russian paper that invoked radiolytic oxidation of sulfides in uranium deposits.[10] This could explain why we were seeing high concentrations of sulfate, but not O_2 or H_2O_2, in our most saline water samples. This was the key to the problem of sustainability of microorganisms in the subsurface over geological time intervals. Not only did we need the electron donor, in this case H_2, but we also needed an electron acceptor. Lisa set about to determine experimentally whether this was a viable chemical pathway and whether it produced a distinctive $\delta^{34}S$ isotope signature.

Although we now felt certain that we were on the right track with the radiolysis energy source in South Africa, the SLiMEs theory proposed by Stevens and McKinley in 1995 had not fared well over the past seven years. Six months after they had published their seminal article in *Science*, Eugene Madsen, Francis Chapelle, and Derek Lovley had published a highly critical commentary in *Science* doubting both its significance and feasibility.[11] This was followed by a report in 1997 based on 16S rRNA gene sequences that showed that methanogens and acetogens, the primary producers of SLiMEs that Stevens had enriched, were present only in trace amounts in the groundwater that he had analyzed. In fact, the microbial community in the Columbia River basalt aquifer appeared to be dominated by diverse and abundant heterotrophs.[12] This did not make any sense, because how could the primary producers, the base of the food pyramid, compose only a tiny proportion of the community? This was pointed out in a paper that followed in 1998 by Anderson, Chapelle, and Lovley, who also showed that the H_2 production from basalt represented only a short-term burst and that the actual production rates were insufficient to maintain a microbial ecosystem.[13] Occasionally this disagreement between two camps would boil over in conferences with accusations that Stevens and McKinley were promoting a fairy tale. But in 2000 Stevens and McKinley countered with a more extensive suite of experimental results that definitively indicated that H_2 production was occurring and at rates which they felt were significant enough, but they still could not explain the relative paucity of primary producers in their groundwater.[14] Then in 2002, Chapelle and Lovley reported discovering a subsurface microbial community dominated by methanogens in a hot spring in Idaho. They also claimed, based on the 16S rRNA gene sequences, that the methanogens were being supported by H_2. They further speculated that the source of their H_2 was not from oxidation of the young basaltic flows of the nearby Snake River Plains but rather from tectonically generated H_2.[15] The latter was a relatively obscure process first reported by Japanese researchers decades earlier during earthquakes and was believed to occur whenever quartz grains were fractured in water.[16] My geophysics colleagues at the time considered the Japanese hypothesis laughable.[17] All this controversy over H_2 generation left many of us in the community scratching our heads as to whether deeply seated life could subsist over geological time without organic carbon derived from the surface. But to me the message was loud and clear that we had to be very certain of our data before we reported that radiolysis was an energy source sufficient to sustain the subsurface microbial communities we were observing across geological time scales.

Meanwhile, Tom Gihring had gathered all the 16S rRNA sequences from the various groups representing two hundred or more samples collected from mines spread across South Africa. When he aligned the sequences and plotted their relationship on a massive genealogical tree, he discovered something quite remarkable. Time and time again, one bacterial species appeared in the clone libraries of the fracture-water samples deeper than a mile and separated by as much as 250 miles. But it was completely absent in the fracture water from the shallower depths, from the mine air, and from the mining water. This true subterranautical species' 16S rRNA sequence was most closely related to that of an autotrophic *Desulfotomaculum* thermophilic isolate,[18] so we nicknamed it *Desulfotomaculum*-like organism (DLO).[19] *Desulfotomaculum*s are SRBs, and this property seemed consistent with our radiolytic-power idea, but the relationship was not close enough to accept that it was capable of sulfate reduction. In many instances the DLO was a minor constituent, but in the case of Lihung's lucky strike, the DLO made up more than 95% of the microbial community. This was more dominant than the methanogens had been in Chapelle and Lovley's Idaho hot spring chemolithoautotrophic community.[20] But was the DLO a chemolithoautotroph? And could we prove that the sulfate was generated from radiolysis?

RADIOLYTICALLY SUPPORTED SUBSURFACE ECOSYSTEMS ON THE EARTH, MARS, AND EUROPA: DECEMBER 2003, CARNEGIE INSTITUTE, WASHINGTON, D.C.

After submitting his Ph.D. dissertation, Lihung moved to the Carnegie Institute in Washington, D.C., to become a postdoc. He enlisted the aid of Doug Rumble and Pei-Ling Wang, two stable-isotope geochemists, to perform sulfur isotope analyses of the minerals and fracture water from his borehole in which the DLO ruled. Doug had recently successfully expanded the repertoire of sulfur isotope analyses in his lab to include the $\delta^{33}S$ and $\delta^{36}S$. These isotopes could be used to look at a phenomenon known as mass-independent isotopic fractionation, which had been found to be widespread in Archean rocks, including the rocks in the Witwatersrand Basin, and therefore could be used to fingerprint the source of the sulfate.[21] These isotopes were our smoking gun for radiolysis. Thanks to Lisa's work on the $\delta^{34}S$ of the dissolved sulfate and sulfide, Lihung was fairly certain that the DLO was reducing the sulfate in situ and fractionating its isotopic composition at the same time. They analyzed the ^{33}S and ^{36}S of the dissolved sulfate and sulfide and compared those values with those of the barite and pyrite found in the fracture minerals. The results clearly showed that the sulfate was coming from the pyrite. Lisa's Ph.D. student Liliana Lefticariu then confirmed experimentally that H_2O_2 did result in pyrite oxidation to sulfate.[22] With all the pieces of the puzzle in place, we had finally confirmed after almost ten years of off-and-on effort that radiolytic oxidation of pyrite was occurring in situ, and we could at last publish that radiolysis was being utilized by a *potential* SRB (figure 9.2).[23] Using a combination of thermodynamic analyses and our radiolysis model, I could now estimate that the environment could support, at a minimum, 50,000 cells per milliliter that turned over their cell mass every 40 to 300 years if the DLO was a chemolithoautotroph, which we still did not know for certain.

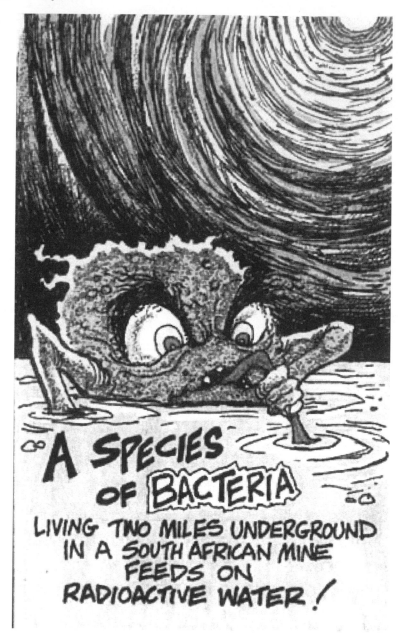

FIGURE 9.2. Cartoon depiction of what a bacterium can do. "*Candidatus Desulforudis audaxviator.*" From Ripley's Believe It or Not website, based on Lihung Lin's *Science* paper (2006) © 2016 Ripley Entertainment Inc.

The radiolytic oxidation of the pyrite did not occur only in the fracture minerals, but throughout the rock matrix wherever pyrite and water were present, and pyrite was pervasive in the Witwatersrand quartzite. I, like most of my colleagues, suspected at the time that no bacteria could exist within these rock units, which all had less than 1% porosity and 0.01% porosity large enough for a bacterium. If true, then the framboidal pyrites within the Witwatersrand quartzite may have formed abiogenically, or if SRBs produced them, they formed billions of years ago when the rock was more porous. The absence of life within the rock matrix would mean that the H_2 and sulfate generated within the matrix was all diffusing into the fractures, where it was being gobbled up by DLO and its compadres. But you always have to check your assumptions, so I sent Mark Davidson on what I thought was a fool's errand. See if you can find sulfate-reducing life inside the rock cores that we collected from Evander, I told him. Mark diligently did what I told him to do, though not without some occasional grumbling. He carefully set up the ^{35}S autoradiography and, to our surprise, found some hot spots sitting inside the interiors of the cores. The rhodamine dye, which we had added to the drilling water, clearly indicated that the drilling water had not penetrated more than a couple of millimeters into the rock cores. Mark went to the University of Western Ontario to use Gordon Southam's SEM, and with help from Greg Wanger spent two days poring over the freshly fractured rock surface in which the ^{35}S had indicated life. Almost ready to give up, they suddenly came upon the bacteria clumped together: clusters of ultrasmall, rod-shaped bacterial cells sitting in the middle of raspberry aggregates of $Ag^{35}S$ and FeS. They were about 0.5 micrometers long by 0.1 micrometer wide and clearly qualified as ultramicrobial cells (figure 9.3). Mark, Greg, Gordon, and I were all blown away. When we compared these SEM images to the mercury porosimetry data, the cell diameters were just smaller than the largest pore-size diameters of 0.15 micrometers.[24] Mark estimated that there should be about 10,000 cells per gram of SRBs, but try as he might, he had no success in isolating and amplifying 16S rRNA from these rock samples to see if they were related to the DLO. It was suggestive, however, that given enough time, micro-

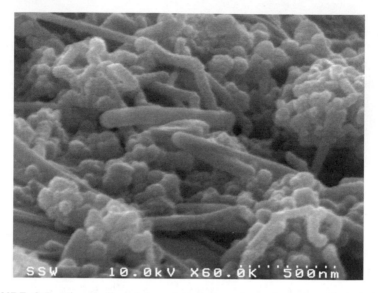

FIGURE 9.3. Ultrasmall rod-shaped bacterium in rock cores collected from Evander Gold Mine at a depth of two kilometers (photo courtesy of Gordon Southam and Mark Davidson).

bial life forms, albeit tiny ones, could penetrate rock in search of sustenance.

Our proof of radiolytic support for chemolithotrophs had an obvious impact on plausible subsurface life on both Mars and Europa, one of Jupiter's moons. Within a week of each other, two papers were published in the summer of 1999, one constraining the potential microbial biomass beneath the Martian cryosphere,[25] and the other constraining the potential biomass in the ocean beneath Europa's icy shell.[26] Both papers pointed out that O_2 was severely limiting in both environments and without replenishing the oxidants of the ecosystem, the Gibbs free energy would quickly reach zero no matter how much H_2 was being produced by the reaction of water with rock beneath the cryosphere of Mars or by deep-sea vents at the base of the Europan ocean. On both planetary bodies, O_2 was being produced by photolytic reactions with the surface ice or atmosphere, but the mechanisms required for that surface O_2 to reach kilometers into the subsurface were complex and geologically slow. Mars exhibited no active tectonics. Although Europa's surface was geologically much younger than that of Mars, Jupiter's tidal stresses seemed to only crack the icy crust of Europa, not recycle it into the underlying ocean. This left CO_2 as the only plausible oxidant remaining, and CO_2 was known to exist on Mars. On Earth, chemolithoautotrophs that utilize H_2 and CO_2 produce either CH_4 (the archaeal methanogens) or acetate (the bacterial acetogens). In 1999, no one had yet detected CH_4 in the Martian atmosphere, and because CH_4 would be destroyed over a period of hundreds of years in the atmosphere by UV radiation, its absence was used as evidence that any subsurface chemolithotrophic activity must be very limited.

Using the known chemical compositions of Martian meteorites and Steve Clifford's porosity estimates, it was a trivial exercise for me to calculate what the radiolytic production rates for O_2 and H_2O_2 would be for Mars as a function of depth.[27] Even though the Martian crust contained far less radioactive elements than that of South Africa, the radiolytic rates were almost as high because of the much higher porosity of the Martian regolith due to the lower gravity and greater impact-induced fracturing. The calculations showed that beneath the icy cryosphere of Mars, radiolysis should be able to sustain a potential biomass as rich and as diverse as what we had seen in the deep fractures of South Africa, and radiolysis could potentially produce biogenic CH_4 at comparable rates. The same argument was applied to the Europan oceans and yielded similar conclusions.[28]

THE DEEP COLD BIOSPHERE: JANUARY 11–12, 2003, HILTON DFW LAKES EXECUTIVE CONFERENCE CENTER, DALLAS, TEXAS

It had taken months before the NAI award arrived at Indiana University for distribution to all of us subcontractors. Apparently the support going to Barbara at the University of Toronto in Canada was much more contentious with NASA than it had been with NSF and had caused much of the delay. I had delays of my own. After learning that we had been funded from NAI, I had contacted Echo Bay Mines in September 2003 only to find out that they had closed Lupin. Echo Bay Mines and another company, TVX Gold, had merged with Kinross Gold USA located in Reno, Nevada. Kinross was now in charge of the Lupin operations, and they had to decide whether or not to reopen that mine.[29] This was the third time I had received funding for doing subsurface microbiology only to find out that I had no subsurface to which to go. Right after the 2003 meeting of the American Geophysical Union, Mike Tansey, who was in charge of the reclamation program at Lupin, contacted me to let me know that they could accommodate us. He also told me to get in touch with Dr. Timo Ruskeeniemi of the Geological Survey of Finland, Geologian tutkimuskeskus, to coordinate field operations with him. Timo was the project leader of an investigation that had been going on for several years at Lupin on the effects of freezing on brine formation in Precambrian shield rocks. Finland had no continuous permafrost deposits, whereas Canada had extensive permafrost deposits, and Lupin was the only operating mine in the world that worked beneath the permafrost. Shaun Frape, a geochemist from the University of Waterloo and an expert on Precambrian shield brines, worked with him providing isotopic data. His support came from Ontario Power Generation, the utility company owned by the Ontario provincial government. All the interest shown by these government entities stemmed from the possibility for storage of high-level nuclear waste in subsurface sites. The concern was whether such long-term storage and isolation would be compromised by a subsequent ice age and permafrost formation or disappearance. Working together, Timo and Shaun had gone underground at Lupin, inspected weeping boreholes with video cameras, and located gas-producing fractures. They then inserted Margot-type packers very similar to the ones we had used in South Africa, and they left them in place. Timo and Shaun research groups had been going back every year since February 2003 to monitor any changes in the borehole chemistry and gas composition. Taking the lead from Mike, I contacted Timo by email at the end of January, telling him that Mike had told me that "Yo de man!" when it came to Lupin hydrogeology and asking whether we could set up a joint expedition. He agreed but explained that we had to move quickly. Kinross was there primarily to mine the pillars and then shut Lupin down. This would give us only eleven to twelve months tops, and Kinross might not be able to handle more than a couple of trips.

This news arrived just as Lisa had called the first all-hands, face-to-face meeting of the NAI team, which was to be in Dallas. During this meeting Terry Hazen proposed that we should attempt to sequence the genome of the DLO to prove definitively whether it was or was not an H_2-utilizing SRB feeding on radiolysis. If PNNL could separate enough DNA from the massive filter that had been attached to Lihung's borehole, he would shepherd it through DOE Joint Genome Institute and work on the assembly and annotation. At this time I had no idea what this entailed or how long it would take. Lisa and I focused on mobilizing what was required for sampling at Lupin in six weeks' time. This was challenging in two respects. We still had no idea exactly how to attach filtering devices to Timo's Margot-type plugs, or packers, what the water pressure and flow rates would be upon opening the valves, and, most important, what the biomass concentration would be.

Very little had been published at that time on the deep, cold biosphere. In 1911, the Russian microbiologist V. L. Omelyansky reported growing bacteria from the soil surrounding one of the wooly mammoths that had been unearthed in 1908. Then in 1942, two Canadian microbiologists discovered viable bacteria in Canadian permafrost deposits. But it was not until David Gilichinsky, a world-renowned Russian microbiologist, began investigations of deep, cold life in million-year-old permafrost located in northeastern Siberia that cryobiology took off in the 1990s.[30] In 1998 he began collaborating with microbiologists to show that the permafrost deposits were not only repositories of extinct life, but also hosts of viable psychrophilic anaerobic bacteria.[31] They had found 10^8 cells per gram in cores down to 60 feet. They speculated that these bacteria could be active in the permafrost even at the −12°C in situ temperatures. A year later, in 1999, a pair of papers published in *Science* had found 10^2–10^4 bacterial cells per gram of ice at 3.6 kilometers (11,800 feet) depth collected by a Russian-supported drilling operation targeting subglacial Lake Vostok in Antarctica.[32] The ice was believed to represent frozen lake water that had been isolated for a million years and was described as a possible Europan analog environment. Compared with the biomass concentrations we had measured in South Africa at comparable depths, these were almost 10–1,000 times less. Would the concentration of microbes in the fracture water at Lupin be like that found in Siberian permafrost or like that found in the Lake Vostok ice?

The year after the Lake Vostok papers, Canadian scientists found evidence of active psychrophilic chemolithotrophy in the rock-rich basal ice of glaciers on Ellesmere Island.[33] The implications were that at the base of the Martian polar ice caps, microbial activity could be taking place. To discover it, however, NASA would need "only" to land on top of one of the ice caps a drilling device that could melt its way through about ten kilometers of ice. A month before our Dallas meeting, a paper was published showing that bacteria were actively taking up isotopically labeled substrates at –15°C in ice. These remarkable results suggested that even at that temperature, thin films of water only a few nanometers thick existed that were capable of slowly transporting nutrients to trapped, living cells.[34] Both of these papers indicated that even within the permafrost layer at Lupin, active microbial metabolism could be occurring and that we needed to be able to perform similar experiments on our core and water samples.

Fortunately for us, another NAI institute located at Michigan State University and led by Jim Teidje, a noted environmental microbiologist, had been working with Gilichinsky for several years on the Siberian permafrost microorganisms. One of Jim's postdocs, Corien Bakermans, had even recently published a paper about potential psychrophiles on Mars.[35] We needed her skills to help isolate any psychrophiles from our Lupin water samples. Lisa phoned her and convinced her to join our expedition to Lupin.

MINE AT THE TOP OF THE WORLD: MAY 13, 2004, LUPIN GOLD MINE, NUNAVUT, CANADA

In May 2004, Lisa and I met Timo for the first time in Edmonton, Alberta. After introductions Lisa and I had to go to the local drug

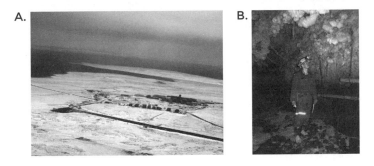

FIGURE 9.4. A. Lupin gold mine from the air (photo courtesy of Lisa Pratt). B. The ice tunnels of Lupin gold mine. Timo Ruskeeniemi for scale (photo courtesy of Lisa Pratt).

store so I could purchase some bleach, rubbing alcohol, and hydrogen peroxide solution that I could use to sterilize equipment at Lupin. Corien then arrived from Michigan and joined us for dinner. Monique Hobbs, who had flown in from Toronto, also joined us. Although Monique worked for Ontario Power Generation and represented the money, she would also be helping us collect samples for noble gas isotope analyses for Ian Clark to determine the age of the fracture fluids. We flew out of Edmonton that evening in a custom-fitted Boeing 727 with about twenty miners. We departed at 11:45 p.m. from a little gate set off from the main terminal that was used by Kinross for their biweekly flights. We sat in the back half of the cabin. The front half was partitioned off and used for cargo and cold storage. After three hours of flying through the night we arrived on an icy, snow-covered airstrip (figure 9.4A). As soon as I stepped through the door of the plane, my nose hairs froze. It must have been –20°C. We boarded a bus for the short trip to the entrance of the living quarters. It was snowing and all I could make out was a huge, dark hulk of a building with a single light over the doorway leading inside. All our bags were thoroughly inspected for alcohol as we stepped through security screening before entering the building. Stacked up inside was the baggage of all the miners who would be departing in an hour for the return flight to Edmonton. They had done their two-week stint and were heading back home for two weeks off.

Lupin had accommodations for 440 workers, and by "accommodations" I mean very nice, Mercure Hotel–style, single rooms with attached bathrooms and comfortable beds. They also had a small medical clinic and doctor on site. A couple of computers we could use to access emails were hidden in a closet-like space. There was even a squash court for those inspired to exercise off shift, a pool table to win some extra money if the salary was not enough, and a sauna to warm one's bones. We were certainly not going outside for a brisk jog with wind chill temperatures of –40°C, and almost perpetual darkness. The highlight of the Lupin operation and center of social life was the cantina. Open 24 hours a day, 7 days a week, 365 days a year, the cantina offered hearty chow and an unlimited array of pastries and other forms of comfort food. In the complete absence of alcoholic beverages, the double-chocolate muffins would be extremely appealing, especially after two weeks. Timo showed us the storeroom in which he kept all of his team's supplies. They had been working there for four years and had accumulated quite an array of tools, pipes, tubing, and just about anything else we might need along the lines of plumbing. We moved our coolers in there. We settled our personal gear in our assigned rooms and rejoined Timo in the cantina for an early breakfast and to work out the plan for the week. One week was all the time we were given on site, and we would need every day to acquire samples from all the boreholes that Timo had packered.

After breakfast Timo introduced us to the mine managers, and we went through our approach. Normally we would have taken a day and a half of safety training before going underground, but working with Timo as our guide, the safety training was reduced to a one-and-a-half hour lecture and, of course, the usual training on the emergency respirators. Timo then took us down to the "drys," where we were given lockers and our thermally insulated coveralls, boots, gloves, hardhats, lamps, and batteries. After we were dressed, and with our sampling gear in our backpacks, we all followed Timo to the cage. On the way we got a tour of the power plant and the mill where they separated the gold and smelted it into ingots. We checked out their Home Depot–size warehouse, in which we could find anything from duct tape to the most obscure of plumbing parts. The five of us walked right up to the cage, which was the size of a normal elevator, not like the jumbo South African cages. Timo punched in the code next to 1100, we climbed into the spacious cage and down we rode to 1100 level by ourselves. I marveled at the convenience of this. As the cage dropped through the 550 meters of permafrost, so did the temperature. When I flicked on my headlamp I could see everyone's breath suspended in front of their faces. At 1,100-meter level (3,600 feet depth), we stepped out of the cage onto a frigid platform from which the usual rusty metal rock bolts protruded overhead. These, however, were draped with yellowish-orange, soda-straw-shaped ministalactites. We walked a couple of hundred yards until we came to the helical tunnel and walked slightly uphill and stepped into another tunnel. The only sound I could hear was that of our steel-toed boots clunking against the tunnel floor. The only light was from our head-lamps, which lit the brilliant fluorescent strips on our dark-brown and blue coveralls. We looked like the fluorescent skeletons that you see on Halloween. From the helical tunnel we walked a few yards into a dark bay known as Merv's shop. Timo reached over and flicked on the fluorescent lights. The bay that lit up was enormous and crammed with all sorts of power tools and building supplies. We followed Timo's eyes down one side of the wall as he located his first borehole. Standing next to it was the largest pile of sulfide-oxidizing biofilm I had ever seen. It was hugging the rock wall, and at the top was a spurting spring of water that would gurgle and occasionally erupt like a mini-geyser. Beneath the layers of cascading filamentous ooze were millimeter-size bubbles of trapped biogenic gas. The rock was also amazing. It was finely layered, like banded iron formation typically is, but then folded again and again, forming sinuous ribbons of light and dark patterns running along the wall. Timo moved to the back of the shop, where he flicked another switch and a ventilation fan turned on. A large yellow tube that had been hanging from the ceiling sprung into life like an enormous python and started wiggling like it was trying to crawl down the tunnel where Timo was standing. "This is the tunnel to the Fountain of Youth," he remarked wryly to us with a wink. We went back to the mountain of biofilm, which I eagerly scooped up into a centrifuge tube and sealed. And Monique screwed a digital pressure gauge onto one of Timo's bore-hole packers.

THE JOURNEY TO THE FOUNTAIN OF YOUTH:
MAY 14, 2004, LUPIN GOLD MINE,
NUNAVUT, CANADA

His packers were the essence of simplicity. He had installed them slightly over a year ago, after their team had finished their first recon-naissance study of the mine. When they first arrived, water was pour-ing from the boreholes and into the Fountain of Youth tunnel, which the mine had dammed at the entrance. The mine would then pump the water out whenever they needed the tunnel for mining purposes and then it would refill. To the mine, the origin of the water was a mystery, hence the name, the Fountain of Youth. Timo and his col-leagues determined that the water originated from an enormous shear zone that intersected the mining operations at the 880-meter and 1,100-meter levels, but not the mining tunnels located deeper. He had convinced the mine managers that he could establish the ori-gin of the water if they would let him seal the boreholes with his packers, and they agreed.

Each packer had a pressure gauge and a ball valve, and nested far inside the borehole were the compression rings that sealed against the rock. Once the boreholes were sealed, the pressure inside them slowly increased as the water, now dammed up in the rock, began to rise in the facture. Eventually the pressure crept up to about 60 bars. Since 1 bar pressure is approximately equal to 10 meters of water, 60 bars indicated that the top of the water level was 600 meters above us. According to Timo, that placed the water level below the bottom of the permafrost, which puzzled him.

Monique opened the ball valve, and we watched the digital pres-sure gauge rise and stabilize at 780 pounds per square inch, which was in fair agreement with the old, rusty pressure gauge above the ball valve. She shut the valve and detached her gauge and Timo reached over to open it again to collect samples in a Styrofoam cup to measure the pH and salinity. The pH was 8–9, similar to that of the fracture water in South Africa. The salinity, however, was much higher, closer to that of seawater, than we had encountered at the same depth in South Africa. The temperature was only 10°C, but it was still warmer than the air temperature. We had pulled off our mitts and donned latex gloves to collect bio samples. We then started collecting samples for gas analysis using our inverted bucket approach. The water pouring into the bottom of the bucket was milky white. As the bubbles began to accumulate in the beaker, the water cleared. Lisa and I had never encountered anything like this before in South Africa. We decided that it must have had something to do with the low temperatures, the high salinity, and the hydrocarbon gases.

Now that I understood the configuration of the packers and the diameter and sex of the pipe outlet, I could set about attaching the filters. First I needed a clean lab bench. Besides, our hands were starting to freeze. I could hardly move my fingers to write in the notebook. I asked Timo for a place to assemble our filters, and he guided us out of the shop and down another tunnel to the refuge bay. It had a heavy steel door set into a cement wall. As Timo pulled it open, it scraped against the cement floor and sounded as if we were opening a crypt. He stepped in and flicked on the fluorescent lights. Inside was a spacious room with several tables and chairs, a small dinette, coffee pot, and refrigerator, with a clean concrete floor and slate-like rock walls. Compared to the refuge bays in South Africa, this was the Ritz of refuge bays and obviously doubled as a miner's lunch room. "The only thing missing is a sow-OO-na,"[36] said Timo, grinning as he cranked up the space heaters. The bay had tubes of sealant that could be squeezed out along the cracks of the door to make the door gas-tight. It also had a rack of high-pressure air cylinders to supply fresh air. The bay could easily handle thirty miners, and you could hold quite a party in there if you had to be stuck underground during a fire.

It was the perfect place for lunch, and after thawing our hands, everyone began unpacking their sandwiches and fruit and taking off their hard hats. I began bleaching the surface of one of the tables. The air was quite still and warmer in the refuge bay, since we were out of the frosty ventilated air of the tunnel coming from the surface. I pulled out the membrane-filter housings, wrapped in their Whirl-Paks, that I was going to use for DNA and for PLFA samples. I would attach the filters to one of the Octopuses I had brought up from South Africa for this trip. But because I had only one Octopus and we were going to sample five boreholes in five days, I had to load the sterile membrane filters into their housings each day. I unbuttoned the top of my coveralls and sat down at the bench. I slipped on my latex gloves and poured alcohol on my hands and on the table. Using a small lighter I set the tabletop aflame to sterilize it. The flames jumped up to my gloves, which quickly caught fire. For a few seconds I thought about what a wonderful way this was of warming my hands and sterilizing my gloves at the same time, but then as the flames started singeing the hairs on my wrist, I began waving my hands madly through the air to put the flames out. While I was distracted saving my hands, the fire on the table was threatening to spread and set the entire table on fire. I quickly squirted bleach on top of it and tamped it out. I looked up to see Lisa pointing a camera at me while everyone else was cracking up over their lunches. I tried to pretend that this was just the normal protocol. I calmly tweezered the filters onto the housings, I sealed them and placed them back into separate Whirl-Pak bags. I pulled the Octopus out of its autoclave bag, attached the filter housings, and then secured the bags to the Octopus to protect them during transport to the borehole.

We packed up and donned our helmets to return to Merv's shop. Borrowing some wire that was readily available in the shop, Lisa and I attached the Octopus, hanging it on Timo's borehole packer. We then gently opened the ball valve and adjusted the pressure to get the maximum flow rates without ejecting the filter housings from their tubing and onto the floor. Having secured the first set of filters on the borehole in Merv's shop, we then followed Timo the into dark tunnel that led to the Fountain of Youth. We crawled over the partially demolished water dam to step into the tunnel entrance. Even though the ventilation had been running for a couple of hours, we could immediately detect a sweet greasy odor that Lisa swore was mercaptan, which is essentially methane with a sulfur attached to it or CH_3SH. With the yellow python quivering above our heads and a steady stream of water flowing at our feet, we followed Timo for several hundred meters past one borehole after another. We would stop to check the pressure of each borehole, and he would open the valve as if he were a bartender opening a brew tap and out would flow fissure water with a frothy head. At the end of the tunnel was a borehole in the floor from which poured the water into the tunnel like a spring. Timo then took us into a smaller side tunnel for another hundred yards to check the final borehole. Along the way we passed a ceiling seep, where Timo had placed Styrofoam cups to collect the steady but slow drip of water. At the end of the tunnel, a small spring of water was flowing out of a borehole in the floor surrounded by a filamentous mat of yellow and orange, much like the ones Lisa and I had encountered in Evander Gold Mine, only it was 30°C colder here. We had found the Fountain of Youth.

It was hard for me to understand why gold miners would compare an artesian spring buried beneath 1,100 meters of frozen rock to a mythical spring located in the Everglades of sunny, tropical Florida. As pointed out to me by my good friend, the astrophysicist Richard Gott, because miners spend a fair amount of time closer to the center of the Earth than most people, they experience slightly greater gravitational acceleration, and, therefore, time passes more slowly for miners than for surface dwellers. He estimated that over a lifetime of a miner that might add up to microseconds. So it's not as if a miner who lived all of his life at the bottom of a mine would reappear on the surface decades later to find out that the world had entered the twenty-third century.

All the boreholes we passed intersected the same shear zone, yet their salinities and pH's were slightly different. Would they yield the same microbial compositions or would there be differences between the boreholes? We would come back tomorrow to set another filter at this borehole. We turned to trek out to Merv's shop and Timo stopped to turn off the lights and ventilation.

We headed back to the cage by a different route, walking past another dark entryway that was past the refuge bay. "What's in there,

Timo?" I asked. Timo turned and led us through the entrance into what our headlamps revealed was an enormous service garage. Other than the fact it was located a kilometer underground in the Arctic, it had all the characteristic features of garages. Parked in it were some of the loaders used to scoop up blasted rock and shovel it into the underground mill. The greasy floors and grimy benchtops were covered with discarded tools and spare parts. Pneumatic hoses and tires were strewn across the floor. The creep factor, however, came from long chains that hung down from the rocky ceiling five meters above us and slowly swayed and rattled with the ventilation. As we walked among the half-cannibalized vehicles, drops of water falling from the ceiling would ding our helmets. Those were the only sounds as we gently crept around. You could not help but feel as if you were walking around in the storage bays of the *Nostromos*, in the movie *Alien*, wondering what might leap out from every shadow as you scan your headlamp over the dilapidated vehicles. "Where are all the miners?" I asked Timo. We had been underground now for six hours and had not seen anyone else. "They are all down on the thirteen hundred level," he replied. Lupin had many levels above us and below us, all with tunnels, refuge bays, machine shops, and motor pools, almost all of which were empty, dark, freezing cold, and silent except for the drip of water and the slightest breeze from the ventilation. It was as if we were in a ghost mine. We turned around and exited the motor pool and headed back to the cage. Timo phoned up to the banks man and a cage appeared within minutes.

Coming up to the surface, we noticed daylight for the first time, because the sun had just risen above the horizon. As I stepped into the warmth of the cage room my glasses fogged. Timo and I removed our wet mining coveralls, socks, gloves, and boots, hung them on wire baskets, and pulled on ropes to raise them up to the ceiling, where the heaters were blasting hot air. "So that's why they call this 'the drys'?" I asked Timo. We then showered, dressed, and joined the women in the cantina. After lunch we donned thick, red Arctic parkas and mitts and went outside to take in the sun and the sights before the sun disappeared below the horizon. All the buildings were red and surrounded by snowbanks. We stood on the edge of ice-covered Contwoyto Lake, and far off in the distance on the opposite side of the lake, we could just make out a miniscule black dot that was an Inuit hut. Other than that, we were in a sea of white. No mountains, just a sea of white. Across the surface of the frozen lake, the occasional flag marked the ice road, although there was no traffic on it today, nor would there be any traffic until next winter. Lupin was the northern terminus of the Winter Road that stretched 430 miles south to Tibbitt and operated from early February to mid-April.[37] At this time of year, the ice would be getting too thin for the eighteen-wheelers to safely traverse.

A razor-sharp wind blew down on us from the north and over our heads on the main building, the Canadian and Nunavut flags both flapping wildly. Lupin mine was in Nunavut and was completely surrounded by Inuit-owned land, but because the original deed had been granted prior to the Nunavut Land Claims Agreement of 1991, Lupin's operations still fell under the jurisdiction of the Ottawa government. The sky was lavender with a romantic orange and pink horizon as the sun began its descent after such a brief appearance. A thin layer of snow covered the windswept ground, which was littered with core boxes, dump trucks, supply crates, all covered with hoarfrost. As we hiked along the perimeter of the red buildings, we came across an outcrop of banded iron formation, much like the one we had seen a kilometer beneath our feet, except that the surface of this outcrop bore the telltale fish-scale texture of glacial scouring. At least ten thousand years ago Lupin was buried with glacial ice. Counterintuitively that meant that the 1,800 feet of permafrost beneath us may not have been present back then and had formed only since deglaciation. If the bottom of the glacial ice had been melting back then, then it was possible that some of that water had penetrated to a depth of one kilometer to mix with the salty water and had carried psychrophilic bacteria with it.

Walking back to the main entrance, we could see the enormous fuel tanks holding the several million liters of diesel that kept the power plant and all the vehicles running. We could also see a plume of steam rising from the gray horn of the fan drift that was choked with ice crystals that had condensed from the steam. As we were looking at one of the snowdrifts draped against the garage, it suddenly moved. Two large white Arctic hares, standing no more than fifty yards from us all the while, abruptly emerged from the white background like shape-shifters, dropped down on all fours, and started hopping away. They were enormous, and I quickly grabbed my camera to photograph them, but when I looked up, they had vanished again into the white background. The first thought that came to me was, how could they be so big? There was nothing to eat up there for nine months out of the year. What could they be possibly feeding on to maintain that body mass? On that thought we decided it would be best to return to the main building before it was dark.

Over the course of the next few days we settled into a comfortable pace like any miner. We would wake up in the morning, check our email, have breakfast, pack a lunch, walk to the drys to change into our mine gear, take the cage down, work in a cold, but not uncomfortable, tunnel all day, come up, take a hot shower, take a walk, have dinner, and go to bed. Dinner was the only chance that we had to interact with the forty or so miners off from their eleven-hour shift. They were all male and covered a broad range of ages. Most tended to stick to themselves, but we were able to draw out some of them in conversation. Having young women as part of your research team always helped in this regard. Most of the miners were Canadian, some were Inuit, and some were American. Like the miners in South Africa, they spent a lot of time away from their families. As remote as Lupin was from civilization, the miners liked working there. It was special among the mines because of its remoteness at the north end of the Ice Road and the long periods of darkness or light. It was about as alien an environment as you could encounter in the mining industry. Most important, it was safe. Some of the miners had just come up from the mines in Nevada and some were going there after they finished at Lupin. The widely accepted consensus was that the American mines were much less safe than the Canadian mines. The infrastructure was poorly maintained and out of date, and there were no refuge bays. None looked forward to going south of the border to work in the American mines, but they often had little choice. Having been underground at Homestake gold mine, I had to agree.

I asked them how they mined the gold from these rock formations. In South Africa, the miners had drilled and blasted a meter-wide stope with a slope of about 30° to follow the ore zone in the sediments. Here, as in South Africa, the gold occurred in sedimentary rocks, but these rocks had been intensely deformed, so much so that the gold layers were vertical. They explained to me that they used the Alimak raise climber, a hydraulically powered giant worm that bored into the rock. It was a rack-and-pinion elevator with a two-meter-square drilling platform that would crawl like a caterpillar up a vertical tunnel, or raise, along an ore zone while they blasted it to smithereens and the rocks fell down to the tunnel below.[38] The raise climber is an impressive piece of hydraulics that enabled them to withdraw rock from just about anywhere with the least amount of disturbance. I could not, however, devise a way of adapting an Alimak to collect uncontaminated biological samples. Besides, Timo had made it too easy for us to get microbial samples by simply turning on a tap. During the week, we visited several different levels, collecting bio and gas samples, inoculating media, and setting up filters.

THE ICE TUNNELS OF LUPIN: MAY 18, 2004, LUPIN GOLD MINE, NUNAVUT, CANADA

On the final day, we planned to retrieve the last filters and check out a potential drill site in the permafrost. Timo decided we should take one of the modified diesel trucks down the helical tunnel. This tunnel descended 4,300 feet in a tight corkscrew path that sloped 15°. One of the miners climbed into the driver's seat as Timo and I climbed into the back of the pickup and the women crammed inside. With little hesitation the driver careened down the snowy track, pulled around a corner, and headed for the entrance of a tunnel covered with big rubber strips. As we passed through, the driver punched the headlights switch and immediately we saw the tunnel ahead and the wall that we were heading straight into. He then jerked the truck into a large tunnel to our right, and Timo and I lurched forward as the driver shifted into low gear and the truck tilted downward. We started corkscrewing our way down into the dark gray, unremarkable rock, but as we continued downward, the headlights and our helmet lights detected sparkling reflections over our head. Suddenly we pulled off the ramp into a tunnel and parked. From the headlights we could see ice crystals and snowflakes of indescribable beauty, size, color, and diversity hanging from the ceiling and decorating the upper portions of the tunnel wall (figure 9.4B).

We left the truck and strolled between ice stalagmites growing up from the floor. We were 300 feet below the surface and at the coldest spot in the mine. At this depth the temperature was always −7°C. The warmer moist air from the depths of the mine naturally ascended the tunnel, but in so doing it passed through the frozen tunnel rock, where the water condensed into these gigantic snowflakes. The air hovered against the ceiling, where it formed one exquisite ice chandelier after another. We were looking at the real manifestation of the vapor diffusion theory that Steve Clifford had proposed was taking place on Mars.[39]

The only sounds in the tunnel were that of the truck and the crackle of ice beneath our feet as we walked toward a plastic sheet lying on the tunnel floor that was Timo's goal. Surprisingly there was a small pool of water in the middle of the sheet, and he reached in with a Styrofoam cup to scoop up the fluid. He stuck his probe into it. The water was −7°C, but highly saline and acidic. Corien collected a little for her incubation experiments. My mind was flooded with questions. From where did this acid water originate? Would we find similar water beneath the surface of Mars? It was tempting to speculate that the salinity and acidity were the result of the fractional crystallization of the water in the permafrost over the past 6,000 years. This was one of the questions in Timo's mind too. But he cautioned us that when Lupin was first excavated in the early 1980s, brine had been used to keep the drills from freezing into the rocks. The salty water in the plastic sheet could represent this drilling brine slow working its way back out of the formation. The acidity could be merely the result of oxidation of the sulfide in the rock formation from the air pumped in with the brine. Timo warned us that the seeps also had very high arsenic concentrations, more than 100 ppm, 10,000 times greater than the permissible levels for drinking water.

As we headed back to the truck, we were struck by how our helmet lights created a golden hue in the crystal chandeliers, almost as if they were lit by electricity. Timo explained how every time they had temporarily closed this mine, the tunnel became completely clogged solid with ice, and they would have to mine through the ice to clear the tunnel in order to reopen the mine. I imagined that something like this must be occurring or had occurred on Mars as well. Satellite images of caves or lava tubes had been reported. An astronaut or robot could descend into those caves, but then they might reach a point where the passages would be sealed by ice. Or they might find a world of ice crystals much like the one I was passing through now. They would certainly encounter saline seeps worming their way through the permafrost and toward the caves and leaking onto the cave floor, perhaps forming a thick biofilm. It reminded me very much of the 1964 sci-fi movie *Robinson Crusoe on Mars*. We climbed back into the truck to head further down the ramp. As we twisted our way down, we would stop occasionally. I got out to look for any signs of biofilms but found nothing. Finally, as we came down to the 380-meter depth, I saw a small biofilm snaking out from a crack in the tunnel wall. We were just above the base of the permafrost and still below freezing, but it was just warm enough for these chemolithotrophs to make a colony.

We pulled into a tunnel at 480 level, just above the base of the permafrost. Timo wanted to show us the boreholes they had drilled the year before and their water collection system. It had several plastic jugs connected with tubing that led into the sealed borehole. There was a steady drip, drip of water into the jugs. This was exactly what Lisa and I had hoped to propose to the NAI, and Timo had the expertise to make it so. We would pick a drill site that was located further away from the helical tunnel, to escape any mining influence, and further up in the permafrost. The surprising result from Timo's boreholes was that at their terminus they did not encounter water. They had encountered a vadose zone beneath the permafrost. This amazed me. Steve Clifford had predicted that vadose zones would also occur on Mars beneath the permafrost at the higher elevations because the subpermafrost water would have drained to the lower elevations in the northern hemisphere.[40] But as Timo described their findings to us, he quickly added that he thought their vadose zone was mine-induced. His reasoning was as follows. Standing in that frosty cave 1,580 feet beneath the surface, we were looking at a borehole that penetrated to 1,800 feet beneath the surface. At that depth we were well below the water level in Lake Contwoyto. Timo described the lake as a through talik.[41] "A what?" I asked. I had never heard this word before. He explained that even though Lake Contwoyto is frozen on its surface for nine months of the year, it is deep enough that it never freezes at depth. Based on the geophysical measurements they had performed during the summer from a boat as they traversed the lake, they were pretty certain that there was no permafrost beneath Lake Contwoyto. This meant that water should be infiltrating the subpermafrost fractures in the Lupin gold mine and that the water should rise all the way to the bottom of the permafrost. But it had not, at least not yet. The low permeability of the fracture systems may explain why it took so long for the mine fractures to reequilibrate with the lake. Even though the Lupin gold mine was located well within the region of Canada where the permafrost is contiguous, it was more like Swiss cheese due to the large number of lakes that acted as through taliks. The same phenomena must have occurred on Mars in its early history when its atmosphere was thicker and the sun was dimmer.[42] Any crater lakes of significant depth would have acted as through taliks, eventually becoming closed taliks, and then finally freezing to the bottom of the crater and sublimating into Mars' thinning atmosphere as Mars grew older and colder. Lupin provided us a perfect analog for studying the early Mars process.

We got back into the truck and continued our journey downward.[43] Within another two turns of the helical tunnel, I could tell we were below the base of the permafrost because the tunnel floor turned to mud and the truck started slipping and sliding. The driver dropped us off at the 880 level and turned around to head back up to the surface. We went into the tunnel to pick up our last remaining filter. Timo's borehole sat behind a deep Olympic-swimming-pool-size pond and was unapproachable, but he had connected thirty yards of teflon tubing to the hole and the valve sat in the tunnel. We had left it running overnight through the filter. So we just shut the valve closed, retrieved the filter, put it in a Whirl-Pak, and then returned to the surface in the cage. That evening over dinner, Lisa, Timo, Corien, and I discussed how we could easily adapt the process of drilling through frozen rock to microbial sampling using the tracers. It would be challenging, however, to get the PFC into a water solution at such cold temperatures. I suggested that we instead try compressed argon gas as our drilling fluid. I thought that given that we would be drilling horizontally through frozen rock, the gas would be enough to carry out the rock chips and cool the bit. Timo figured that we would have to drill horizontally at least 100 meters to escape the contamination from the tunnel drilling, and most likely 300 meters if we were going to have a chance of intersecting saline water pockets. We calculated that we needed to drill two holes, one to intersect the fracture and the other to core through the facture so we could cleanly capture both the bacteria in the water and any on the fracture surfaces. Not using drilling water was advantageous for collecting uncontaminated saline fracture fluid. Besides, most proposed drilling strategies on Mars would use compressed Martian atmosphere for the drilling fluid, given the lack of water. The discussion went on into the late evening accompanied by copious napkin notes. We decided that we would set up an anaerobic glove bag and autoclave in the mine's chemistry lab. We had all been enchanted by Lupin gold mine, the staff's support, and what Timo had accomplished. In our field expedition report to NAI headquarters, we would boast how Lupin was the perfect site for NASA to test technologies that would probe for potential life beneath the cryosphere of Mars. "Don't forget to tell them how you almost burned the mine down," said Timo, looking across the table at me with a wide grin that his mustache could not conceal. We all started cracking up. The next day we flew back to Edmonton with all our samples in coolers and from there dispersed them to laboratories in the United States and Canada.

We were now in a race to prepare for a spring 2005 drilling campaign at Lupin. But what drill rig would we use? If we did not use the mine's, we would probably have to ship one and the rest of our equipment on the Ice Road from Yellowknife. That meant having all the gear deployed to Yellowknife and ready to load on a truck by January 2005. And we had to have our NASA funding in place two months prior to that in order to ship the equipment from the United States to Yellowknife through Canadian customs. We also needed to request an additional $100,000 from NAI to cover the costs of drilling at Lupin, and that request had to be submitted very soon. Timo had contacted Kinross about the proposed drilling campaign to get cost estimates for rehabilitation of a site at the 300-meter level. They were supportive of the idea, if we could drill into one of the areas that they were considering mining. Timo found out that Boart/Longyear had an LM37 drill rig on site that was capable of doing this as well as a "bazooka" drill for shorter, 30- to 40-foot boreholes, that we could use for surface drilling.[44] I had contacted Geoff Briggs at NASA Ames, who mentioned they had a drill rig they had used over the summer on Ellesmere Island to drill permafrost. I thought this would be a good opportunity to test the NASA drill rig's capability of drilling hundreds of meters. I was stunned, however, to find out that in their successful test on Ellesmere Island near Eureka, they had only penetrated a couple of meters. Rather than relying on NASA's drill to obtain our samples, we opted for the LM37 and started preparing the proposal. But in October we received word from Timo's contacts at Lupin that Kinross was deciding that it might have to abandon trying to mine the pillars because of poor ground conditions. Then, Lisa and I had just finished the budget for drilling at Lupin when an email arrived in December, right before Christmas, saying that Lupin was shutting down. Mike Tansey told us that everyone was either dumbstruck, in shock, looking for a new job, or just rolling up their sleeves and starting the evacuation. The equipment would be pulled out and packed for shipment on the Ice Road that January. They were asking for Timo to get up there to remove his equipment. As if that news was not bad enough, we had failed to successfully amplify any DNA from the filters we collected on our first trip. The flow cytometer had failed to detect cells in the water samples, suggesting that the biomass concentrations in these fracture waters had to be at least a hundred times less than that for our South African fracture water. We had nothing to show for our efforts. After Christmas, we regrouped, and I began ordering enough filters to set up six samplers. We planned our return trip to Lupin for March, right before the mine would shut its doors for good.

When we arrived, Lupin had truly turned into a ghost mine. The cantina was empty except for us. This time we went through and set up hollow fiber filters with high-pressure connections that could handle much higher flow rates of about six liters an hour, and we left them on the flowing borehole for three days. One week later we had a day flight out to Edmonton. As we began our ascent, we could see the lonely red buildings of Lupin standing in stark contrast to the surrounding sea of whiteness and then slowly disappear as if it had been a dream. As we rose into the cloud deck overhead, the whiteness of the sky merged imperceptibly with the whiteness of the rolling tundra beneath. As we pierced the haze, we rolled on top of an ocean of clouds extending to the horizon. For the next two hours we could see the shadow of our plane dancing across its surface outlined by a rainbow ring. Finally we reached the edge of the cloud ocean, and the tundra emerged again from beneath, but this time it was covered with evergreens. After we departed, Lisa went up a few weeks later, one year to the day since our first visit to Lupin, to collect samples and brought a videographer with her to capture the last time man would see the Fountain of Youth at Lupin. Timo shut off the lights at Merv's shop for the last time before they headed for the cage. Lupin settled in for a very long winter's hibernation with no spring in sight, and the water level slowly began to flood the tunnels.

JULES VERNE HELPS US NAME OUR NEW SUBSURFACE CHEMOLITHOAUTOTROPH: SPRING 2004, LAWRENCE BERKELEY NATIONAL LABORATORY, BERKELEY, CALIFORNIA

While Lisa, Timo, and I struggled with the Arctic, PNNL had been quite successful in extracting large-molecular-weight DNA from Li-hung's massive filter after several attempts. They had shipped the

DNA to the DOE Joint Genome Institute, where it had been sheared into 3,000-kilobase chunks for sequencing. Over a period of two years they had woven each of the 3-kilobase sequences to each other to produce segments of the genome that they eventually were able to stitch into one long continuous and complete 2.3-megabase loop of chromosomal DNA of the DLO.[45] There were a few extra bits of sequence that didn't seem to belong to the DLO, but these represented less than 1% of the sequences. The DNA loop had only one 16S rRNA sequence, and it matched Tom Gihring's DLO 16S rRNA sequences.

The task of annotating the remainder of the sequence fell to Dylan Chivian, an aspiring bioinformaticist working on our project. He quickly identified several protein coding genes for H_2 oxidation and the complete sulfate reduction pathway. The sulfate reduction pathway was the only ATP-producing metabolic pathway. The DLO was definitively an SRB. We had nailed it. The only way this microorganism could have survived was by living off the radiolytically produced H_2 and sulfate. But the information Dylan was able to decipher from the first genome of a deep-terrestrial subsurface microorganism was full of surprises (figure 9.5). The DLO had a complete N_2 fixation pathway. That was a shock to us. N_2 fixation was energetically very costly, requiring lots of ATP. The existing dogma about the subsurface was that it was energy limited. Was there enough energy from radiolysis to fix N_2 to make ammonia for the synthesis of amino acids for proteins? He also found that the DLO was autotrophic, which was no surprise, but that it used an acetyl-CoA pathway identical to that found in methanogenic Archaea, not similar to those found in bacteria. So parts of its genome had been transferred horizontally, or by HGT, from a methanogen, one that most likely coexisted with it in the deep subsurface.[46] This was consistent with what Duane and Gihring had found in the Dream Borehole. He also found several short sections of the genome with scrambled nonsense coding, called CRISPR sequences, which were designed to defend the organism from viral infestation. Was the DLO under attack from viruses 9,000 feet beneath the surface? Could they be controlling how many DLO

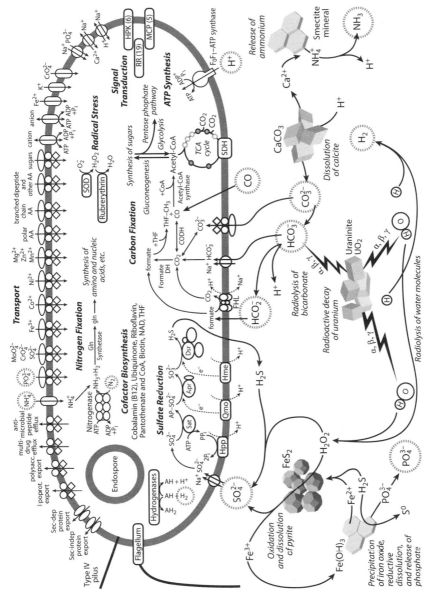

FIGURE 9.5. Cartoon depicting the different metabolic and anabolic processes coded in the genome of "*Candidatus* Desulforudis audaxviator*" (courtesy of author).

there were by lysing the DLO? Could they be responsible for the HGT and helping the DLO to evolve to new conditions? By examining multiple copies of the genome, each from a distinct cell, Dylan discovered that the genome was almost exactly duplicated between each cell. It was almost clonal. This seemed to suggest that the 10 billion cells represented by the DNA had originated from one progenitor with very little mutation. This did not seem consistent with the idea that each cell division occurred once every 300 years as we had estimated, but that it had grown more recently.

The more Dylan deciphered, the more excited we became. The DLO had genes for flagella and for gas vacuoles. This meant that the DLO could propel itself through the water while adjusting its buoyancy like a submarine, processes critical for negotiating its way through fractures. Dylan found genes for proteins that involved signal transduction. In other words it was a hunter. It was looking for something, perhaps a key nutrient or signal from other microorganisms, that said, "Food over here!!" All of this activity required energy, and the only energy available was radiolysis. The moribund state of the deep biosphere that I had accepted until this point was now out the window. It was clear that in these deep, hot environments chemical gradients existed and the microorganisms were exploring them.

The DLO became the poster child for the Indiana-Princeton-Tennessee Astrobiology Institute. The only problem was that the organism needed a real name. I had never named a living species in my life, so we formed an email group trying to come up with names for the DLO that had not already been used. Dylan first suggested "Desulfotomaculum zarathustra." Zarathustra was named after Nietzsche's hermit and was also in the title of Richard Strauss's *Thus Spake Zarathustra*, which is the music for the opening dawn sequence in *2001: A Space Odyssey*, which was set in Africa. This was certainly more creative than my "Desulfotomaculum Mponeng." Fortunately for me, Nora had taken a philosophy course as an undergrad and had a copy of the book *Thus Spake Zarathustra* on her shelf. We sat down and skimmed through it. Heavy stuff, all about Übermensch, tend to your garden, and God is dead, with a tiny bit on Darwinian evolution, but all in all, no obvious connections to subsurface microbiology. Others from the group regarded Nietzsche as elitist and neo-Nazi at the core, a pre–Ayn Rand in a Wagnerian sort of way. So we decided that "Zarathustra" was perhaps not the most politically correct choice for the species name. That last thing we wanted was to be accused of promoting a Nazi bacterium from hell. I contacted Esta and Derek to see if they could suggest a Xhosa or Zulu word that might best describe our DLO. Esta suggested "Tokolishe," which in Zulu mythology was an eyeless, dwarf-like water sprite that created all sorts of mischief. DLO would be like a Zulu zombie bacterium from hell. "Desulfotomaculum tokolishe" had a nice ring to it, though. But it was promptly pointed out to me that we needed a Latinized version for the species name, which ruled out our Zulu mischief maker. In the mean time, Chivian had created a higher-resolution phylogenetic tree from the concatenation of about thirty-one genes that were related to the ribosome, and the tree clearly indicated that the DLO was not only a new species but also a new genus.[47] So now we had to come up with a Latin genus name as well. Clearly it would start with "Desulfo," since that is what the DLO did, reduce sulfate. Gordon Southam had sent us a beautiful SEM portrait of the DLO. It was shaped like a rod, less than 0.2 micrometers in diameter but more than 2 micrometers long (figure 9.6). I translated "rod" to Latin using an Internet translator, and it came up with "rudis." "Desulforudis" sounded right and no one had ever used it before. But we still needed the species name. On a plane ride back from a conference in Europe, Dylan was thumbing through a copy of Jules Verne's *Journey to the Center of the Earth* when he came upon the message written on parchment in Icelandic runes by a sixteenth-century alchemist, Arne Saknussemm, that Professor Lidenbrock translated into Latin and reversed the order.

In Sneffels Yoculis craterem kem
Delibat umbra Scartaris Julii intra
Calendas descende, **audas viator**,
Et terestre centrum attinges . . .
Kod feci. Arne Saknussemm

Or in English,

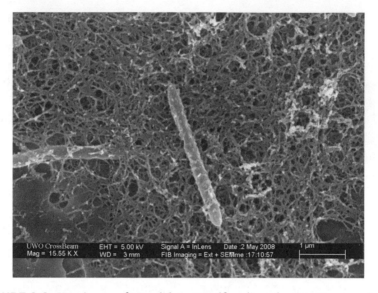

FIGURE 9.6. SEM image of "*Candidatus* Desulforudis audaxviator" captured on a filter sample of Mponeng borehole 104 along with abundant filamentous material (courtesy of Gordon Southam).

Go down into the crater of Snaefells Jökull, which Sartaris's shadow caresses just before the clends of July, O daring traveler, and you'll make it to the center of the Earth. I've done so. Arne Saknussemm.

Properly rendered, *audax viator* stood for "bold traveler." This fit in perfectly with the genome's identification of its motile, chemotrophic capability and its apparent distribution across the subsurface of South Africa. "*Candidatus* Desulforudis audaxviator" was boldly exploring for underworld realms where it would survive and even flourish. We submitted the paper to *Science* in June 2006.[48]

THE JOURNEY TO HIGH LAKE: JANUARY 2005, ONTARIO POWER GENERATION HEADQUARTERS, TORONTO, CANADA

We had not given up hope on the Arctic field campaigns. Back in December, Timo had told Lisa and me that he had been looking at two other mining prospects, which were operated by Wolfden Resources, that were further north of Lupin. In January we met in Toronto with Timo, Shaun Frape, Monique Hobbs, and the others to discuss the drilling options. Wolfden was undertaking a massive drilling program at High Lake and reopening an underground ramp at Ulu. We might be able to piggyback on their drilling campaigns. Both were primitive camps that were only open during the summer. The soonest we could start drilling would be summer 2006. Ulu looked the most promising, but Timo had to find a drill rig that we could use underground, and the closest appeared to be in Yellowknife. Because Ulu was beyond the Ice Road, we would have to fly the rig in on a C130 Hercules and fly it out after we had finished, at a cost of $60,000 Canadian. Lisa and I proceeded with the drilling proposal to the NAI, although we were not sure which property would be the final target. We would need the funding very soon if we were going to secure the necessary agreements and deploy supplies to the site in time. The Ulu camp opened in late April, but by early May Wolfden told us that they had underestimated the amount of ice that clogged the old underground ramp, and they could not rehabilitate the deepest level in time for us. So we changed the target at the last minute to High Lake, which already had a drill on site. Wolfden's geologist in charge of the property, Ian Neil, amended the contract to include an extra borehole for the Indiana-Princeton-Tennessee Astrobiology Institute at about $150,000. With Tommy's and Susan's help, I finished the SIP by the first week of June. We then had six weeks to order and ship all the equipment we needed for coring to Edmonton, to clear Canadian customs, to put it on a truck to Discovery Mining Services in Yellowknife. They would load the supplies on the biweekly Twin Otter flights to High Lake.

The flight from Edmonton to Yellowknife, the capital of the Northwest Territories Province, is two hours. Within twenty minutes of departure we had left behind the patchwork quilt of farm fields and arboreal forests sliced by a rectilinear grid of roads which gave way to an uncharted expanse of forests, lakes, and tightly meandering streams. To our left the snowy tips of the Rocky Mountains loomed in the distance. Straight ahead a dense white cloud bank stretched from horizon to horizon. Soon we were engulfed by the haze that strangely refracted the sunlight from above, lighting the terrain unrolling below like a spotlight. Kettle ponds partially covered with green algae, a thinning, receding timberline, and hummocky meadows laced by meandering streams passed beneath us. This was classic postglacial tundra. No roads, power lines, or fences. No visible sign of humankind's footprints until we reached the outskirts of Yellowknife.

Yellowknife sits astride Archean granite and greenstone rock poking up through the glacial cover from the last ice age. An oasis of civilization parked on the northeastern edge of the massive Great Slave Lake, Yellowknife was the jumping off point for the great ice road that ran 430 miles north to the shuttered Lupin mine. The gold mines Con and Giant, which used to sustain Yellowknife, were mines in name only, now having extracted the most profitable ore before shutting down. Their headgears stood over the town like silent guard towers. But the entire region was still buzzing and blasting with diamond, gold, and semiprecious metal exploration and mining, and Yellowknife remained the hub of their supply lines. Its population fluctuated seasonally with ecotourists flocking there during the summer for tours of the Northwest Passage. During the winter months, Japanese couples discreetly visited specialty camps like Aurora Village to copulate in Indian-style tepees beneath the northern lights. We stayed overnight at the Explorer Hotel, which was just on the edge of "downtown" Yellowknife. We picked up our field gear from the airport and drove it to Back Bay, which was used as the local floatplane base. We squeezed our gear on board a Twin Otter floatplane, and the five of us—Timo, a fellow Finnish geologist, Matti Talikka, Susan, Corien, and I—climbed in and strapped ourselves to the canvas foldout seats. We were sitting right behind the pilot and copilot; nothing separated us from their cockpit. We had been given earplugs to soften the drone of the engines that permeated the cabin. After the steps were folded up, the doors shut, and the fuel hose disconnected, one of the ground crew unhitched the mooring line, tossed it onto the float, and waved us off. After starting the engines, the pilot gently gunned them, and we started moving out into the bay like a gargantuan water strider. Once we were in the middle of the channel, not more than a hundred meters from shore, the pilots slammed the engine throttles full forward, and the plane shook from the vibrations as we were pressed back into our seats. I quickly grabbed my earplugs. The plane surged forward in pulses, as the pontoons hit one wave after another, each time with greater impact. We were heading for some wooden cabins on a hill across the bay not more than a few hundred yards. And then as soon as the pontoons cleared the drag of water, we launched into the air like a missile.

As the plane arced over Great Slave Lake, I could see the southern shore decorated with little cabins and the occasional boat. We then headed north, leaving the lake behind us. It was July 21, 2006, a year after Lupin closed. It had been March 2005 when we last flew up to Lupin. At that time the lakes had been completely frozen, and a steady stream of semi-tractor trailers could be seen migrating north through the middle of the lakes. Now we could more readily see the true nature of the landscape. As we flew further north, the trees that engulfed Yellowknife retreated into the long, linear rock crevasses formed by Archean fault zones and Proterozoic dike swarms. Still further north, the green meadows that had replaced the trees began to thin out into smooth white granite surfaces. Nunavut was more like a lake-bound archipelago than a continental shield. All the freshwater, whether in lakes or in rivers, flowed across the surfaces. Every little hollow had a lake and an outflow channel from which water poured during the summer melt. In this country, the water never penetrated underground, and rivers rarely developed the intricate lace work tributaries that we're accustomed to further south. In western Nunavut, the land of mostly freeze and occasional thaw, the streams rushed down the hill and into the ocean and never lingered for long to meander their way. Occasionally we would come across a cliff face with a recently formed landslide. These thermokarst features were becoming more frequent as the climate warmed and the permafrost underneath thawed,[49] leading to the collapse of the soil that it had held in place by the ice. As we flew overhead, the shape of the landscape reminded me very much of the valley networks and collapse features on the oldest surfaces of Mars, the origins of which were subject to much speculation. But from this Arctic perspective, it seemed that early Mars, like the current Arctic, must have been frozen most of the time. The occasional thawing from changes in its orbital inclination or warming from meteorite impacts or volcanic eruptions would produce short-term runoff focused within the uppermost few feet of regolith.

High Lake was only a kilometer long by a couple of hundred meters wide, looking just like any one of a thousand lakes spread across western Nunavut. It was well north of the Arctic Circle and only forty kilometers south of Coronation Gulf, which connects Amundsen Gulf to the Beaufort Sea. But High Lake is easy to spot from orbital photographs or even Google Earth because of the two square kilometers of orangish smudge bordering its western shore. This orangish smudge was an incredibly huge gossan as big as the lake itself and easily the biggest in Nunavut.[50] Gossans form when rocks rich in sulfide ore are exposed to O_2 at the Earth's surface. Sulfide-oxidizing bacteria quickly colonize the sulfide exposures and transform the fool's gold into ferric iron hydroxides, which produce the orange color, and sulfuric acid in the water, which runs off to acidify the creeks and lakes.[51] Wolfden's geologists had filed a claim based on that photoimagery. Now Wolfden was at High Lake coring the deposit to map its vertical depth and the richness of the massive copper and zinc sulfide deposits. After having logged 260,000 feet of core, they had decided that the deposit was rich enough. High Lake would soon become an open pit mine, initially. The ore would be transported along a road to a deep-water port forty miles further north, where it would bc loaded into ships to carry the ore to China. The road did not exist yet, nor did the deep water port. But with global warming clearing the Northern Passage of ice for a good portion of the year, it was a plan that would soon become economically feasible. When that happened, the Arctic mines would no longer need Hercules flights from Lupin or Yellowknife; they could bring supplies in on ships and truck their supplies down a few tens of miles from the north.

HOW TO CORE THROUGH PERMAFROST AND COLLECT WATER, STEP 1: JULY 23, 2006, WOLFDEN MINING CAMP, HIGH LAKE, NUNAVUT, CANADA

That was the vision at least, but for now Wolfden's mining camp was tentatively perched on the southwest shore of High Lake. As it came into view, the camp looked to be no more than a dozen wooden shacks and tents and a huge pile of fuel drums. We swooped down over a bare rock ridge from the south and dropped into the lake. As soon as the pontoons hit, we were thrown forward in our seats as the pilots pulled back on the throttle. The Twin Otter came to a stop within a few hundred yards. We motored over to a rickety wooden dock that reached into the water just far enough for the Twin Otter to moor. It had been a clear sunny day as we flew up with the wind from the south. When we opened the doors and began to climb out, we were first greeted by the mosquitos, who were quite happy to see us. A ground crew of Wolfden employees also met us and helped unload our gear from the back. We moved the Ar and N_2 high-pressure cylinders to one of the storage sheds and carried our personal gear up to one of the white, four-person tents heated by a kerosene stove. We then walked up to the main office, where the first call of business was camp orientation. Trish Toole, who was the camp manager, handed us two battery-powered walkie-talkies and showed us how to use them. Since we would be working away from the mining camp, this would be our only means of calling for help. One channel was for Major Drilling Group International and the other was for Great Slave Helicopters. We were all handed the bear spray and bear bangers and shown how to use them. "Whatever you do, point the banger up into the air, don't point the banger at the grizzlies. That will just piss them off," Trish warned us, as the other Wolfden employees gathered in the office started chuckling. If we did see grizzlies anywhere near our drill site or the core shack, then we were to call in, and the office would dispatch one the helicopters to shoo them away. She also instructed us on the use of the air horns

and their location just in case a grizzly entered the main camp looking for food. Of course, feeding the curious bears, wolves, foxes, and
wolverines that would occasionally drop in for a friendly visit was as
forbidden as it was to shoot them. She also discouraged feeding the
sik-siks, the Arctic version of prairie dogs.

After the orientation concluded, we met the helicopter pilot over
lunch in the mess hall and arranged the transport of our field supplies from the storage area up to the drill site and to the core shack,
where we would process the cores.[52] All transport on site was either
done on one's back or by helicopter as no wheeled vehicles were
permitted. Susan, Corien, and I were in charge of getting the core
shack and tracers ready, and Timo and Matti were in charge of preparing the drill site. After lunch, we met at the storage compound
and gathered the boxes and gas cylinders destined for the core shack
into a big rope net that we hooked onto a big chain, which was then
attached to a helicopter. Once that was done, Susan, Corien, and I
then turned and dashed up the hill away from the camp and over a
ridge top about a hundred yards to reach the core shack. Just as we
arrived, so did the helicopter with our sling load. The pilot gently set
our load down right next to the entrance of the core shack, where I
disconnected the sling and fanned my hand over my helmet to indicate to the pilot it was clear to take off. The pilot then returned to the
supply area, where Timo and Matti had gathered a load to be hauled
to the drill site. A few minutes later we saw the helicopter with a load
slung beneath it head up to a dark gray ridge of Archean volcanic
rock a half-mile south of us. It hovered near the tip of the ridge,
where we could see a drill rig tilted at a 65° angle. After a few minutes
of hovering it pulled away with a sling load of fuel drums and returned to the storage area to deliver empty drums.

After Terry showed us how to operate the diesel generator in the
shed next door to the core shack we turned on the power to the latter.
The core shack was the second-largest building on the site and had
large windows to let in the twenty-four-hour daylight. A single kerosene stove sat at one end to heat the shed, and Terry demonstrated
how to negotiate with the stove in order to actually get the thing to
work. Susan, Corien, and I set about assembling the anaerobic glove

bag and gas cylinders. Once the bag was assembled and inflated with argon gas, I donned the gloves and started sterilizing the bag's interior. Susan and Corien used bleach and ethanol flamed by propane torches to sterilize the various surgical implements we would use to split and crush the cores into heaps of pea-sized rock and then coax them into Whirl-Pak bags. They wrapped the tools with aluminum foil and passed them into the glove bag, where I laid them out on top of autoclaved brown paper. The core surgery was now open for business.

To set up the tracers for drilling, we needed to hike to the drill site. We sprayed on DEET and donned our head nets before stepping outside. Although it was a short hike, we had to first cross a hummocky ice-wedge polygon bog without sinking up to our shins.[53] Having passed the "ankle-twisting" bog, we then climbed three hundred feet up a "knee-banging" gorge to a ledge, which we then traversed for about five hundred feet to get to the drill rig. Timo and Matti were organizing the core boxes and their equipment in a small two-person shed they would use to log the core and collect the drilling water samples. Timo had spent the previous six weeks plotting the several hundred boreholes that Wolfden had already drilled over the past years to identify the best borehole for us. Since our budget was too tiny to support drilling from the surface, we chose to deepen a shallow borehole so we could capture the transition from the permafrost to the water beneath it. This borehole had to be close enough to camp that we could carry the cores to the core lab. It had to be hydraulically up gradient from the other boreholes that had penetrated the permafrost to avoid contamination. Finally, the borehole, when extended, would need to penetrate a fault zone that almost certainly contained water-bearing fractures beneath the permafrost. Only one borehole met these qualifications: borehole 03-28. It was located on the southwest corner of the property on top of the ridge overlooking the camp, was only 1,000 feet deep, and was angled toward a major north-south fault zone that ran along the western edge of High Lake. The future mining plan placed this borehole south of what would become an enormous open pit. The borehole, once we completed it, could then be used by the mine in the future for environmental monitoring of any potential groundwater contamination. Since the mean annual temperature at High Lake was 5°C colder than at Lupin, Timo estimated that the bottom of the permafrost was at approximately 350 to 450 meters (1,150 to 1,500 feet) depth, depending upon the thermal conductivity of the rock. Given the 65° slant of the borehole, this depth meant that we would be collecting almost 650 feet of core.

We met with Bruno, who was in charge of the Major Drilling team. He seemed extremely capable of adapting to our requests, particularly our request not to use salt in the drilling water. The week before, he had had the rig moved up to the site. He also had his guys install a water pump at the small lake just west of the core shack to pump the fresh lake water up to the drill rig. He had set up two diesel-powered water heaters in series that heated the water to 80°C. Although we could not filter the water through 0.2-micrometer filters to remove the lake microorganisms, such a high temperature was sufficient to melt their lipid membranes and thus kill them. If we had tried to filter the water and the filters clogged, the hot water might shut off, at which point the drill rods would freeze to the borehole. That was too risky.

The drillers had already drilled out the ice that was plugging the borehole and had circulated the hot freshwater through the borehole for five days before we arrived to flush out any calcium chloride ($CaCl_2$) brine that had penetrated the formation when the borehole had been drilled the previous year. Upon returning to the surface, the much cooler water traveled downhill through a hose into a sludge collector and back out to a sump pond located well away from any of the adjacent lakes. Under the strict environmental guidelines granted by the Nunavut government in Iqaluit on Baffin Island, the mine could not discharge any drilling fluid of any kind anywhere near standing bodies of water.

Bruno, true to his word, had installed a T-valve into the drilling-water line just beneath the drill rig. Susan and I then set to work installing the HPLC pump for the PFCs and connecting it to the plumbing. I had never worked with PFCs before and was mostly taking notes while assisting Susan with connecting the HPLC line that would carry the PFC solution to the pipe thread through various bushings. She finished by installing a second valve that we could use to sample the PFC solution and measure its concentration before it was diluted with the drilling water. Knowing the pump rate that the drillers liked to use when coring and the maximum pump rate of Susan's HPLC pump, we estimated the concentration of the PVC solution we would need during the complete coring operation. Bruno had promised that after the drillers had drilled through the ice blockage of the old borehole they would switch to the drill bit used with the wireline coring tool, which we double-checked to ensure that our tubes would fit inside the coring tube. They did. Bruno was three for three.

Susan had brought a bottle of latex fluorescent microspheres and little Whirl-Pak bags, and I had brought the temperature strips. We played around with Major Drilling's wireline coring tool until we worked out how to attach the bags and where to insert the temperature strips. We needed to monitor the temperature on the core down hole as it was being cored to make sure that the core was not heated by the friction of the bit to more than 25°C. Such a high temperature would kill the psychrophilic microorganisms that Corien and Susan were hoping to cultivate and isolate. Killing the psychrophilic SRBs would also invalidate the ^{35}S activity measurements I wanted to make. Basically we had to keep the drill rods between 0 and 25°C using hope and prayer and adjusting the penetration rate of the bit. Susan hooked up the HPLC pump to the generator we had borrowed from the mine and switched it on to test the pump. I heard a loud "zzziitt" and "pop" and then "dammit!!!" The first two sounds came from the HPLC pump, the last from Susan. It looked like we needed to rewire the generator, so we called Bruno on our walkie-talkie. A half-hour later Bruno came up huffing and puffing to the drill site. He reached into the generator and did a little rewiring. Susan had replaced the fuse on the HLPC pump. They switched everything on, and after a bit of knob twiddling, the piston of the HPLC pump began its slow "ka-chunka, chunka" cycle. "Thanks, Bruno," she smiled and gave him a big hug.

By the time we were finished, the next shift of drillers had just arrived. It had been a full day since we had flown in that morning. Susan and I hiked up to the top of the ridge behind us to take in the

FIGURE 9.7. Wolfden mining camp at High Lake, Nunavut, during summer 2006; taken from ridge looking to the east. Rock shed in lower left core is where the cores were processed. Helicopter pad is in center, near pilot's quarters. Fuel drums are on the right. Main building and workers quarters are closer to the lake (photo from author).

view (figure 9.7). It was 7 p.m. and the sun had swung around to the northwest. The ridge on which we were standing cast a shadow over the valley directly beneath us. No more than two hundred yards distance to the southeast was a belt of orange-hued rocks and dirt that stretched between two small ponds. We were going to intersect that belt at about 1,600 feet deep, and the drill rig was pointed straight at it. Beyond that, a green-bespeckled, gray hummocky surface stretched to the horizon. I could easily make out a grayish-black mesa on the horizon, which seemed only a few miles away because the air was so clear, but it must have been at least fifteen miles. Fluffy and flat clouds drifted overhead as their shadows undulated across the ridges and streams. No human contrivance was visible, or likely existed, as far as our eyes could see. It seemed so friendly and inviting. We could easily imagine camping out here during the summertime, living on fish and without fear of any deadly poisonous reptiles or insects. It was just those pesky mosquitoes that were off-putting. In a way it was reassuring that parts of the planet still existed that had not been completely overrun by the resource extractors. But in the century ahead, when the Northwest Passage would be ice-free almost year-round, would all of this become prime real estate? Would the countryside be carved up into ranches connected to small mining towns by asphalt roads that reached down to beachfront resorts with casino gambling on the Beaufort Sea? Certainly we would be overlooking an enormous pit, hundreds of meters deep with an inclined ramp traversed by twenty-ton dump trucks. The wind was shifting behind our backs toward the north. We looked in that direction, and just north of the blue-roofed white huts that marked our camp was a broad alluvial plain. It started at a bright yellow hill to our left and followed an orange-colored valley to turn into red ochre gorges that dumped their contents on the western shore of the tranquil metallic blue-green lake. The only things growing on this barren, Martian-looking wasteland were white stacks of core boxes. The green shrubs that decorated the surrounding gray rocks stopped at the edges of orange landscape, not daring to set root in it. When we had finished our coring, we would have time to investigate this terrain much more closely. The rover *Opportunity* had landed at Eagle Crater on Mars two years earlier. Within a year it had identified hematite concretions called blue berries, gypsum, and a ferric iron salt mineral called jarosite. This salt was typically found in gossan deposits on Earth and suggested that the Martian plains of Terra Merridiana had once contained acid lakes, not unlike High Lake, or at least acid groundwater, 3 billion years ago. Papers had been written suggesting that such an environment was prohibitive to life. Would we find the same types of minerals in these Arctic gossan deposits as were found on Mars? Certainly we would find life.

Then we both heard "sssick, sssick." Susan and I looked down at a couple of sik-siks looking up at us and chirping their dual syllabic call. "I think they are telling us it's dinner time," I said to Susan. We started gently climbing down to the mining camp, ignoring the sik-

siks, much to their consternation. After dinner and a shower, which we took after the drillers had taken theirs, following proper camp etiquette, Timo, Matti, and I settled into our tent, where we found a toasty warm, but smelly, kerosene heater waiting.

HOW TO CORE THROUGH PERMAFROST AND COLLECT WATER, STEP 2: JULY 24, 2006, WOLFDEN MINING CAMP, HIGH LAKE, NUNAVUT, CANADA

The next morning, over breakfast, Timo updated us on the drilling after talking to the night crew. They had just taken the first core and were now tripping out the pipes to attach a bit that had the larger inner diameter we needed for our polycarbonate tubes. They would be ready to take our first microbial core by noon. This gave us just enough time to finish preparations and start adding the tracers to the drilling fluid. Susan and Corien went to the drill site, while I stayed behind and checked on the core vacuuming station I had set up in Bruno's machine shop. We wanted to see if we could get enough gas from the pore water to measure its composition and, we hoped, estimate its age using the same approach that Johanna had used in South Africa. After testing the generator, pump, argon gas regulator, and valves, I dashed up the hill to the drill site. Susan had started pumping the PFC solution into the line as the drill water was pumped into the drill rods. We decided to keep pumping the PFC tracer throughout the entire coring operation in case we intersected any cryopegs and needed to evaluate to what extent the drill water had mixed with them.[54]

The drillers were almost finished lowering the pipes with the drill bit, so Susan and I prepped the core barrel. Susan slowly poured methanol along the inside of the three-yard-long polycarbonate tube while I rotated it to sterilize the inside and then we slid it into the core barrel. She then sat down in the dirt at the end of the core tube, took one Whirl-Pak bag filled with yellowish latex fluorescent microspheres and stuffed it into the core catcher at the bottom of the core tube (figure 9.8A). We then tightened the core catcher so that the

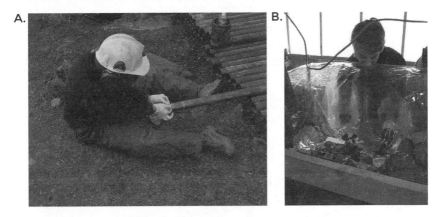

FIGURE 9.8. A. Susan Pfiffner loading bag of fluorescent microspheres into NQ3 coring barrel (photo by author). B. Corien Bakermans subsampling cores in the anaerobic glove bag at High Lake during summer 2006 (photo courtesy of Susan Pfiffner).

wires of the Whirl-Pak bag were wedged between the tube and core catcher. I slid the temperature monitoring strips into the tiny gap between the core liner and the core barrel. Grasping each end of the core barrel, we hoisted it up and walked it over to the driller, who hooked one end up to the quick-release of the wireline cable and raised the core barrel up until the lower end hovered above the open hole. He then lowered it down the hole to the bottom and locked it into place. The drillers started the drill motor, and the whine of the drill in such close proximity forced us to pull down the ear protectors attached to our hard hats. The nice thing about that was that we didn't have to put up with the whine of the mosquitoes buzzing around the perimeter of our head nets.

The drillers were penetrating the metavolcanics at about 100 feet per day, which meant that we should reach the fault zone in about seven days. We had enough fluorescent microspheres to collect ten cores, so we would be collecting a microbial core every 70 feet, or about every seventh core. Based on the drillers' rate of penetration, that would mean collecting a microbial core approximately every sixteen hours. The microbial core they were coring right now would be ready in approximately two hours. This gave me time to collect samples of drill water from the sump can at the base of the drill rig for conductivity measurements and PFC analyses. I also needed to collect filtered drill-water samples for cation and anion analyses, as well as samples for organic acid analyses and total and dissolved organic carbon. Soon after coring began, I started collecting samples of water from the sump can for DNA, RNA, and lipid analyses. Susan and I also collected sump-water samples for enrichments in sterilized serum vials that had been filled with N_2 gas. Finally we collected water from the sump can for fluorescent microsphere and cell-count analyses. I placed all these samples into the cooler with Blue Ice that I had carried up from the main camp. Susan went down to the core shack to help Corien prepare the core cooler. They arrived carrying the cooler just as the coring run was completed and the wireline core barrel had been brought to the surface. The driller handed us the microbial coring tube, while Timo handed him another core barrel. Susan and I pushed the core with its liner from the core tube. The squished Whirl-Pak and temperature strip fell out, which I immediately read. Good, the temperature was well below 25°C. We quickly carried the tube over to the small bench we had set up and sawed it into four 2.5-foot-long sections, bleaching the blade between each cut. Susan capped the ends of the tube with bleached caps. Susan and Corien then placed the tubes gently into the cooler, which they had filled with argon gas and Blue Ice to keep them as anaerobic and cold as possible. They then carried the cooler down the gorge and across the bog to the core shack, where the anaerobic glove bag was waiting.

Because the core segment that was wedged in the core catcher was the last part of the core to be drilled and potentially the most microbially contaminated, because it was outside the liner, I selected it for the pore-gas analyses. After sawing up the microbial core, I quickly cut a four-inch-long slice of this segment and placed into a waiting stainless steel vacuum canister in the core-logging shack. I then ran this downhill to the vacuum pumping station where I evacuated and flushed it with high purity argon several times, after which, I sealed the copper tube on top of the canister. Now any gas diffusing from the rock would be trapped in the canister for analysis.

I returned to the core shack where Susan and Corien had already transferred the four cores into the glove bag and onto sterile brown paper. We reached into the glove bag arms and pulled sterile latex gloves over our glove-bag gloves like doctors preparing for surgery. We wiped our hands with ethanol and then used a small rotary tool to remove the core liner from each core one at a time. After photographing and logging each core, we used a sterile hammer and chisel to subsample the cores (figure 9.8B). We then placed each subcore into separate Whirl-Pak bags. By the time we finished, we had filled seventy different orders for samples in eight hours.[55] It was now eleven hours after the microbial coring had begun, and we had five hours left to store the samples, completely sterilize the tools and glove bag, prepare another microbial coring barrel, get something to eat, and catch some sleep. We were seriously undermanned. We carried the coolers with the dry and Blue Ice down to the dock and lowered them into the icy waters of High Lake, tying them up so they would not float out into the lake. It was the best we could do to keep the samples as cold as possible until the next shift arrived the following week with more dry ice.

The sky had become gray and completely overcast by the second day of coring. The wind from the north stirred up whitecaps on the lake, and there was just enough chill in the air to keep the mosquitos at bay for the first time since we had arrived. The rate of penetration had been averaging a little less than a hundred feet per day. Earlier that morning, Timo and Matte had recovered and processed core no. 8. Up to that point, the boys of Master Drilling had achieved 100% core recovery, even in the fractured volcanic rock that transitioned into a sulfide-rich hydrothermal zone. Unfortunately core no. 9, our second microbial core, proved more challenging. The core barrel jammed six feet into the run. After pulling up the drill string, the drillers flailed away at the core barrel for half an hour with hammers until they were finally able to pull out a partially melted and thoroughly cork-screwed polycarbonate liner. We were perplexed, but rather than process the dubious core, we prepped another core barrel and sent it back down again for another attempt to obtain our second microbial core. This time we completed the run without jamming, but the wireline snapped on retrieval. Fortunately the drillers were able to fish the core barrel from the bottom of the hole. By 10 p.m. we finally were able to begin processing the second microbial core that we had started at noon. By 6 a.m. we had finished processing the core and in three hours would have to start another core run. Fortunately, other than this slight snafu, the coring went off smoothly, and over the next few days we settled into a routine of twenty-one hours on and three hours off, snatching cat naps and grabbing food between runs. The three of us had turned into zombies, using our muscle memories to process the cores. Throughout the long Arctic day and night, we would stalk up and down the ridge and through the bogs in search of fresh microbial cores to take back to our lab and chop up. By day 4, Susan was completing my sentences and every time she reminded me of what to do, I replied, "Yes dear."

HOW TO CORE THROUGH PERMAFROST AND NOT COLLECT WATER, STEP 3: JULY 29, 2006, WOLFDEN MINING CAMP, HIGH LAKE, NUNAVUT, CANADA

By day 5 the drillers told us they thought they were below the base of the permafrost. They discovered this by turning off the hot-water circulation for several hours, raising the bit a few feet off the base of the borehole, and then lowering it back down. While in the permafrost, this motion would encounter the resistance of partially frozen drill water, but once below the permafrost zone, no resistance would be felt. They were basically ice fishing for the bottom of the permafrost with a 1,600-foot-long drill rod. After consulting with Bruno, Timo decided that it was time to remove the wireline coring tube and drop his temperature probe down the drill string. It read 7–8°C at the bottom of the hole. We were definitely through the permafrost. Bruno told his drillers to trip out the drill string, so the drillers slid a plug down to the bottom of the drill string and pulled up the drill string with the water it contained. This was a "wet" pull and was done quickly to remove drill water from the borehole before it froze to the rock surface. The drillers immediately began inserting casing into the borehole to a depth of 320 meters to maintain borehole stability in the permafrost region. The casing would also keep any residual drilling brine in the rock formation from flowing into the borehole, and also prevent melted ice in the permafrost zone from flowing out and partially sealing the hole. After that was done, we started removing the drilling water using a deviously clever eighteen-liter bailer that Bruno had made in his shop. We dropped it down the hole and raised it nineteen times during the subsequent twelve-hour shift. The temperature of the water had decreased to about 3°C, and the salinity was increasing to well above that of the freshwater we were using. We succeeded in opening a borehole through the permafrost and into the subpermafrost water-bearing fault zone. The drilling plan with freshwater had worked. Clearly, if we ever decided to send a human mission to Mars to drill through the cryosphere, Bruno should be on it.

With the coring completed and satisfied that the water sampling was going smoothly, the three of us decided we needed some well-earned sleep. Corien headed down to the women's tent while Susan crashed on the floor of the core shack, and I passed out on top of one of the tables. During the next day, while we slept, four additional bailer samples were taken. Timo used his conductivity detector to measure the water level in the borehole, the temperature of the water, and its salinity. The salinity had risen to about 4,500 ppm total dissolved solids, more than a hundred times higher than that of the lake water used for drilling. Timo and Matti filtered this bailed water for lipid and DNA analyses even though the bailer was not sterile. In addition, they took samples for PFC and chemical analyses. Over the course of twenty-four hours, they removed about 400 liters of water from the hole in order to lower the water table below the permafrost so that it would not freeze and to remove the contaminated drilling water. During this process, the borehole water was cleared of the rock dust from drilling, and its rising salinity appeared to indicate that true subpermafrost formation water was steadily pouring into the borehole through the fractures in the fault zone. The down-hole conductivity probe detected the top of the water at 1,460 feet below the surface. Timo estimated that the saline water was entering the borehole at a rate of about one liter per hour.

When Susan and I awoke the next day, Timo and Matti were sitting at the picnic table over breakfast with glum visages and fell looks. "The borehole ate my probe," Timo said as we walked in and sat down with our pancakes and eggs. Their probe had gotten stuck in the borehole, and they could not retrieve it. It was sitting at a depth of 400 feet and must have been trapped in ice. Earlier that morning the drillers had attempted to drill out the ice without using any water, but they failed. We thought that water vapor, either from the subpermafrost water and/or from atmospheric moisture, had condensed at the coldest section of the borehole and formed an ice plug. We could not collect any more water samples, and the samples that we did have were heavily contaminated by the drilling water. The only thing we could do now was to plan for a drilling campaign in the early spring of 2007 to attempt removal of the ice plug before Major Drilling had to move the drill rig to a new location. If the ice drilling was successful, then we could isolate the borehole with a packer and perform long-term monitoring of the subpermafrost microbial community.

Dan McGowan and Adam Johnson, graduate students of mine and Lisa's, had just flown in the day before to begin a week's worth of water sampling, but now it seemed they would be packing up their gear and returning to the United States much earlier than anticipated. Dan updated me on the DNA analyses of the second set of Lupin samples. He had detected a microbial community dominated by bacteria, but no methanogens. This finding seemed consistent with the isotopic data on the CH_4 supplied by Shaun Frape's lab.[56] The dominant phyla were quite similar to those found in South Africa at a comparable depth. The 16S rRNA gene libraries suggested that anaerobic sulfide oxidizers and SRBs dominated the community, implying a subpermafrost sulfur-cycling environment similar to that Lihung had seen in South Africa. But what stunned us was that there were only about 100 cells per milliliter in the fracture water,[57] a concentration a thousand times less than we had typically found in the

South African fracture water at the same depth and that Karsten Pedersen had found in the Swedish subsurface fracture water. This concentration was comparable, however, to that reported for the Lake Vostok ice cores.[58] Could it be that thick glacial ice and permafrost was limiting important nutrient or electron-acceptor fluxes? If so, then the estimated size of any potential biosphere beneath the cryosphere on early Mars would have to be reevaluated. Alternatively, perhaps most of the biomass was attached to the mineral surfaces, and more so than at higher temperatures. After all, the Siberian permafrost sediments had 10^8 bacterial cells per gram,[59] and 10^7 bacterial cells per gram had been found in the silty ice at the base of the Greenland ice sheet, more than three kilometers from the surface.[60]

To answer this question, Susan focused on analyzing the cores we had collected from High Lake. Thanks to the heating, the drilling water contained fewer than 10,000 cells per milliliter when normally we would have expected a million cells or more per milliliter. The fluorescent microsphere and the PFC tracers had also worked and definitely indicated a thousandfold reduction in contamination of the core surface from the drilling water. Because the core diameter was too small for paring, the initial PLFA analyses were performed by crushing whole cores. The results seemed to indicate that these cores had the equivalent of approximately 100,000 to 800,000 bacterial cells per gram of rock. To deduce how much of that was surface contaminant, Susan briefly immersed the cores in the same methanol-chloroform solution she used to extract the cell lipids and then crushed the cores and extracted them again. The results were staggering. All the cores were below the detection limit of about 10,000 cells per gram with one exception, and even that core had only 30,000 cells per gram.[61] From the $^{35}SO_4$ autoradiography we could definitively detect that sulfate reduction was occurring in the core interiors even after Susan's surface sterilization approach. We could not, however, amplify any DNA from the cores.[62] It did not matter if the cores were from the permafrost or from beneath the subpermafrost. These results suggested that the biomass in the permafrost-impacted rocks was low and that probably the biomass in the subpermafrost water would be low as well.

We probably should have stopped there. But Timo had wanted to get his probe out of that borehole, and we still had not given up on the idea of slipping a packer down the hole and pulling water up to the surface. After all, a human mission to Mars would have to do the same thing. Down-hole electric pumps powerful enough to pump water up from 1,500 feet depth did not seem to exist. Even if they did, it was not clear to us how we could keep the water from freezing and sealing the tubes on the way up. Shaun Frape contacted the consulting firms that supplied the mining companies with packers to find out if they had any solutions to this problem, but it turned out that no one had been that interested in collecting water samples from a borehole beneath permafrost over time. I contacted the U.S. Army's Cold Regions Research and Engineering Laboratory for advice.[63] They also had nothing to offer but wanted to know if I ever found a solution. After many fruitless hours of emails and Internet searches, I did what I should have done in the first place: I talked to Susan and Tommy. They suggested that I contact Barry Freifeld, a research scientist at LBNL who had worked with Tommy and Susan on a CO_2-injection experiment in Texas. He had developed a deep borehole water-sampling tool called the U-tube. This clever device was like two very long soda straws that merged into one soda straw at the bottom of the borehole (figure 9.9). Where the two straws merged was a small check valve. He would lower this assembly until it was immersed below the water table or the packer. The water would then flow up into the two straws until it reached its maximum height. For many formations this was well below the surface, beyond the range of down-hole pumps. He had two valves at the top of each of the soda straws, one of which was connected to a high-pressure gas cylinder. The other was the outlet valve. He would then open both valves and turn on the gas. The gas pressure on the one straw, which he called the drive leg, forced the check valve to close at the bottom. The water in the two straws would then flow up and out of the one straw leading to the open outlet valve, which he called the sample leg, until nothing but gas was flowing through the two straws. As soon as Barry turned off the gas, the check valve beneath opened up and the water flowed up into the two straws ready to be pumped out again. He

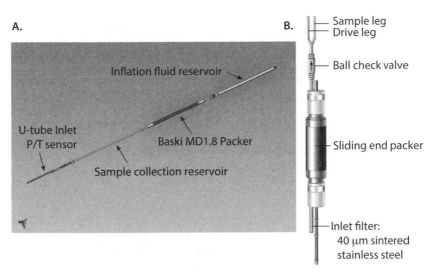

A.

Inflation fluid reservoir

U-tube Inlet
P/T sensor

Baski MD1.8 Packer

Sample collection reservoir

B.

Sample leg
Drive leg

Ball check valve

Sliding end packer

Inlet filter:
40 μm sintered
stainless steel

FIGURE 9.9. A. High Lake bottom-hole assembly showing the sequence of components from bottom to top (left to right on the schematic). B. Schematic layout of a U-tube sampler. The sample collection reservoir formed the "U" portion of the U-tube at High Lake. It is still sitting there, frozen in its borehole (from Freifeld 2008).

described this to me on the phone, sent me his paper, and walked me through it. It was simple, elegant, and required no power.

Barry had the solution to a problem that no one who had worked in the Arctic had ever been able to solve. He merely had to make two modifications to his U-tube. First, he had to add heat tracer tape to the outside of the two straws to keep the water from freezing and clogging them. Second, he had to run a jacketed fiber-optic cable along the length of one of the straws. This cable was sensitive to changes in temperature and could measure the temperature with a one-yard spatial resolution by detecting the laser light that was back-scattered up the cable as the laser light was transmitted down the cable. This distributed temperature sensor (DTS) was critical for our application. We had to supply enough power to the heat tracer tapes to keep the tubes from freezing, but not so much power that the temperature exceeded 25°C. We had no idea beforehand where along the deployed U-tube the hot spots and cold spots would be located, so we were threading a very fine temperature "needle." Finally, we would add a pressure transducer, a temperature sensor, and a conductivity probe to the bottom of the U-tube. The TROLL was a downhole sensor that monitored temperature, conductivity, and pressure; its pressure transducer would tell us how quickly water was flowing into the borehole and thus the permeability of the fault zone we had drilled. It was up to Lisa to go back to NAI and request enough money for Barry to construct and deploy the U-tube and turn our borehole into the first subpermafrost observatory.

THE U-TUBE VERSUS THE PERMAFROST: JULY 23, 2007, EXPLORER INN, YELLOWKNIFE, NORTHWEST TERRITORIES, CANADA

It was July 23, 2007, almost a year to the day since my last trip up to High Lake. During that time, Lisa had secured the funding for Barry, who then had battled with the LBNL purchasing department to get the supplies he needed for the first permafrost U-tube. After that, he had them shipped up to High Lake. Barry and I met at the Explorer Inn in Yellowknife. Along with several students, we were a full planeload. The next day was sunny and clear as we packed our gear into the Twin Otter and flew north to High Lake. About an hour out of Yellowknife, I looked up from my laptop to see that we were flying straight into a shroud of clouds that stretched to the horizon on both sides. Rather than climbing, the pilot surprisingly dove down to find better visibility. His strategy worked for a while, but soon we were flying so low that, as Slim Pickens would have said, "we'd need sleigh bells on this thing."[64] The pilot was hugging the topography as we flew in and out of shrubby dells and lakes and over the tops of rocky ridges. He would bank to the left and then right to dodge a rock ledge here and then there. It almost appeared as if the tips of the wings were about to clip the ground and hurtle us like a pinwheel to a fiery conclusion. But this pilot was reeeeally good, as he kept dipping in and out of one vale after another for over two hours. I had no idea where we were, nor how the pilot knew where we were. Suddenly, we passed over a ridge with a drill rig sticking up and High Lake appeared beneath us. The pilot dropped the plane onto the lake. We had made it! After we had pulled up to the dock and the pilot had killed the engines, I pulled out my earplugs and stepped outside onto the dock. I then turned to profusely thank the pilot, who was clearly not a happy pilot, a bit shaken, if not stirred.

The new camp manager, who worked for Zinifex, the company that had taken over Wolfden that year, met us. I thanked him for allowing us to stay and mentioned how lucky we had been flying up from Yellowknife. The manager explained to me that it was never a matter of luck running this camp. "We're in the @##@!! coastal zone so we get all this @##@!! coastal shit weather. So I have to make sure that they can @##@!! land before they take off from Yellowknife, otherwise I make a @##@!! $10,000 mistake." I got his point; clearly being a mining camp manager in the Arctic was a stressful job. He was under a lot of pressure to get the camp up and running and was falling behind schedule, and now he had to deal with scientists.

The silver lining to the coastal weather was that it kept the mosquitoes at bay, and we didn't have to wear the head net and gloves. We went down to the mess hall to talk to Chris, the helicopter pilot, about slinging Barry's equipment and a generator up to the drill site. Chris called us the Naysaw guys,[65] a nickname that stuck with us for the rest of our stay. He told us that his helicopter was sick, but he should have it repaired by the next morning. We walked up to the drill site. Not much had changed on the ridge since we left it the year before. The drill pad was still there with its wooden shack perched on its deck and the drill tower leaning over toward the ridge. The little wooden shack we had used the year before for logging core would now be used by Barry to monitor the U-tube bottom-hole array (BHA). The next morning Chris started slinging Barry's equipment up to the ridge while the drillers connected the power to the rig and put in the pumps and hose from the lake to the drill rig. Unfortunately, Bruno was not there to offer his sage advice with the drilling, so I struck up a friendship with the new driller. Barry and I started unpacking the BHA from its wooden crate and laying it on the ground. One section was a chrome-colored, bullet-shaped probe that was the TROLL. Another section was a six-foot-long rubber-coated pipe that was the inflatable packer. Another section was a nine-foot-long silver-colored cylinder that contained a reservoir for propylene glycol, essentially antifreeze, which would be used to inflate the packer. Finally, there was a silver-colored pipe that would cover the entire assembly to protect the wiring. While Barry and I started assembling the BHA, the others in the team set up five spools on a pipe that straddled two saw horses.[66] Three spools were for the 6,000 feet of stainless steel tubing, two spools of which were for the U-tube, the sample, and driver legs, and the third of which was for the inflatable packer that would sit above the TROLL. The fourth and fifth spools were for the heat tracer tape and the cable for the TROLL. At the same time, Chris was flying over our heads slinging three silver-colored, 50-gallon heater tanks so the drillers could start heating the drill water. I added some fluorescein dye to the water to track the contamination of the subpermafrost water by the drill water once we had punched through the ice. I checked the water temperature, which was about 90°C, and the steam for the inlet trough rose above the drill pad. By 2 p.m. the drillers had nineteen of the ten-foot rods down the hole but had not encountered any ice with the bit yet. Barry now had the TROLL connected to the U-tube assembly. He had placed a water filter at the tip of the TROLL to make sure that no suspended dirt would gum up the U-tube. Adam and Randy paced out the heat tracer tape and put colored tape every 82 feet so we would know the depth of the BHA when we finally lowered it into the borehole. We finished the BHA by 6 p.m. and then went to dinner.

Afterward we hiked back to find the drillers in a bit of a fluster. They had broken through some ice at 420 feet down the hole and lost water circulation. That was not a good thing because it was the circulating hot water that kept the drill rods from freezing to the casing. While awaiting the outcome, Barry connected the tubing on the BHA to the packer and inflated the packer. He partially inflated the packer to make sure everything was leak tight. He then spliced the connection between the TROLL and the TROLL power/transmission cable and checked the output with his laptop. The TROLL woke up and started sending data. The BHA looked like it had survived its long journey from LBNL. Meanwhile, the drillers had restored water circulation and resumed adding more drill rods to the string. We finally went down to get some sleep.

I woke up early, about 3:30 a.m., as usual. Apparently my South African circadian rhythm was not affected by Arctic twenty-four-hour summer daylight. It was way too early for coffee. The new camp cook would not put on the morning joe until 7 a.m., right before the night shift arrived for breakfast. So I strode up to the drill site to meet the night shift and see how they were doing. They had stopped again. Apparently they had been drilling ice all night until they hit something that seemed to jam the bit. They had started pulling up the rods to see what it was. "That would be Timo's conductivity probe," I said to one of the night shift drillers. At 7 a.m., I was having my first cup of coffee in the mess hall when the nighttime drillers showed up and tossed the black cable onto the table in front of me. I looked it over. There was no probe and this was not all of the wire. There must still be more cable down the hole. Somehow the probe must have gotten trapped way down the hole and the cable froze to the side along the entire length of the hole. That was not good. They told me the day shift had started putting the rods back down the hole.

After breakfast, Adam and I went up to the core shack to unpack the lab supplies for collecting the water samples we had left behind last year. The walkie-talkie started crackling. It was the chief driller. They had a problem. We hiked up to the drill site in a northerly rain/sleet storm to have a look. He explained that after they had started drilling into the ice again, the drill pipes froze in place. They were at 700 feet, about half the way through the permafrost and still well within the casing. How could they be frozen? It was Barry who noticed that one of the heaters had failed sometime in the previous twelve hours, when they had been raising and lowering the drill string, and the water circulating through the hole had not been fully heated. The temperature had not been high enough to keep the bit and rods from freezing to the casing. The hot water could not circulate through the rods now, and with every passing minute, the freezing front would advance up the rods and seal the drill string solidly to the permafrost. The chief driller quickly tried to nudge the drill bit loose. He ramped up the rpm's on the drill motor and then engaged the rods. The drill pad shuddered and shook, but the rods didn't move. He tried it again, and this time the drill pad shifted beneath my feet like a mini-earthquake. I thought I had seen the rods move. "Let's try again," I said. He revved up the engine even more and then engaged the rods. This time the rig lifted and bounced, but the rods still didn't budge. The others had repaired the broken heater and starting shoving a hose down into the drill rod. The drill chief thought that they could put enough heat down the rods to free the drill string. But I could tell from his look that he was not optimistic. Barry and I continued working on the BHA while they flushed hot water down the drill rods. Occasionally, the drillers would fire up the drill motor to try to nudge the rods, but they seemed stuck fast.

Six hours later we walked down to meet with the group from Frape's lab for dinner. They had been spending their time trying to melt through the ice in another borehole but so far had not succeeded. They sketched out their design before us on the table. They were heating a copper pipe attached to a pump with 850 watts, and as the ice melted, the pump would withdraw the water and sink further down the borehole. Barry looked the schematic and shook his head. He knew it wouldn't work. The thing about Barry is that he is competence and confidence to the nth power. The last three days working in the field with him convinced me that he knew exactly what he was doing and what he needed to do it. "There is no way that will work with that amount of power down the hole"; he didn't mince his words. I explained the situation at the drill rig and told them that our only hope for getting to the subpermafrost water was to free the drill string and that would require adding $CaCl_2$ to the drill water. None of the students from Frape's lab wanted salt down the hole because the contamination meant that it would screw up their $\delta^{37}Cl$ data. I explained that it wouldn't be great for the microbiology either, but we had just spent well over $150,000 getting Barry's

U-tube up here and I was not about to walk away from our one shot to use it. Besides, I had almost 700 feet of drill rods permanently frozen in the hole. I wasn't sure how much Zinifex would charge me for them, but I suspected it easily exceeded the limit on my credit card. I tried to assuage them by explaining that once the U-tube was in place and functioning, the $CaCl_2$ brine could be pumped out given enough time. It might require coming back the following summer. We were out of options, so I called the chief driller on the walkie-talkie and told him to start adding the salt, under the glare of the five disappointed faces in front of me.

The next morning the wind was warm and coming from the south. It was sunny and the mosquitoes were out with a vengeance, having been deprived of blood for three days. After an early breakfast we visited the site. They had been pumping saltwater for twelve hours and had only advanced about a couple of feet an hour. It amazed me how quickly the ice had frozen the drill rod in place. We built some big handles for the spools and then discussed with the chief driller how we would use the wireline to haul up the U-tube assembly, which had grown to 40 feet. While we were waiting for the drillers to free the drill string, I decided to do a survey of the gossan deposits. One thing that was clear early on in this survey was that the gossan did not extend beneath the active layer into the permafrost. All the mineral alteration so visible from satellite imagery was restricted to the top three feet of the surface. Beneath that one yard of acidic soil, the minerals of the mafic metavolcanic rock were fresh and unaltered. The gossan formed an alluvial channel that began up by core shack lake and ran down to High Lake. The channel had coarse chunks of fresh rock mixed with a red, muddy matrix of iron hydroxides. I did not find any "blue berries," the spherical hematite concretions that had been reported by *Opportunity* at Eagle Crater, on Mars, but I did see plenty of coarse-grained hematite and magnetite on the weathered rinds of the rocks. The conceptual model for the formation of the blue berries involved groundwater migration, and hematite concretions with similar morphology had been reported in the Navajo Sandstone in southern Utah. So perhaps the *Opportunity* team was right about how the blue berries required groundwater. But my instinctive understanding of the Martian cryosphere model of Steve Clifford permitted only subcryosphere groundwater.

But at High Lake there was no groundwater associated with the gossans, just energetic surface water from the spring melt and summer rain interspersed with dry days, like today. Mud cracks were everywhere in the main channels. Occasionally I could see a tiny patch of green algae, but otherwise the gossan was devoid of vegetation. On the edge of the channels, a thick, black carpet of lichens and fungi was woven several millimeters into the soil. It had a ropey, crenulated structure that actually shaped the depositional pattern of the channel, and there were no mud cracks where this carpet existed. I carefully took my knife and cut out pieces of the fungal mat and shoved them into a centrifuge tube. I wondered whether the surface of our Earth had looked much like this in the Archean. As the cyanobacteria started pumping O_2 into the atmosphere, the sulfide- and iron-oxidizing bacteria must have started producing deposits like this all over the planet. Perhaps Mars at one time had similar biomats.

At 8:30 p.m. the night drillers called me on the walkie-talkie. They had freed the drill rods, so I went up to the drill site to check. They were still trying to get the hose out of the drill pipes using the wireline. When they finally pulled it out, they told me they would circulate the hot, now salty, water down the drill rods for another two hours before they started drilling again. I returned at 4:30 a.m., and they had drilled through about a hundred feet of ice. By 10 a.m. they were at 935 feet depth and had not seen any more of Timo's conductivity meter. I estimated that they should reach the bottom of the permafrost before midnight, but at 9 p.m., at 1,150 feet depth, both the water pump and hydraulic pump failed. The night shift worked feverishly on the repairs, although with the brine in the hole, they didn't have to worry about the rods freezing.[67] At 7 a.m. on a sunny Sunday morning, the drillers were at 1,300 feet depth and still encountering solid ice. By noon they were at 1,453 feet depth and had been fishing for the bottom of the permafrost by first raising the rods a yard, holding for an hour and then letting them drop to see if they encountered any ice. By the time they reached 1,600 feet early that afternoon, they could not detect any resistance. We had finally penetrated the ice.

We decided to circulate the water for a couple of hours to bring up any wire shreds from Timo's probe that may have fallen to the bottom of the hole and then started tripping out the rods. I added one more dose of fluorescein dye as an independent tracer of the brine contamination. By 10 p.m. the rods were out of the hole. Now we had to scramble. Even though we had not bailed the water out of the hole, the water level had dropped significantly after the rods were withdrawn, and we worried that any ice blockage at 390 feet formed from air moisture condensation (as had happened last summer) would stop our deployment of the U-tube. Seven of us plus one of the drillers gently picked up the U-tube, carried it to the front of the rig, and started feeding the top end toward the rig, where another driller clipped it to the wireline. I scrambled up the ladder to the crow's nest and grabbed the top end of the U-tube as Barry, Adam, and Randy pushed it up from below. The chief driller pulled in the slack on the wireline. Barry had the honor of launching the U-tube down the open end of the borehole. The drillers played out the wireline until the top end of the U-tube reached the borehole. Holding onto the U-tube with a pipe wrench, we connected the stainless steel tubes, the TROLL, and the heat tracer wires being fed to us over the back wall of the drill rig from the spools. We then disconnected the wireline and released the pipe wrench. The U-tube was now hanging from the stainless steel tubing, which we used to gently lower the U-tube down to its target. Adam and Randy spooled out the tubes and wire from behind the rig. When the colored tape told us that the packer was below 1,600 feet, we stopped. Using a pipe and some steel wire, we tied the U-tube to the top of the borehole.

It was now the moment of truth. I connected the N_2 high-pressure gas cylinder to the packer stainless steel tube. Barry turned on the generator and started heating the heat tracer tape. He then stepped into the wooden logging shed to monitor the temperature of the permafrost layer with the DTS and the pressure beneath the packer with the TROLL. I turned on the gas to the packer. Hopefully it had inflated down there. I then connected the second N_2 high-pressure gas cylinder to the drive leg of the stainless steel tube. I looked to Barry through the window of the shed. He gave me the thumbs up, so I opened the valve to the drive leg and could hear the N_2 gas coursing into the tubing. The TROLL temperature was reading $-0.6°C$, which was cold, but not below freezing, for this salty water, and Barry shouted out that the pressure had dropped in the TROLL from 2,400 kilopascals to 150, a sign that the check valve had closed. I held onto the outlet valve, pointing it to a huge plastic bottle, when all of a sudden, fluorescein green turbid water started shooting out of the valve. We filled one bottle after another. After we had collected eleven liters of the green stuff, the sample tubing started sputtering, and then a whoosh of dry N_2 pushed out the last few drops. Barry came and shut off the N_2 gas to the drive leg, and I closed the sample valve. Now we just needed water in the borehole to fill up the U-tube again. It had worked, just as Barry had said it would. We could even leave the U-tube here over the winter without fear of the tubes icing up because we could fill them with N_2 gas under high pressure to keep them dry.

Over the next three days the crew would collect five more samples, sixty liters total, and the green hue of the fluorescein dye was starting to turn a pale yellow-green when all of a sudden they were not able to get any water. The TROLL temperature had climbed to 6°C, close to where it had been last year, but the temperature of the tubes in the top 800 feet was still below $-2°C$. Barry turned the power from the generator up to the maximum output. He now had 20 watts per meter coursing down the wire, and it nudged the temperature up ever so slightly to $-1.8°C$. But the TROLL told us that salinity of the water had declined as the formation water diluted the $CaCl_2$ brine and that its freezing point was now only $-1.3°C$. Barry later determined from modeling the temperature profiles that at the depth of 140 to 200 meters, the thermal conductivity was five times that of the surrounding rock.[68] According to the core logs, that is the depth at which the mine had drilled through the sulfide zone. The sulfides had much greater thermal conductivity and transmitted the heat we were generating away from the tubes and into the rock. We had not prepared for that.

FIGURE 9.10. The Marty the Martian sign left at the Zinifex High Lake mining camp, summer 2007 (photo courtesy of Lisa Pratt).

If we had put insulation around the heater tape, our approach would have worked. But we had no insulation with us, nor any way of getting insulation to the site before Major Drilling removed the drill rig. But Barry had downloaded an impressive amount of data from the DTS, enough to determine the thermal conductivity structure and the paleotemperature history of the site. It was the best borehole temperature data set that had ever been reported for the Arctic. More important, we learned what did not work when it came to drilling through permafrost and what was required to make it work on Mars. During the next two days, we packed up all our gear and took the next Twin Otter back to Yellowknife, unfortunately without decent microbial samples for a second time. Before we left, the videographer made an arrow sign with Marty the Martian on it and hung it on the post outside the mess hall. "MARS 78–378 million miles" (figure 9.10).

During the winter, Shaun Frape would report that the analyses of the $\delta^{37}Cl$ of chlorine indicated that, despite our best precautions, all the samples, even the ones collected in 2006, were contaminated by the drilling brincs used to make the original borehole. Still I could not give up as long as Barry's U-tube was up there with nothing but the grizzlies to guard it. I thought we could pull it up (very difficult because the borehole was probably clogged with ice), insulate the tubing (easy), and gently lower it back down (not so easy). We would not need a drilling crew for that, just a lot of muscle and some room at the camp. In January I flew up to the Zinifex office in Thunder Bay and pleaded with them to grant us some space for the summer of 2008, but it was no use. They were not as inclined to help us "Naysaw guys" as Wolfden had been.

After four summers of trying to uncover the universe hidden beneath the permafrost of the high Arctic, we had developed a healthy respect for frozen rock when it came to drilling, but we also gained some significant insights as well. The first was that the potential abundance of biomass beneath permafrost was much less than we had seen at comparable depths in nonpermafrost locations. Whether this was due to reduced in situ energy and nutrient fluxes in ice-saturated rock remained an unanswered question. The second was that, to borrow a Wobberism, "there was a snowball's chance in hell" that a robotic, or even a human, mission to Mars could drill through several kilometers of permafrost and collect enough fluid samples containing any Martian organisms to detect them. The combination of the difficulty of moving fluids through frozen rock and the low biomass concentrations had not been on our radar when we attended the first NASA Mars drilling workshop back in 1996.

CHAPTER 10

THE WORM FROM HELL

All at once an enormous head shoots back up, the plesiosaur's head. The
monster is fatally wounded. I don't see its immense carapace anymore.
Only its long neck rises, falls, lifts again, droops once more, lashes the
waves like a gigantic whip, writhes like a worm cut in half. Water spatters
to a considerable distance. It blinds us. But soon the reptile's death throes
come to an end, its movements get weaker, its contortions grow calmer,
and that long piece of snake lies like an inanimate object on the quiet
waves.

—*Jules Verne, Journey to the Center of the Earth*[1]

NO DEEP DRILLING ON MARS: MARCH 23, 2008, NASA AMES RESEARCH CENTER, CALIFORNIA

In March 2008, NAI hosted a workshop at NASA Ames on accessing
the potential deep biosphere on Mars. There I summarized our frus-
trations with drilling through hundreds of meters of frozen rock. It
amazed even me how the demeanor of this workshop was completely
different from the one hosted by Geoff Briggs twelve years earlier.
Gone was the naïve optimism that innovative technology like cable
plasma drilling would enable us to drill through the cryosphere. Instead each presentation focused on simple auger, percussion, and rotary systems, none of which achieved depths greater than two yards.
This depth would be a minimum requirement to get beyond the
destructive impact of cosmic radiation on any organic biomarkers.[2]
The newer, cheaper, faster, better option being proposed was to send
a rover to a recently formed impact crater. New meteorite impacts
had been documented on the surface of Mars by the *Mars Global
Surveyor* orbiter during its past ten years of operation.[3] Perhaps they

could find one that reached a depth of ten, maybe one hundred, yards. But nothing short of using a nuclear warhead, however, could get beneath the kilometers-thick cryosphere to search for extant life as we know it.[4]

Following the discussion of deep-drilling approaches, the potential PP issues were presented by Cassie Conley, who had taken over from John Rummel as the head of NASA's Office of Planetary Protection. The tenor of these PP issues had also shifted significantly from our discussions back in the 1996 meeting. At that time the emphasis had been a concern with back contamination of Earth from Mars by some Andromeda-strain-like Martian pathogen. But in 2006, the National Research Council produced a report concerning the forward contamination of Mars by Earth, specifically by NASA spacecraft.[5] The report was highly critical of the protocols that NASA had been using to sterilize its spacecraft and of its assay for microbial contamination. The report placed the Office of Planetary Protection squarely in the path of the Mars Exploration Program, which had many more ambitious robotic surface missions in the works but which also lacked sufficient funding to implement the degree of sterilization that had been used for the Viking missions. The Mars Exploration Program Analysis Group responded to the report by identifying more specifically, based on the best understanding of terrestrial extremophilic microorganism physiology, where on Mars the "special regions" were in which terrestrial microbial stowaways could survive or even propagate. Also because the search for life on Mars was a search for life "as we know it," it meant that the "special regions" were also places where potential Martian habitats might exist today.[6] The NRC report recommended that those special regions effectively be off bounds to spacecraft unless the spacecraft met very high requirements for sterilization. The subcryosphere of Mars was clearly a special region, and even if a human or robotic drilling mission could penetrate the cryosphere, the danger of contaminating, or worse, destroying a potential Martian habitat by invasive Earth microorganisms, was and still is viewed as too great.[7]

Finally, Penny Boston, an astrobiologist from NMT, suggested that mini robotic surveyors could be used to characterize caves on Mars.

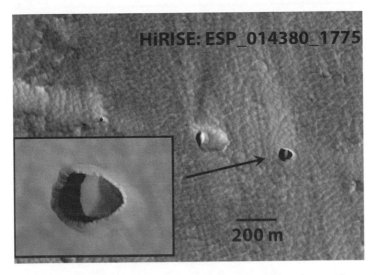

FIGURE 10.1. Collapsed pits associated with extensional tectonics, northeast of the volcano Arsia Mons on Mars at –2.27°S, 241.90°E. HiRISE image ESP_014380_1775. Image credit: NASA/JPL/University of Arizona (from Cushing 2012).

Certainly caves existed, because they had been detected in some of the volcanic calderas (figure 10.1). In fact, the deepest caves in the United States are lava tubes on the island of Hawaii. Mars had the largest extinct volcano in the solar system, Olympus Mons, and because of the low Martian gravity, a stable lava tube could extend to great depths. Planetary protection issues aside, caves provided astronauts with unique opportunities. Listening to her presentation, I thought wistfully about the ice tunnels of Lupin Gold Mine. Given the rational pessimism about deep drilling on Mars, I had to admit that Penny might be right.

MARTIAN ASTROBIOLOGISTS ENTER LECHUGUILLA CAVE: FEBRUARY 1994, 20 MILES SOUTHWEST OF CARLSBAD, NEW MEXICO

Penny Boston had attended the first International Conference on Microbiology of the Deep Subsurface, in Orlando, Florida, in January 1990.[8] She was looking for funding and projects involving subsurface microbiology. Chris McKay and she had been working on a theory about a potential subcryosphere biosphere on Mars (they had been grad students together at the University of Colorado, Boulder). Chris had moved on to NASA Ames to pursue his passion in exobiology. Penny had become the research director of Complex Systems Research, a NASA contractor located in Boulder, Colorado. At the meeting, she met with Mikhail Ivanov, who had long been studying microorganisms from Russian oil and gas fields. The Orlando meeting proved seminal in advancing her own thinking about subsurface microbial life, and she went on to publish a paper with Mikhail and Chris on potential subsurface Martian life in 1992.[9] She had become completely entranced with the idea of subsurface life but chose an alternative path to the deep biosphere.

Caves exist on Earth and would have undoubtedly formed on Noachian-age Mars.[10] The standard theory of cave formation is that the carbonic acid in rain seeps underground into the limestone and slowly dissolves it away. Noachian-age Mars is believed to have had a significant CO_2-rich atmosphere. This was required in order to provide sufficient greenhouse heating to warm the Martian atmosphere to the point that surface water could exist. The carbonic-rich surface water would lead to carbonic-rich groundwater and thus caves. Another type of cave is a lava tube. This forms as basaltic lava cools at the top and bottom but remains molten enough in the interior to continue to flow downhill. If the upper crust doesn't collapse, the void left behind is a lava tube. With the low Martian gravity these caves and lava tubes could be quite big and pervasive. Why not look at the microbiology of caves, then?

Penny undertook her first serious cave expedition in February 1994,[11] a five-day trip into the Lechuguilla Cave. Lechuguilla Cave was and still is the deepest limestone cave in the United States. It is located in the Carlsbad Caverns National Park, just twenty miles southwest of Carlsbad. Lechuguilla Cave was unknown until a team of spelunkers dug through a pile of rubble in 1986 in what was then a relatively dry minor cave accessed by a twenty-seven-meter deep pit. In the early twentieth century, it had been mined for bat guano to use as fertilizer. These mined caves were innumerable along the canyons that cut into the southward- and eastward-facing slopes of the Guadalupe Mountains. My family was familiar with these caves. During World War II my dad had disassembled a Ford tractor and reassembled it inside a cave near Sitting Bull Falls to mine the bat guano in there. Not all of these caves were connected to underground caverns, but the spelunkers noticed that a wind was blowing the rubble in and out at the end of Lechuguilla Cave. When they cleared the rubble, they found a tunnel that eventually led to more than 140 miles of huge rooms, ponds, and passageways that reached down 1,600 feet below the surface. Lechuguilla has never been open to the public even though the grandeur of the gypsum chandeliers and the calcitic moonmilk falls matched and sometimes surpassed that of the Carlsbad Caverns.[12] I had seen many of these intricate formations draped around multicolored pools revealed by strategically placed lights when I visited the Carlsbad Caverns as a kid. The caverns were so beautiful that 20th Century Fox filmed part of their adaptation of Jules Verne's *Journey to the Center of the Earth* in them. They had amazed me even though I really did not know then how they were formed. We used to look up at the giant stalagmites rising to meet the roof of the Big Room, which stood higher than Niagara Falls, while the ranger who guided our tours would always tell us that it took many millions of years for these caverns to form. According to the standard theory of cave formation, back then was that the carbonic acid in rain had seeped underground into the limestone and slowly dissolve it away. Lechuguilla and the Carlsbad Caverns both sat on the northern edge of the Chihuahuan Desert and received minimal rain. When it did rain, it came in torrents that would charge down the canyons into Carlsbad and flood its streets. Both caverns resided inside the Permian Period Capitan Reef complex that is comparable to the modern-day Great Barrier Reef off the coast of Australia. If you stand on the ridge at the entrance to the caverns and look south over the flat, desolate plains of southeastern Texas with all its Permian Period salt and carbonate deposits, it is easy to imagine you are standing on a reef and looking down into the Permian seabed.

Geologists recognized the importance of the enormous Lechuguilla cavern system almost immediately after its discovery. Lechuguilla sat in the back-reef facies beneath fine-grained impermeable sedimentary rocks.[13] These sedimentary rocks blocked what little groundwater infiltration there was from penetrating into the underlying carbonate. The depth to the water table was 1,500 feet, which was not unusual for the more arid portions of the southwestern United States and meant that air penetrated to that depth. All the oil drilling to the south and east revealed that an extensive sulfidic-rich oil reservoir existed at a greater depth beneath the Capitan Reef. Oil-degrading SRBs were reducing the Permian evaporitic sulfate at still greater depths and creating sulfide gas,[14] which was rising into the overlying Capitan Reef, where it met air. The sulfide gas combined with O_2 to produce sulfuric acid, which dissolved the carbonate from below. The impermeable cap of sedimentary rocks protected the ornate and delicate speleothem gypsum chandeliers and sulfur deposits of Lechuguilla.[15] This was the evidence that carbonic groundwater penetrating from above had not formed the caves. Gypsum is a highly soluble mineral that dissolves in fresh groundwater almost as readily as salt does. The theory that sulfuric acid was the main corrosive agent of carbonate caves, so-called sulfuric acid speleogenesis, was first proposed by David Morehouse in 1968. He had studied Level Crevice Cave in Dubuque, Iowa, one of the many carbonate caves in the region that were mined for their lead and zinc sulfides. Although the water in the cave had a neutral pH, Morehouse observed that the natural cavern sizes bore a direct relationship to the amount of sulfide in the rock and that the rooms and passages were bigger at depth and smaller near the surface. This was completely inconsistent with the standard carbonic acid theory. The water had high concentrations of sulfate, and the walls of the cave were coated with ferric iron hydroxides. Morehouse even proposed that the large rust-colored mats of ferrous-iron-oxidizing *Gallionella* that grew in the cave were responsible for oxidizing the sulfides and generating the sulfuric acid. In 1981, a speleologist named Egemeier, who studied the Lower Kane Cave in Wyoming,[16] arrived at a similar conclusion, but in his case sulfidic groundwater, not sulfide minerals, was the source of the sulfate. Egemeier did not go so far as to suggest that bacteria were involved in producing the sulfuric acid, but he did point out the similarities between the Kane Cave and the Carlsbad Caverns.

If Lechuguilla was a magnificent example of just how biogenesis could construct an underground world, then what roles did the microorganisms play in its formation? In the early 1980s, algae, fungi, and bacteria were considered to contribute to small-scale depositional features in caves, such as the moonmilk. The term "biokarst" was defined to classify these small-scale phenomena.[17] To form carbonate, these organisms had to oxidize organic matter into CO_2, which would then form carbonate. Bacteria can enter the cave through groundwater flow or hitch a ride as parasites or endosymbionts on any number of troglophilic springtails,[18] beetles, spiders, and crickets. They could even be carried in by bats and deposited on the surface in the guano deposits. Additionally, the wind coursing into a cave could carry fungal and bacterial spores, because caves can breathe. As the barometric pressure outside decreases or increases with weather fronts passing overhead, the cave exhales and inhales, respectively. It was this wind that led to the discovery of Lechuguilla. So how far down into Lechuguilla Caves did life go?

This was the question uppermost in Penny's mind when she accompanied Chris McKay and Kim Cunningham on that life-altering trip in February 1994. To prepare for the trip, she did three hours of vertical ascent/descent training in Boulder. Penny did not have the slightest inkling of how difficult exploring Lechuguilla was going to be, or else she never would have attempted it with so little skill or experience. During her five days underground she kept thinking that she just had to live long enough to get out of there and then she would never have to go back. She managed to get out under her own steam but with a broken rib, an eye swollen shut with a lump of Fe/Mn speleosol fluff in it (her first clue that there were microbes in the fluffy speleosol),[19] a sprained ankle, and numerous pizza-sized bruises. But as her wounds were healing, she replayed in her mind the beauty and amazing nature of what she had seen and she became hooked. She realized that she had to learn to do caving safely because she wanted to work in those places. Three years later she moved to Socorro to join the faculty in the Department of Biology at NMT. There she met Tom Kieft for the first time, even though both had attended and given talks at that pivotal subsurface microbiology symposium in Orlando in 1990. The move to Socorro also gave her the opportunity to explore the microbiology of the many lava tubes that existed in the young volcanic flows distributed along the north-south Rio Grande Rift near Socorro.

Penny and her colleagues Diane Northrup of the University of New Mexico, Albuquerque, and Kim Cunningham of the USGS in Denver discovered that many of the iron-, manganese-, and sulfur-bearing carbonate walls of Lechuguilla were covered with filamentous microbial mats and fungi. They formed a ubiquitous patchwork quilt of yellow, red, and chestnut brown mats across the walls as deep as 1,500 feet and many miles away from the entrance. The mats were most abundant where the warm, moist air rising from the lakes at the base of the cave system condensed on the cold ceilings of the upper passages and halls. All of this was taking place well beyond the reach of bats, insects, or even the cave's breath from the entrance. Lechuguilla had its own internal microclimate. The microbiologists found chemolithoautotrophs, suggesting they were fixing CO_2 and N_2 from the cave's atmosphere as they oxidized the reduced iron and manganese of the calcite and dolomite into a metallic oxide fluff and the sulfur into sulfuric acid. The sulfuric acid, in turn, dissolved more of the metal-rich carbonate, providing more nutrients for the chemolithoautotrophs. These corrosion-formed mats had fallen to the floor of the cave. The thickness and extent of the floor fluff was orders of magnitude greater than had been found in other caves in the Guadalupe Mountains, the United States, Australia, and Turkmenistan. The researchers discovered that the heterotrophic bacteria and fungi were not living on dissolved organic carbon infiltrating from the surface, as the standard model for biokarst would have predicted; instead, the hctcrotrophs were feasting on the organic matter provided by the chemolithoautotrophs. Finally, the scientists found evidence that many of the ancient stalactites had formed around microbial filaments. Lechuguilla had its own internal subterranean ecosystem sustained by chemoautotrophy, and it had been around long before the entrance was discovered, perhaps for millions of years. The spelunking microbiologists published their findings a few months before Stevens and McKinley published their discovery of chemolithoautotrophic SLiMEs.[20]

SPELEOPHILIC LITHOAUTOTROPHIC SUPPORTED ECOSYSTEMS? MOVILE CAVE, ROMANIA, AND CUEVAS DE LOS CRISTALES AND DE VILLA LUZ, MEXICO

Soon after, other papers began appearing that reported chemoautotrophically supported cave ecosystems. In the same year that Lechuguilla was discovered, a construction project in Dobrogea, Romania, accidentally penetrated a hidden cavern lying just 60 feet below the surface. Its discoverer named it Movile Cave. It wasn't until the overthrow of Romania's ruler, Nicolae Ceausescu, in 1989, however, that microbiologists began to explore the full extent of Movile Cave. The unique aspect of this cave was immediately apparent when they encountered a highly diverse community of troglophilic animals that seemed to have evolved for millions of years in isolation from the surface. At the greatest depth of the cave the scientists encountered a lake. On the surface floated small crustaceans that were the top of a food pyramid supported at the bottom by chemolithotrophic, sulfide-oxidizing bacteria with names such as *Beggiatoa, Thioploca, Thiotrix, Thiospira,* and*Thiobaccillus.*[21] To explore the submerged extent of the cave, scuba-diving microbiologists entered the lake and discovered isolated air pockets, or air bells, deeper in the Movile Cave system. The air in the air bells was low in O_2 and rich in CO_2 and CH_4. The water in them was also covered with thick mats of sulfide-oxidizing bacteria. Because of the low O_2, crustaceans could not live there and feed on the bacteria. But in those mats the microbiologists found hermaphroditic female, bacteria-eating nematodes feasting on a cornucopia of bacterial filaments.[22] There were more than a million nematodes per square meter of microbial mat. The possibility that aerobic protozoans and metazoans could live in anoxic or sulfidic environments had only been reported in the late 1980s. Movile Cave's air bells provided a living example of such an ecosystem.[23]

Although we had never encountered caves in South Africa, we had heard plenty of stories from the South African miners about tunneling into giant fissures in the rock at great depths, fissures large enough for a human to enter and crawl around. Then in 2000 we heard that miners had tunneled into a cathedral-size cavern 600 feet beneath the surface in Mexico. The miners were working for a small silver, lead, and zinc mine in the town of Naica, which is in the Sierra Madre Mountains on the south side of the Chihuahuan Desert, about 350 miles south of Lechuguilla. The limestone cavity was choked with giant selenite crystals up to 40 feet long. One of the discoverers had commented that the crystals were so brilliantly reflective that they looked like moonbeams that had taken on weight and substance. Penny Boston, Tom Kieft, and Chris McKay joined an expedition to that cavern, the Cueva de los Cristales, just to see if they could detect bacteria trapped in fluid inclusions in the selenite giants and see any signs of the type of bacteria Penny had been encountering in the caves further north. The big difference between the Naica crystal cave and the ones in which she had worked before were that the air temperature of the Naica cave was 136°F and its humidity 95%. To enter the cave even for short periods, the researchers had to don bodysuits cooled with air circulating across ice blocks. The Cave of the Crystal Giants, however, appeared to have been formed by a strictly inorganic process, like a gargantuan geode.[24]

All these observations raised the question, was it possible that the chemolithoautotrophic sulfide-oxidizing bacteria were the principal agents in forming the caves and the karstic topography that results when the caves collapse? The best place to get an answer to this question was a cave in which the process was ongoing. Cueva de Villa Luz, or Cave of the Lighted House, located in the state of Tabasco, in southernmost Mexico, was just such an active sulfidic cave. It was easily accessible to the public, being only a mile and a half south of the pueblo of Tapijulapa. Biologists had been documenting the fish species there since the 1960s, but it wasn't until the arrival in 1997 of a microbiologist from Chapman University, Louise Hose, that microbial sampling began. The samples revealed that hyperacidophilic bacteria were living at pH levels as low as 0.1 and had formed intricate lacelike filaments, subsequently termed snotites. Snotites were concentrated at the end of the cave in an alcove appropriately named snotite heaven. Working with an increasingly multidisciplinary team of scientists, Hose made the case in 2000 that significant corrosion of the cavern was occurring through microbial oxidation of H_2S.[25] The light isotopic value for the H_2S suggested that SRB had produced it, perhaps those found within an oil reservoir that was some thirty-five miles to the northwest.

In 2005 Gaetan Borgonie, a nematologist from Ghent University, was attending a nematology convention in Kansas. He and other nematologists had gathered to sort out the evolutionary tree for all nematodes that had been sequenced and to reconcile it with their understanding of the evolutionary development of these tiny round-worms. It was a mess. Gaetan proposed that one approach to resolving the phenological development of nematodes with their phylogenetic evolutionary patterns might be to look at nematodes from more extreme environments. Presumably these nematodes would have originated from or adapted to these extreme niches. Gaetan had watched a segment of David Attenborough's *Planet Earth* about Cueva de Villa Luz in Mexico and its complex biodiversity of vertebrates and invertebrates and wondered whether an acidophilic nematode could be found there.[26] Although the cave had very high H_2S concentrations, so did Movile Cave, and nematodes flourished there. In reply to Gaetan's inquiries about obtaining samples, Louise had sent a photo of an organism that looked like a very long, knotted-up nematode. She suggested that Gaetan contact Laura Rosales Lagarde. Laura was a geologist who spent a lot of time in the Cueva de Villa Luz and knew the caves well. She agreed to let Gaetan tag along to several sites in and around the caves with her. Within four months of seeing the Cueva de Villa Luz on television, Gaetan traveled to Mexico to collect samples. He was traveling alone, because his colleagues at Ghent said it was an insane idea to think one would find worms at a pH of almost zero. That dismissive judgment just egged Gaetan onward.

Walking on the trail through the jungle toward the rising plume of steam pouring out of the entrance to the Cueva de Villa Luz was like stepping onto the movie set of an Indiana Jones movie. A milky white river issued from the cave and collided with the turbid brown, humic-acid-rich water of the Rio Azufre. Gaetan and Laura followed a stairway down to an entrance to the cave from which emanated the intense rotten-egg odor of H_2S. From there they ventured in, traversing about three-quarters of a mile and passing beneath numerous skylights. Looking up, Gaetan could see they were only about a hundred feet beneath the floor of the jungle. Looking down, he was astonished at the sheer number of fish swimming in the milky white river running through the cave. Laura explained that the milky color was due to colloidal elemental sulfur particles that formed when the H_2S from the springs was oxidized by the O_2 in the air. On the cave walls bacteria oxidized the H_2S all the way to sulfuric acid, which dissolved the limestone. Everywhere the current crossed a channel of limestone, Gaetan could see the long white filaments of sulfide-oxidizing bacteria. The walls of the cave were not like anything he had ever seen. It was more like the skin of a giant diseased dinosaur. It was pockmarked with white puslike patches interspersed between black and gray anastomosing stripes. The striped pattern occasionally formed a spongelike growth. Snakelike filamentous streams now and then emerged from the carbonate skin and wound their way to the floor. Laura had taken him to one of the several mineral springs from which the warm water and H_2S was emanating. The lizardlike skin was very well developed around these springs. The black stripes looked as if they were a kind of fungus that was growing between the gray gelatinous goo covering the limestone (figure 10.2). When he looked closely at the white patches, he could see they comprised clusters of millimeter-size prismatic crystals that were almost gem quality.[27] Some of the clusters were completely coated with white mucus, as if the minerals were fighting off an infection. But the most exquisite features were the long loops of hairlike strands that hung from the crystals. These were the snotites. They were decorated with drops of water, like pearls on a necklace. Small mites were flying about and landing on the snotites to feed. Laura used some pH strips to measure the pH of the water drops, and they were between 0 and 1. She had brought Gaetan to one of the most acidic sites in the cave. He carefully pulled off his backpack and pulled out a pair of latex gloves and centrifuge tubes to collect some of the bacteria-rich slimy snotites. "Come to me my little worms," he thought as he coaxed the

FIGURE 10.2. Gaetan Borgonie in Cueva Villa Luz, summer 2006 (courtesy of Gaetan Borgonie).

slime into the tube and sealed it with a little water from the spring. Gaetan joyously harvested newly formed snotites and larger older snotites with his centrifuge tubes. Satisfied that he had collected about every possible type of acidic biofilm, he followed Laura back out of the Cueva de Villa Luz.

Back at Ghent University in the Nematology Section of the Department of Biology, Gaetan immediately set to work analyzing his snotites. As a senior full professor, he could bring to bear an amazing array of instruments available to him, including a CT scanner. The outer layers of the snotites were covered in a new nematode species, which he named *Mesorhabditis acidophila*.[28] The snotites turned out to be hollow tubes, the walls of which were composed of the filamentous sulfide-oxidizing bacteria, *Acidothiobacillus*. Juvenile and adult nematodes, about a half millimeter long and 15 to 20 micrometers in diameter covered the outer and inner layers of the tubes. The central fluid in the tube was highly acidic, and Gaetan found only "dauer" stage nematodes present in the acid core.[29] The flying mites covered only the outer layer of the snotites. Intriguingly, the snotites that did not have any nematodes also did not have any mites. From this observation, Gaetan deduced that the mites were feeding on the nematodes and that the nematodes, in turn, were feeding on the bacteria. The mites were not feeding on the bacteria, as was previously thought. The nematodes could hide out in the acid core where the mites dare not go and then come out to rejuvenate in the outer layers. They were hiding in an acidic refuge from predators. Cueva de Villa Luz had whet Gaetan's intellectual appetite, and he was now set on pursuing the topic of extreme nematodes to much greater depths. Literally.

In October 2006, after the summer of coring up at High Lake, I received an email from Gaetan Borgonie. "Dear Professor Onstott, I am a nematologist . . . been reading your work . . . nematodes are extremely resistant animals . . . *how could I get samples of soil from that deep?*" My pupils dilated upon seeing the last phrase. Most of the time I got emails only from journalists or videographers who wanted an exciting trip down a South African gold mine, but this was the first email I had received from any scientist, least of all a biologist, who wanted to go down and collect samples. I recalled the sample of worm-bearing biofilm that Lihung had collected five years earlier from Evander Gold Mine no. 8 shaft, 11 level—or was it 13? Upon thawing it, I could see that there were some worm-shaped things in the biofilm. I replied to Dr. Borgonie that we might have something like nematodes. Gaetan's replies were peppered with more questions than I could answer. By December he had started preparing nematode growth media, and our plan was to meet up with Esta in April 2007. Esta cautioned us about the biodiversity treaty, and we spent the time prior to the trip drafting the necessary memo of understanding about what would be done with any nematodes we found. Gaetan's lab was well equipped in terms of international permits to carry nematodes in and out of the country, but we needed to make sure he left cultures behind in South Africa for their nematologists. I contacted Walter Seymore, the chief geologist at the no. 8 shaft at Evander Gold Mine, and asked him if he could take us down to 11 or 13 level. Walter had been of enormous help in getting us core samples years earlier and was very keen on the idea of looking for worms.

THE HUNT FOR THE WORM FROM HELL BEGINS: APRIL 22, 2007, BLOEMFONTEIN, SOUTH AFRICA

I picked up Gaetan at the small Bloemfontein airport on a pleasant austral autumn Saturday. We then swung by the microbiology department at the University of the Free State to get him outfitted with coveralls, boots, helmet, etc. and then called Esta. The next day we drove up to Fochville and stayed overnight at a local B and B nearby. We had already planned a trip down to 18 level of Tau Tona Gold Mine, where we took the opportunity to introduce Gaetan to a burping, stinking borehole at 12,000 feet depth.

As hard as it was to imagine, the security at Tau Tona was much stricter than it had been in the past. We were required to sit through a couple of hours of orientation given by the mine manager himself, which made it increasingly difficult to schedule underground trips. Part of the reason for the heightened security and orientation was the boom in illegal mining during the past decade by the zama zamas. Signs were posted everywhere in Tau Tona showing the cartoonish miner "lion" mascot of Tau Tona warning of the signs left behind by zama zamas. They would live underground, hiding in ventilation ducts during the day shift and then come out at night after blasting. They would perform crude gold extraction in the tunnels, collecting enough gold to pass out of the mine to their coconspirators, who would in turn smuggle in food. As a result, we were no longer allowed to carry food down with us.

After the briefing we grabbed our gear and headed toward the cage. It took us three cage rides to get down to 16 level, with each cage becoming progressively smaller. After getting out on 16 we then followed a tunnel for one-and-a-half miles and arrived at a steel door in the rock wall. We opened the door and stepped onto the landing of a steel stairwell that went straight down for 300 feet. As I followed the others down the stairs, the only thing I could think was that this must be the deepest stairwell on the planet. At the bottom we opened another steel door onto 18 level and were blasted by the heat pouring from the tunnel. After hiking another third of a mile, we came to a tiny side door that opened into a room. At the end of a stinking, foul pool of water that stretched across the hundred-foot-long room was the burping borehole. While the rest of us were sweating away, Gaetan was eagerly collecting water samples from the borehole. The pH was about 8 and the sulfide concentration was enormous, but the temperature was 48°C. I thought that temperature was beyond the limit of eukaryotes, but as it turns out, nematodes had been reported from hot springs before. When he had finished collecting sufficient water, Gaetan insisted that we collect multiple samples of the soil from the room and the tunnel floor as we headed back to the stairs. Every ounce of strength I had was gone by the time I had hiked and crawled up the thirty stories of steps to 16 level. I had been curious to see how Gaetan reacted to the trek, but he seemed quite at home in the dark, dank recesses of Tau Tona, and his height and heft gave him an advantage in the tussle that usually occurred in the cage rides down and up from the mines.[30]

The next day we departed for Evander Gold Mine, while Esta and her crew returned to Bloemfontein. On the drive over, Gaetan explained to me about the hardiness of nematodes. His favorite story was the breakup of the shuttle Columbia 61 kilometers above the Earth's surface on February 1, 2003. Columbia had been carrying a biological experiment that included many different species. As Gaetan explained to me, all the members of that experiment perished along with the shuttle crew, except for the nematodes. The NASA investigation teams had found the canisters that contained *Caenorhabditis elegans* littered across the ground, and several months later, with the permission of the investigation oversight committee, scientists at NASA Ames were able to open the canisters. To their delight they found the little nematodes alive and doing quite well, according to Gaetan.[31] He told me stories of nematodes that had gone dormant for forty years and then woke up to feed. I wasn't sure which to marvel at the most, the nematode or the nematologist who would perform a forty-year-long experiment. Gaetan also described the ability of the nematodes to survive low O_2 concentrations. According to Gaetan, nematodes were everywhere in Antarctica, where they enter a state of anhydrobiosis to survive the extreme desiccation of Antarctic tundra. After listening to Gaetan for two hours talking about nematodes, I was becoming quite convinced that nematodes could survive just about any environmental insult. After all, if nematodes could survive reentry from space, then why wouldn't we expect to find them in the bowels of our planet? We had arrived in Evander and I turned right onto the road to the no. 8 shaft. There we met Walter in his office, and he laid out the plan for the next morning.

At 9 a.m. the three of us went down to 11 level. We stepped out onto the well-lit platform and had no trouble finding the sites that Lihung had sampled eight years before, because we could see the bright pink biofilms hanging down the black rock walls of the Ventersdorp metabasalt. Overhead was a spectacular array of foot-long soda-straw-shaped carbonate stalactites. Upon closer inspection we could see that each biofilm hung below a weeping borehole of water and that at its core was a white filamentous wad of sulfide-oxidizing bacteria. Each biofilm was surrounded by an orangish gelatin, which in turn was encircled by a thick, black patina. It was the orangish gel that contained tiny, black, sticklike things that I thought were the worms. "I've got worms!" I shouted to Gaetan and waved him and Walter over. Gaetan didn't seem quite as excited as I expected. "No, those are fly larvae," he said with distain. But maybe as at Cueva de Villa Luz, where there were flies, perhaps there were nematodes here. The biofilms had a peculiar iridescence to them I had not seen in biofilms before. Walter turned off the ceiling light at our request, and the biofilms lit up with sparkles that must have been some type of mineral precipitates in the biofilm. We found several more of the biofilms, and Gaetan set about collecting them in centrifuge tubes while I tried to collect the water dribbling from the borehole using a water bottle. A black-and-orange-striped cockroach appeared from a crack in the wall above me and crawled down toward the borehole to see what all the excitement was about. I had seen plenty of cock-roaches before, down to about 1,700 feet in the Evander Gold Mines. I had also heard all the "giant cockroach carries off lunch pail" sto-ries from the miners, but now I began to wonder whether the cock-roaches had been feeding on the biofilms that were so abundant in Evander Gold Mines. I proceeded to measure the pH, salinity, and temperature as best I could. The pH was its usual 8.5–9, and the salinity was very low, as expected. All of this was carried out under the cockroach's gaze. I said to the cockroach, "Do you realize that you are wearing my university's colors?" It twiddled its antennae at me and then scurried down to the tunnel floor somewhere beneath my feet. Meanwhile, Walter had managed to get us some of the soda straws from above as well. We finished our safari in a short time then returned with Walter to the surface and shared lunch in his office.

Afterward we went back to the Highveld Inn. Gaetan went imme-diately to his room to set up a microscope. Thirty minutes hadn't passed before Gaetan was knocking on my door. "T. C., you've got to come see this." I stepped into his room, where he already had slides laid out on the table. He pointed to the eyepiece of his microscope. I readjusted the focus and there they were, the nematodes. They looked like nothing more than tiny black commas in an orangish, bubbly gel, but they would flex and move, and there were a lot of them. Oc-casionally their shapes would expand briefly as if they were eating. I had been so used to looking at the tiny Brownian motion of bacterial rods and cocci at very high magnification that I was struck by the size of these behemoths in comparison. "And that's not all." He wiggled a centrifuge tube of water in front of me, and I noticed that a tiny bubble of water that went downward, instead of floating to the top. "You've got tardigrades in your water," Gaetan cheerfully pronounced. I had never heard of a tardigrade before. "What is that?" I asked. "They also call it a water bear, but it is like the tiniest insect. It will actually grab an air bubble and then swim down to eat bacteria while breath-ing the air from the bubble. They are also very hardy creatures, as well, and can be found in all sorts of extreme environments. "We'll call this one the Esta beetle," said Gaetan gleefully. Gaetan was like a

kid in a candy shop, and I was greatly relieved to see he was pleased
with his haul.

Over dinner that evening at the Highveld Inn restaurant, Gaetan
regaled me with more stories about nematodes. Some of them, about
nematodes infesting animals and humans, were not the best dinner-
time conversation, but the more horrific the details the more Gaetan's
eyes would glimmer and his smile would grow bigger. We eventually
got down to a more serious business of understanding what we saw.
The nematodes were confined to the sulfide-oxidizing bacterial biofilms, just as he had found in Cueva de Villa Luz. Only this time, the
pH was very high, high for nematodes that is. I had been assuming
all along that the nematodes had been carried in by the miners with
the equipment, like the spiders and cockroaches had been. We had
pretty much ignored characterizing the bacteria of the biofilms despite their inherent beauty simply because they could be contaminated by microorganisms carried by the ventilation air. I explained to
Gaetan that even though we had been at a depth of 4,300 feet, the
water we were seeing on 11 level was coming down from the Transvaal dolomite aquifer through the metabasalt of the Venterdorp Supergroup and that the dolomite aquifer was very rich in bicarbonate.
That explained why we saw so many carbonate stalactites at that location and why the pH was so high. The water in the aquifer obviously
was sulfidic and therefore must be low in O_2. Perhaps aerobic chemolithotrophs do travel underground with the groundwater. Then when
the mine intersects that groundwater with a borehole, the rare, surviving, aerobic chemolithotrophs explode into growth wherever the
sulfidic water hits the mine air upon exiting the borehole. This provides a very energy-rich redox gradient. We had previously witnessed
very rapid growth of biofilms in newly mined stopes at 10,000 feet.
Gaetan replied that the same might be possible with a nematode,
perhaps one in a dauer stage. When it arrives at the borehole and
encounters this magnificent banquet of bacteria and a little O_2, it
resuscitates, begins to feed and reproduce, and before you know it,
you have a colony of nematodes. "So what you are saying is that it
might be surviving back inside the aquifer, but how long could it
live? Doesn't it have to reproduce?" Gaetan got very excited about this

part, explaining how some nematodes, parthenogenetic females, can reproduce asexually. But some nematodes can only reproduce sexually, with a male partner, and some nematodes can do both. I could see the advantage of not requiring sexual reproduction in the deep biosphere. Finding a mate down there would be a life-consuming quest, and the chances of species propagation before dying would be nil.

The next morning we arranged for a return trip to collect additional samples of the soil around the boreholes and along the tunnel and of the service water. This was the only way of determining whether the nematodes could survive outside the biofilm. Over the course of the summer while I was wrestling with the U-tube in High Lake, I was getting regular updates from Gaetan on his worms. He had not found any nematodes in the soil samples from Evander or Tau Tona, and he had not found any living nematode in the hot, sulfide-stinking water at Tau Tona. He had tested the pH range of the Evander biofilm nematodes, and they were quite happy at pH 8.5, but when he tested lab cultures of surface nematodes, they died at that pH level. To test the eating preferences of his biofilm nematodes, he made an agar plate, half of which was covered with a bacterial lawn of the biofilm bacteria and half with the common laboratory bacterium, *E. coli*. Without a doubt, time and again, the nematodes turned up their noses at the *E. coli* and instead grazed solely on the biofilm bacteria. Meanwhile we had received ^{3}H and ^{14}C analyses on the Evander seep water, and they indicated that it was not modern, but quite old. If the nematodes did arrive with the water, then they truly weren't mining contamination from other mining levels. By September Gaetan was ready to return to South Africa for more underground safaris in other mines in search of nematodes. He had already completed SEM and TEM imaging, 18S rRNA sequencing, and complete morphological descriptions. He had even dissected some of the nematodes he had preserved on-site to extract the bacteria from their tube-shaped stomachs and determine by 16S rRNA what type of bacteria the nematodes had been eating before he had captured and killed them.

"FROM THERE TO HERE AND HERE TO THERE, FUNNY THINGS ARE EVERYWHERE": JANUARY 2009, NORTHAM PLATINUM MINE, TRANSVAAL PROVINCE, SOUTH AFRICA

Gaetan had applied for sabbatical leave for the 2008–9 academic year, a request that was begrudgingly granted by Ghent University. With help from Esta's graduate students,[32] Gaetan set about designing a filter that could be attached to boreholes to collect worm-size organisms separate from the bacteria. In January 2009 Gaetan decided to visit the Northam platinum mine with Esta, Derek Litthauer, one of Esta's students, and a science writer by the name of Marc Kaufman.[33] There Gaetan collected nematode-bearing stalactites from the ceiling and discovered it was the same species he had found at Evander mine, more than 180 miles away. This perplexed us all and prompted Esta to send me an email quoting Dr. Seuss, "From there to here, from here to there, funny things are everywhere."[34] Admittedly we were still in shock at the zoo Gaetan was discovering down there. We had encountered "*Candidatus* Desulforudis audaxviator" in deep fractures separated by hundreds of miles of fractured rock, but it was motile and could swim. Gaetan had told me that nematodes do not swim. Was it reasonable to expect a nematode to crawl almost 200 miles? I wasn't sure how fast a nematode could crawl, but I was certain that there was a nematologist somewhere who had probably measured it. I still felt that mining contamination was the most likely source of the nematodes. Gaetan and the rest of us agreed that the only way to resolve this was to find a nematode in the fracture water itself. We had to adapt the approach we had used before, filtering thousands of liters of water to catch the worms.

By February 2009 Gaetan informed me that the new filters seemed to be working great, and he had started installing them on several boreholes for filtering over months. By July 2009, Gaetan had finally found his dream worm. It had been filtered from borehole water, and he was able to resuscitate her in Esta's lab. Up until that point he had succeeded in finding the occasional nematode in some boreholes, but despite his best emergency room protocols could not get them to grow. But in a filter sample from a one-mile-deep borehole in Beatrix Gold Mine, he had found a nematode that responded to his patient ministrations, feeding her a little bit at a time until she was finally able to produce an egg, and then one miraculous day the egg hatched, giving birth to a beautiful baby nematode!

Gaetan flew me down to South Africa that July to introduce me to *Halicephalobus mephisto*, his Worm from Hell, and to go over the draft of a paper he wanted to submit to *Nature* describing her. First he showed her to me underneath a binocular microscope in one of Esta's labs that Gaetan had transformed into a worm ER. She looked like a wiggly black comma. Then Gaetan showed me the SEM images of her. She was gorgeous to behold, having a long tail stretching down to a curved tip. She had a narrow isthmus and a ten-micrometer median bulb. She was a perfect 1:3:4 in terms of the lengths of her cheilostom, gymnostom, and stegostom, making her mouth the perfect shape for eating bacteria. And then there were her lips—all ten of them, six labial and four cephalic—which stretched around the irregular circular mouth and down to her neck. As photogenic as *H. mephisto* was, Gaetan warned me that other members of her genus were deadly. "*Halicephalobus* infects horses, and several people who have cleaned the stalls of horses with open cuts on their skin became infected with the nematode. The nematode multiplied exponentially and penetrated every organ of the body in three days. The people died. There is no cure," Gaetan told me, beaming like a proud father. Being an owner of horses myself, this description sent shivers up my spine. I stepped away from the binocular scope and started checking my fingers for any open cuts. He then proceeded to describe at what temperature she could grow and what she liked to eat. He saved the best for last. She was parthenogenetic, that is, she reproduced asexually! He had borrowed the name "mephisto" from the medieval Faust legend referring to the Devil as "he who loves not the light" (or in this case, she who loves not the light). Gaetan and I sat down to go over their draft manuscript and the data they

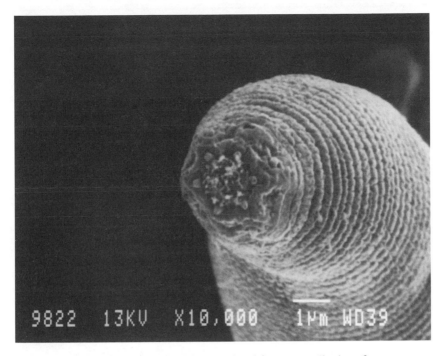

9822   13KV   X10,000   1μm WD39

FIGURE 10.3. *Halicephalobus mephisto*, isolated from one-mile-deep fracture water in the Beatrix Gold mine in 2009 (SEM image from Gaetan Borgonie).

had used to convince themselves that *H. mephisto* was not a mining contaminant.[35]

When the paper was finally published in *Nature* two years later, the editors had utilized the term "The Worm from Hell" on the cover to describe *H. mephisto* along with an SEM image of her face (figure 10.3), much to our and the mine's amusement. The public media had a field day reporting on the result, which they quickly compared to the movie *Tremors*. Although the movie involved much larger worms systematically decimating a small Nevada town, as the online media and blogopshere articles circled the globe during the subsequent seventy-two-hour news cycle, the reported size of our *H. mephisto* kept getting larger and larger, eventually reaching five meters in length!! Although granted that some nematodes in whale placenta reach eight meters in length.

But even discovering the 0.5-millimeter-long *H. mephisto* was like finding Moby Dick swimming around in Lake Ontario. The ability of multicellular organisms, like nematodes, to survive on so little O_2 had enormous implications for subsurface life on Mars and particularly Europa, where astrobiologists had claimed the photolytically and radiolytically produced O_2 concentrations would be too low for complex organisms. The biocomplexity of the subsurface ecosystems was potentially much more complex than we had believed in the days of SSP. In the case of karstic aquifers, the space for these complex ecosystems and their larger inhabitants was in itself created by H_2S produced by some microorganisms and then oxidized by other microorganisms.

EPILOGUE

Prior to the SSP coring campaign of 1986, the published results of terrestrial subsurface biomass relied upon what could be grown on media, and these values were typically fewer than 1–10 cells per gram of rock or per milliliter of water. These numbers had to be considered minimum estimates of the terrestrial subsurface biomass, with two notable exceptions. The first was from Olson et al.,[1] who published cell counts in water samples pumped from boreholes penetrating the Madison Limestone Aquifer at a depth of about one mile. The second was from Wilson et al.,[2] who reported cell counts from sediment cores collected to a depth of 25–30 feet in Lula, Oklahoma. Other reports on cell counts for the terrestrial subsurface had been published earlier by Ekzertsev,[3] Mishkov,[4] and Kuznetsov et al.,[5] all in the Russian scientific literature. These counts were based on both water and sediment cores and ranged from 10^5 to 10^9 cells per gram or milliliter (figure 11.1A). During the ten years that the SSP was active, 300 microbial cores were collected, along with several dozen water samples, and a similar number of total cell counts and PLFA concentrations were measured and published (figure 11.1B). This data defined a trend consistent with the decreasing biomass abundances reported for subseafloor sediments by Parkes et al in 1994.[6]

Since the end of SSP program in 1996, approximately 164 published studies of the deep terrestrial subsurface, which have included cell counts of sediments, ice, and water, have added an additional 1,300 data points filling in the picture of the terrestrial biomass abundance (figure 11.2). These studies are spread across all six continents. What they reveal is a highly variable biomass concentration that ranges from 10 to 10^9 cells per gram or milliliter, with a decline in biomass concentration in rock as a function of depth. Surprisingly,

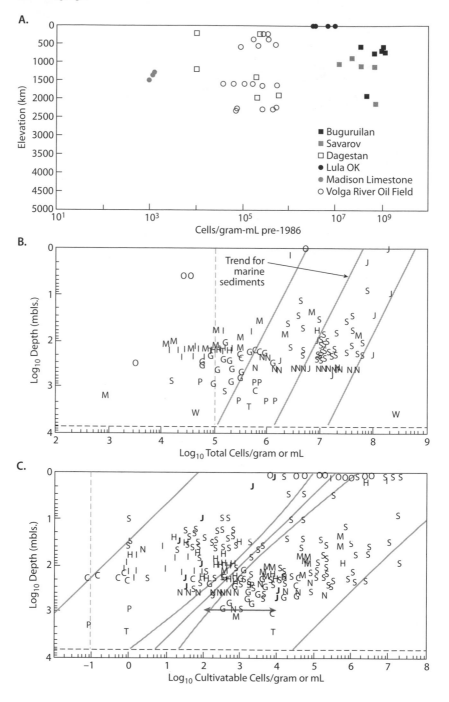

the data set still defines an array that is consistent with the decreasing biomass abundances of subseafloor sediments reported more recently by Parkes et al.[7] despite the much higher porosity of seafloor sediments. The cellular concentrations in groundwater, however, do not exhibit any such decline with depth. Taken together, the data confirm that the subsurface biomass represents a substantial fraction of living carbon on this planet and the domain of the subsurface biosphere is becoming clearer (figure 11.3).

The discovery of *H. mephisto* in 2009, the time I chose to end this book, demonstrated that even after twenty-five years of exploration, the subsurface biosphere still had many surprises left for us. Since then, advances in sequencing technology have enabled a more thorough characterization of the biodiversity of the deep subsurface. These techniques have revealed that the subsurface biosphere contains a large proportion of undefined phyla that represent new branches on the tree of life, if you will. This undiscovered life has recently been referred to as microbial dark matter, borrowing a term from the astrophysicists.[8] Since 2000, the exploration of the marine subsurface biosphere has greatly intensified, has become an integral part of the ocean drilling program, and is supported by many marine science programs.[9]

During the SSP campaign it was clear that a fraction of the microbial biomass was active using FISH probes. Now we are able to compare the DNA with the rRNA and determine which organisms are active and which are not. The new technologies have also permitted us to locate the specific genes being expressed by profiling the mRNA, in a process called metatranscriptomics,[10] and the proteins being expressed by the microbial communities in the deep subsurface via a process called metaproteomics. Other single-celled eukaryotes, such as active fungi, have also been found in the deep subsurface.[11] Viruses have also been identified in the deep subsurface, and

---

FIGURE 11.1. A. Cell counts as a function of depth prior to 1986. B. Total cell counts versus depth determined by acridine orange direct counts (fig. 1 from Onstott et al. 1999). C. Cultivatable cell numbers versus determined from serial dilution assays on growth media. Solid letters refer to rock samples and open letters to water samples.

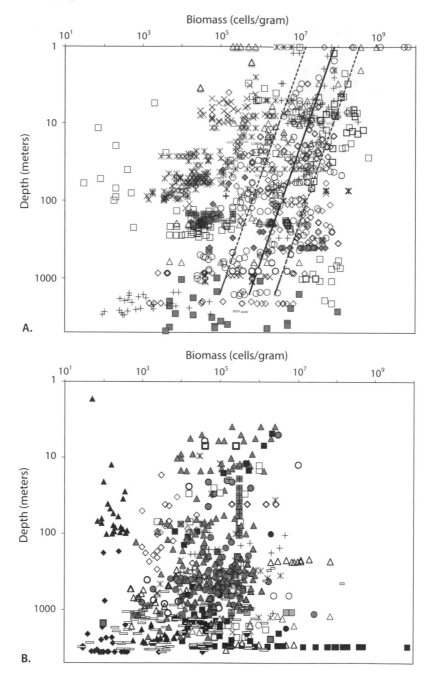

in the years to come a more thorough understanding of the role they play in the ecology of the deep subsurface will be revealed.[12] Rather than inferring the rate of cell turnover from geochemical arguments, we can now directly measure the rate of biomass turnover in the subsurface with the use of amino acid racemization.[13] We are finally at the point where we can determine whether the geochemical and isotopic signals being measured in subsurface environments are being produced by present-day microbial metabolism, even if at a slow rate. The application of these techniques has revealed many surprises and more riddles that remain to be solved. Two questions left over from the days of SSP, however, have not been resolved as of yet: (1) the age of subsurface microorganisms, and (2) whether life can originate in the subsurface.

To answer the first question one needs to establish reliable phylogenetic relationships between subsurface microorganisms and their closest surface cousins and then estimate the time since divergence. This was the holy grail sought at the SSP Clocks Workshop in 1992. Since that time, molecular clock models have become much more sophisticated, the genomes of many organisms in the three domains of life have been sequenced, and geological calibrations of evolutionary events have been established that constrain the rate of the clock.[14] The goal of much of the research in molecular clocks is to determine the geological age of prokaryotic evolutionary events deep in the Archean Age, where the extrapolation of mutation rates is tenuous at best. The application of such clocks to dating subsurface microorganisms over much shorter time scales, however, has not been revisited. A much more extensive database of 16S rRNA sequences now exists than could have even been imagined back in 1992. It is time to return to this question to determine whether the dates derived from the application of molecular clocks to accepted deep subsurface bacteria

---

FIGURE 11.2. A. Cell counts from rock cores as a function of depth based upon a compilation from the literature through 2014. B. Cell counts from water samples as a function of depth based upon a compilation from the literature through 2014.

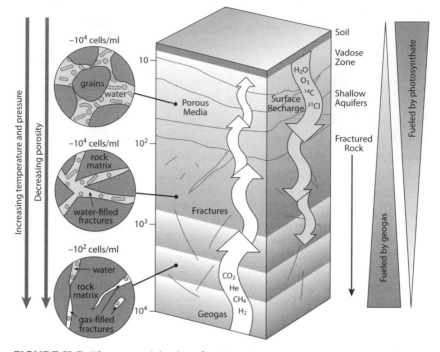

FIGURE 11.3. The terrestrial subsurface biosphere. Porosity and permeability decrease with increasing depth. Geogas produced by abiological process migrates upward but diminishes with decreasing depth. Fresh meteoric water, O_2, and rhizozone-generated nutrients migrate downward, diminishing with depth. These two fluxes combine with water-gas-mineral interactions to sustain microbial communities. But because the permeability and geological structure is heterogeneous as a function of depth, the microbial abundance and diversity do not diminish uniformly with depth. In this figure the biosphere is subdivided into the mesozone ($15°C < $ Temp $< 40°C$), thermozone ($40°C < $ Temp $< 80°C$) and hyperthermozone ($80°C < $ Temp $< 120°C$). The hydrothermalzone is the region in which complex organic molecules are stable but no known life forms can exist (based upon original figure created by author and published in "Earthlab," a report to the National Science Foundation.)

are consistent with what is known about their potential time of residence in the subsurface.

The second and more important question is whether life can originate in the subsurface, as Tommy Gold once proposed. The expeditions to South Africa began with the search for abiogenesis of life in the deep subsurface, but we failed to find it due to our lack of preparation. But this search continues today wherever ancient water exists deep in the Precambrian crust. The answer to this question will have the greatest impact on the search for life in our solar system. Even without having answered this question, subsurface microbiology has already provided insights into how life could conceivably exist beneath the surfaces of Mars and the larger moons within our solar system, such as Europa. The mere possibility of detecting extant subsurface life-forms remains the hope of many scientists, but the inaccessibility of these environments challenges the approaches available to NASA's planetary exploration program and its limited budget.

Even though the diverse communities found deep beneath our planet's surface in the last few decades do not compare to the scale of Jules Verne's fictional subterranean imagination, their discovery has altered our perception of life that lies hidden from sunlight. Today the exploration of subsurface life is not about how big the subsurface biosphere is or how far it extends, although these are still significant questions in and of themselves. It is more about discovering the boundaries of our imagination. Removing the blinders that hobble us and our perceptions of the deep biosphere and of life in general is a never-ending challenge. We must always be prepared to be surprised, because many wonders still await us down there in our hunt for deep life.

APPENDIX A: CHRONOLOGY OF THE EXPLORATION OF SUBSURFACE LIFE[1]

1793   Alexander von Humboldt (1793) describes algae and fungal species in gold mines located in the Fichtel Mountains of Germany.

1910–   Galle (1910) grows bacteria from coal samples that produce
1911 CH_4 and CO_2.

1911   Omelyansky (1911) grows bacteria from permafrost soil associated with the Sanga-Yurakh mammoth, unearthed in 1908.

1914   Schröder (1914) also grows bacteria from coal samples, but they do not produce CH_4.

1922   Von Wolzogen Kühr (1922) discovers microbial sulfate reduction in "deeper layers of the earth," which were 10- to 37-meter-deep cores of Pleistocene sediment near Amsterdam.

1926   Bastin et al. (1926) grow sulfate-reducing bacteria from water samples collected from soured oil wells in Illinois and California down to 942 meters. They speculate whether the bacteria's ancestors were deposited with the sediment during the Ordovician and Pennsylvanian epochs or were introduced during drilling.

1926   Ginsburg-Karagitscheva (1933) claims that in 1926 she reported growing fermenting anaerobes from water samples collected from springs and wells as deep as 1,007 meters near Baku.

1928   Lieske and Hoffman (1928) grow bacteria from coal collected at 1,089 meters depth in the Westfalen coal mine,
claiming they were transported to depth by groundwater.

1928   Lipman (1928) grows bacteria from rock and Pliocene sediment collected from "a depth of several hundred feet."

1930   Bastin and Greer (1930) grow bacteria from oil-well water
from Illinois collected from 275 to 600 meters depth.

1931   Lipman (1931) grows bacteria from coal samples collected
at a depth of 550 meters from the Locust Summit coal
mine in Pennsylvania and claims they were deposited with
the sediments during the Pennsylvanian Epoch, 252–290
million years ago.

1932   Farrell and Turner (1932) collect their own coal samples
from Locust Summit coal mine from a depth of only 120
meters and grow bacteria only from those coal seams intersected by water-bearing fractures and not from the dry, unfractured coal seams. They dismiss Lipman's claim of entombed ancient bacteria.

1935–   Issatchenko (1940) reports on the pink water from 1,500
1940   to 2,000 meters depth in the oil fields in the Aspheron
Peninsula northeast of Baku. He reports the occurrence
as due to cells of *Chromatium*, anoxygenic photosynthesizers, and refers to the possibility of O_2 generation by
radiolysis.

1942   James and Sutherland (1942) report growing bacteria from
permafrost in Canada.

1943   ZoBell (1943) isolates halophiles from limestone and gyp-
sum block collected at a depth of 450 meters from the
Grande Ecaille sulfur mine on the Mississippi River delta in
Louisiana.

1944   ZoBell fails to grow any bacteria from water obtained at
2,134 meters depth from an "aseptically" drilled oil well in

Los Angles, California. This raises doubts about the indigenous nature of SRBs in oil fields.

1946 Cox (1946) is critical of the claims by ZoBell of bacteria making petroleum and challenges the evidence that the bacteria might be indigenous.

1949 Miller (1949) reports finding SRBs in the interior of salt cores collected between 213 and 247 meters depth from Tertiary-age, sulfur-bearing salt domes in Texas and proposes that bacteria were responsible for the formation of sulfur deposits.

1950 Kuznetsov (1950) reports microbial cell counts from oil-field water collected in the oil fields of the Cheleken Peninsula in Turkmenistan and from the Terek-Dagastan oil field in Azerbaijan, south of Baku. No depths were provided.

1951 Ekzertsev (1951) reports the first cell counts from drill cores of sedimentary rocks collected from 560 to 2,090 meters depth in the Saratova oil fields of the Volga River basin.

1951 ZoBell (1951) references SRBs from the production water of Ventura Avenue oil wells from 1,300 to 4,400 meters depth but cannot determine whether they might be indigenous or contaminants.

1955 Morita and ZoBell (1955) document the concentration of viable bacteria in Pacific pelagic sediment from 0.4 to 7.5 meters depth beneath the seafloor.

1957 Smirnova (1957) performs first tracer experiment on drill cores collected down to 400 meters depth using cultures of *Serratia marcescens*, a red-pigmented bacterium.

1958 Meshkov (1958) performs first cell counts on water from oil-bearing formations in the Paleozoic strata underlying the basin north of the Caucasus. Several other reports in the Russian literature on the microbiology of oil-field water occur between 1957 and 1958.

1960   Naumova (1960) isolates hydrocarbon-degrading bacteria in the Dagestan oil field north of the Caucasus.

1960   Reiser and Tasch (1960) isolate bacteria from salt collected at 193 meters depth in the Wellington salt mine in Kansas. Like Dombrowski, they believe the bacteria may have been trapped in the salt since its deposition in the Permian.

1961   Lindblom and Lupton (1961) culture anaerobic bacteria from Orinoco delta sediments from a depth of 50 meters and relate their presence of the organic-matter characteristics.

1962   Telegina (1961) isolates hydrocarbon-degrading bacteria from water associated with gas wells down to 228 meters deep in the Azov peninsula.

1961– Al'tovskii et al. (1961) are the first to examine the change in
1962   the microbial community composition of groundwater as a function of location on the hydrological path in sandstone aquifers that recharge in the northern slopes of the Caucasus and discharge as springs further north. The depths of wells sampled range from 210 to 2,123 meters.

1962   Mekhtieva (1962) grows bacteria from the water associated with the Middle Volga River oil fields.

1962   Kuznetsova (1962) also reports cell counts from groundwater as a function of location on the hydrological path in sandstone aquifers that recharge in the northern slopes of the Caucasus and discharge as springs further north.

1963   Dombrowski et al. (1963) report the isolation of bacteria from salt collected at 230 to 4,300 meters depth, mostly from salt mines in Europe, the United States, and Canada, and in formations ranging in age from the Permian to the Cambrian. Dombrowski claims that these bacteria were entombed in the salt during its deposition.

1965   Iizuka and Komagata (1965) report on bacteria and fungi in sediment and water samples collected from the Higashiyama oil mine at 180–270 meters depth and from oil-brine samples pumped from the Yabase oil field.

1965   Bogdanova (1965) presents the first comparison of results of enrichments from both pared and flamed cores and from borehole water samples from the same formations. The wells were drilled in Tertiary Period sediments along the Lower Volga River basin in the Caspian lowlands. The sample depths ranged from 87 to 504 meters.

1966   Sugiyama (1966) pares and analyzes cores of Miocene sediments collected during geological drilling to depths of 600 to 3,200 meters on the southern part of the Niigata Plain in Japan. The study is the first to examine the distribution of fungi in the deep subsurface.

1966   Zinger (1966) reports results of bacterial growth on hydrocarbon from groundwater in the Lower Volga River basin with respect to oil and gas deposits.

1967   In the first subsurface microbial investigation in the United States since ZoBell's efforts in the late 1940s, Davis (1967) investigates hydrocarbon biodegradation by reporting growth of SRBs in a freshwater oil reservoir in the Carrizo Aquifer as a function of distance from recharge and depths ranging from 65 to 1,277 meters.

1968   Morehouse (1968) proposes sulfuric acid speleogenesis for carbonate caves in Iowa and suggests that iron-oxidizing bacteria may have been responsible.

1970   Rozanova and Khudyakova (1970) dismiss the possibility of indigenous subsurface bacteria in oil fields and attribute their presence to oil-field flooding. (Until this time, the Russian geomicrobiologists have been the last holdouts for the theory of indigenous subsurface bacteria in oil reservoirs, but this paper closes the chapter on that period.)

1980   Dockins et al. (1980) report on bacteria in the groundwater of the Fort Union Formation in Montana, as interest in the relationship between groundwater chemistry and bacterial activity begins to grow again in the United States. On the basis of $\delta^{34}S$ isotope analyses, they also argue that the sulfide in the water is biogenic.

1980   In the first study of subsurface bacteria in the United Kingdom, Whitelaw and Rees (1980) report on nitrogen-cycling enrichments from two cores collected from the surface to 30 and 50 meters depth in the surficial Chalk Aquifer.

1981   Olson et al. (1981) report direct counts of bacteria and enrichments of SRBs and methanogens from 1,247 to 1,790 meters depth in the Madison Limestone Aquifer in Montana. Within the USGS, the idea that bacteria change the groundwater chemistry of aquifers is becoming more accepted.

1981   First Greenland ice core is completed to 2,037 meters through the ice sheet. Although the goal of the coring is for climate research, the cores will provide material for follow-up microbial studies of the deep cold biosphere beginning in 2004.

1982   Oremland et al. (1982) at the USGS begin investigations of the microbial activity of seafloor sediment on Leg 64 of the Deep Sea Drilling Program in the Gulf of California, picking up where ZoBell left off thirty years earlier.

1983   Wilson et al. (1983) collect core of Quaternary Period sediment to a depth of 5 meters and examine the microbial community with support from the EPA.

1983   Belyaev et al. (1983) isolate methanogens from a sandstone core collected at 1,675 meters depth in the Bondyuzhskoe oil field, and argue, based on the isotopic composition of methane, that the methane gas in the oil reservoir is produced by methanogens rather than being thermogenic.

1980– In 1980, the Studiecentrum voor Kernenergie and Centre
1986 d'Étude de l'énergie Nucléaire begin construction of an underground laboratory located at a depth of 240 meters in
the 34-million-year-old Boom Clay in the town of Mol, Belgium, the first URL in Europe. The Canadian government
begins construction of the Whiteshell Underground Research Laboratory in Pinawa, Manitoba, in 1983. The Swedish Nuclear Fuel and Waste Management Company breaks
ground on an underground laboratory on the Baltic coast
island of Äspö in 1986. The Äspö URL becomes by far the
most successful in terms of characterization of subsurface
microbial community composition and activity in subsequent years thanks to Karsten Pedersen.

1986 The P24, P28, and P29 cores are collected from within the
Savannah River Plant as the first subsurface microbiology
study launched by the SSP of the U.S. DOE. Clays and
sands are sampled from 8 to 265 meters, covering the lower
Tertiary Period and upper Cretaceous Epoch. The results of
this campaign are published in a special issue of *Geomicrobiology Journal* (vol. 7, 1989).

1986 Drilling begins at the Siljan Ring structure in Sweden as a
wildcat oil/gas prospect advocated by Tommy Gold. No microbiology is undertaken as part of the drilling project, but
thermophilic isolates will eventually be recovered from the
6,700-meter-deep borehole.

1986 Spelunkers break through rubble in southern New Mexico
to discover 200 kilometers of caves reaching 489 meters beneath the surface that will be named the Lechuguilla Caves.

1986 A construction project in Dobrogea, Romania, accidentally
penetrates a hidden cavern lying just 20 meters below the
surface. Its discoverer names it Movile Cave.

1987 Chapelle et al. (1987) of the USGS collect cores from 25 to
170 meters depth in the Patapsco Aquifer of Maryland and
analyze the microbial abundance.

1988   The SSP cores the Cretaceous Epoch Cape Fear Formation in South Carolina at 365 to 470 meters depth at site C10 in the first coring campaign dedicated to deep microbiology. For the first time, full tracer technology, developed by the SSP, and protocols are implemented for the collection of subsurface microbial samples.

1988   Kolbel-Boelke et al. (1988) report on microbial enrichments from Pleistocene Epoch sands from core collected to a depth of 30 meters.

1989   Colwell (1989) reports microbial results on sedimentary interbeds of the Snake River basalts of Pleistocene Epoch at 70 meters depth within the INEL property drilled in 1988.

1990   First international meeting on subsurface microbiology is held in Florida and supported by the SSP.

1990   Pedersen and Ekendahl (1990) report microbial results and abundance from groundwater wells in Precambrian granites in Sweden to depths of 860 meters.

1990   The SSP cores the Snake River Plains basalt down to a depth of 187 meters and develops new coring and tracer approaches using argon gas. In the same year the SSP begins coring at Yakima on the PNL site, eventually collecting cores from Miocene Epoch lake sediment and paleosols at depths to 190 meters.

1990   Parkes et al. (1990) begin measuring the microbial biomass of subseafloor sediments during Leg 112 of the JOIDES cruise to Peruvian margin.

1991   Amy, with SSP support, investigates microbial abundance and composition in Miocene volcanic ash within the tunnels underneath Ranier Mesa on Nevada Test Site to depths of 450 meters. The first results are reported in Amy et al. 1992.

1992 SSP piggybacks on Texaco drilling operation at Thorn Hill Farm in Virginia, obtaining microbial cores from 2,700 meters depth.

1992 Gold (1992) publishes "The Deep, Hot Biosphere," citing Karsten Pedersen's unpublished work.

1992 Pedersen and Ekendahl (1992) report microbial results from Stripa iron mine groundwater to depths of 1,040 meters.

1992 Thorseth et al. (1992) attribute complex dissolution features in volcanic glass to borings by unknown microorganisms and begin finding them in subseafloor basalt at the Juan de Fuca spreading center.

1993–
1994 Stevens et al. (1993) and Zheng and Kellogg (1994) begin investigations of subsurface microbiology of the Miocene Epoch Columbia River basalts in Washington and Idaho.

1993 Norton et al. (1993) begin isolating halophilic Archaea from the Permian salt in UK mines to depths of 1,200 meters.

1993 Stetter et al. (1993) isolate hyperthermophilic Archaea from 3,000-meter-deep oil wells in Prudhoe Bay, Alaska, setting the stage for the renaissance in petroleum microbiology.

1994 The SSP supports deep subsurface microbiology coring campaigns at Cerro Negro, New Mexico, to depths of 240 meters in Cretaceous Epoch shales and sandstones and at Parachute, Colorado, to depths of 2,100 meters in Paleocene and Cretaceous sandstones.

1994 Szewzyk et al. (1994) isolate a thermophile from more than 3,500-meter-deep water in the Siljan Ring borehole.

1995 Stevens and McKinley (1995) publish their SLiMEs paper on autotrophs in the Columbia River basalt.

1995 Cunningham et al. (1995) publish their discovery in Lechu-guilla Caverns, New Mexico, of autotroph-driven speleogenesis.

1995– L'Haridon and Jeanthon report isolating more thermo-
1996 philes from oil reservoirs in the Paris Basin and Cameroon. Microbial surveys of oil reservoirs around the globe, includ-ing studies of the western Siberian oil fields, are reported and certain phyla that seem indigenous to deep oil reser-voirs are found.

1996 McKay et al. (1996) publish the ALH84001 results. NASA Ames holds its first workshop on drilling into a Martian subsurface biosphere.

1996 Onstott et al. collect and analyze their first microbial sam-ples at 3,100 meters depth in the ultradeep gold mines of South Africa.

1996– Reports on the microbial abundance and community com-
1997 position are published from water and clay samples col-lected at depths of 200 to 450 meters from the Whiteshell, Äspö, and HADES Underground Research Laboratories.

1998 Whitman et al. (1998) publish on the global subsurface bio-sphere, establishing the significance of the deep biosphere.

After this time, so many deep biosphere studies are published that only a few are highlighted here.

1998 U.S. Science Advisory Committee, upon the recommenda-tions of Tommy Phelps, begins implementing U.S. DOE tracer protocols for future deep microbiology ocean drill-ing expeditions.

1998 Rivkina et al. (1998) report on the microbiology of core col-lected from depths down to 30 meters into ancient perma-frost deposits in Siberia.

1998– Coring of the icy cover over Lake Vostok down to 3,400
1999 meters is completed, and Priscu et al. (1999) and Karl et al.

(1999) report the microbial contents of the ice. (Collection of the underlying Lake Vostok water would not occur until 2014, however.)

2000–
2011   ODP Leg 201 becomes first deep microbiology-driven ocean drilling expedition, this one off the coast of Peru, and produces a legacy of papers that lasts a decade.

2000   Vreeland et al. (2000) report isolating a *Bacillus* from WIPP salt that they believe has been isolated for 250 million years.

2002   Chapelle et al. (2002) identify a subsurface ecosystem at a hot springs in Idaho dominated by methanogens supported by H_2 produced by another mechanism, silicate cataclastic reactions.

2002   Orphan et al. (2002) discover clusters of cohabitating SRBs and methanogens in seafloor sediment that account for the anaerobic oxidation of methane, in which the methanogen operates its metabolic pathway in reverse.

2005   Microbial studies of two deep drilling projects are initiated. One is centered on the Chesapeake Bay Impact Structure, from which Cockell et al. (2012) analyze sedimentary cores from 125 to 1,660 meters depth. The second is the Chelungpu Drilling Project, which recovers core from metamorphic rock down to 4,800 meters deep and probably represents the deepest core samples ever analyzed for microbiology (Zhang et al., 2005).

2006   Lin et al. (2006) prove the existence of a radiolytically supported subsurface ecosystem.

2008   Chivian et al. (2008) publish the first metagenome of a deep subsurface, radiolytically supported, single-species biome.

2009   Borgonie discovers the first deep subsurface meiofauna (Borgonie et al. 2011).

APPENDIX B: CHRONOLOGY OF THE U.S. DOE'S SSP MEETINGS

1984 Frank Wobber creates the SSP with $40,000 from the U.S. DOE. He envisions a three-phase program beginning with Phase I, characterization of the subsurface microbiology at site using coring.

1985 May. Pennsylvania State University. Frank Wobber, Eugene Madsen, and Jean Marc Bollag, soil microbiologists from the university, host the workshop that leads to the formation of Frank's nascent Microbiology of Subsurface Environments group.

1986 February 24–25. Augusta, Georgia. Frank Wobber and Jack Corey along with Frank's SSP microbiologists and Jack's SRP environmental geologists meet to discuss collecting cores for microbiology by piggybacking on Jack's drilling campaign to evaluate the TCE contamination of the drinking water aquifers beneath the SRP.

1986 June. DOE Savannah River Laboratory. By the first week of June all recipients have been assigned their tasks and they converge on SRP for a one-day meeting on the drilling project, now named Deep Probe by Carl, before the cores are collected and shipped the following week.

1986 September 22–23. DOE Headquarters, Germantown Maryland. Frank holds a second stakeholder's meeting for Deep Probe to go over the results from the P-series drilling campaign and to present his long-term strategic plan for the SSP.

1987   February 17–18. Augusta, Georgia. Frank invites outside experts from different universities and DOE laboratories to critically evaluate the unpublished results of the P wells and the methods used to acquire them and to make recommendations on how to improve those methods in preparation for the "fourth hole." By the end of this planning workshop/ review, the recommendations are clear. They must invest more time, money and thought in "tracers," the term now coined to describe additives to the drilling system that serve as an indicator of potential microbial contamination in the samples.

1989   January. Tallahassee, Florida. Frank gathers the Phase I investigators to present their final reports on the C10 samples less than six months after they receive their samples from the field. Frank lays out the plans for new coring operations at DOE sites out West.

1989   March. Salt Lake City, Utah. Frank introduces the Transitional Program at a planning meeting with the stated goal of transferring drilling technologies to industry and developing new sites for subsurface microbiology that provide a scientific basis for remediation of DOE's Cold War legacy of radionuclide and toxic metal contamination.

1989   June. Las Vegas, Nevada. Frank holds a meeting to explore possible microbial sampling of tunnels at the Nevada Test Site.

1989   August. Redmond, Washington. Frank carries out an internal review of the Transitional Program strategy.

1989   September. Idaho Falls, Idaho. Frank gathers investigators at the Idaho Research Center for a meeting of the Master Strategy Steering Committee to decide which of the two sites, GeMHEx site at PNL or the Snake River basalt site at INEL, will be drilled during the Transitional Phase of Frank's program. In the end he decides to do both. After the meeting, Frank puts the request for proposals, the RFP, out on the street, so to speak. Any investigators, those from universities included, who want to work on these sites must submit full proposals to Frank by Jan. 1, 1990.

1989 October. Manteo, North Carolina. Frank holds a second meeting to finalize the Transitional Program strategy and funding prior to submission of the plan to DOE headquarters.

1990 January. Tallahassee, Florida. First International Symposium on the Microbiology of the Deep Subsurface.

1992 January 14–16. Las Vegas, Nevada. Frank holds a Deep Microbiology Investigators stocktaking meeting during which all his investigators report on the results of the samples collected, including experiments performed on the samples just collected from INL, PNL, and NTS, and they present hypotheses concerning subsurface bacteria and their origin. At the end of this meeting Frank tells the investigators about the upcoming Texaco drilling opportunity and recruits investigators to go to Thorn Hill no. 1 to collect samples.

1992 September 30. Annapolis, Maryland. Evolutionary Clocks Workshop. Frank brings together a group of geologists, geochemists, and hydrologists and a group of molecular biologists to determine what needs to be done in order to use sequence information on subsurface microorganisms to determine how long they have been underground .

1992 December 2. Annapolis, Maryland. Second Taylorsville Basin stocktaking meeting. Frank gathers the investigators involved in the analyses of the core samples collected from the Triassic sediments nine months ago. At this meeting David Boone reveals a new anaerobic bacillus that he wants to name *Bacillus infernus*.

1993 June. U.S. DOE Headquarters, Germantown, Maryland. Frank gathers a group of his SSP investigators to present their latest results, including the new findings from the Triassic Taylorsville Basin, in person to the director of OHER, Galas.

1993   September. Bath, United Kingdom. The second International Subsurface Microbiology Meeting. This is the first truly international meeting devoted to not only the techniques but also the new discoveries in deep subsurface microbiology of continents and seafloors.

1993   October 27. Gaithersburg, Maryland. Frank holds the third Taylorsville stocktaking meeting.

1993   December. Annapolis, Maryland. Deep Microbiology Meeting. Three-day conference to welcome new PIs, to finalize the report on the deep microbiology program, and to lay out the work schedule in preparation for drilling at Cerro Negro in June.

1994   April 7–8. Lewes, Delaware. Frank organizes a review of the ESRC.

1994   June, La Jolla, California. Meeting of JASON to evaluate Frank's SSP.

1995   February 25–28. Salt Lake City, Utah. Origins stocktaking meeting.

1995   March. University of Tennessee-Knoxville. Stocktaking meeting for other SSP subprograms.

1995   July. Portland, Oregon. Final Origins deep subsurface microbiology stocktaking meeting for DOE SSP.

1997   January. Idaho Falls. Post-SSP meeting to formulate the Test Area North Project to remediate the Snake River Plains Aquifer from TCE pollution generated by the atomic plant.

NOTES

INTRODUCTION

1. Lipman 1931, 183.
2. Kuznetsov et al. 1962.
3. Alexander 1977, 23.

CHAPTER 1: TRIASSIC PARK

1. The neutron density tool, which is used to measure the neutrons, is very sensitive to atomic hydrogen because of its large neutron-scattering cross-section. As a result, the neutron log provides estimates of the gas and water content in the rock. The seismic tool provides formation density and, hence, porosity. The resistivity provides an estimate of the clay content. The gamma logging tool measures the gamma rays emitted by the radioactive elements potassium, thorium, and uranium, thus determining the rock composition, and also the gamma rays that it sends into the rock formation and that are inelastically scattered by the Compton effect, thus determining the rock formation's density. A self-potential log was also used; it is sensitive to the salinity of the water in the rock formation. And last of all, the engineers had a temperature probe to measure the present-day ambient temperature.
2. "Gas play" is the term used to describe high concentrations of volatile hydrocarbons, such as methane, ethane, propane, and butane, which are measured from the drilling mud as it degasses into the circulation system upon reaching the surface. With a known flow rate of the drilling mud, the depth at which the gas plays originated are deduced and correlated with the lithology of the drill cuttings originating from the same depth. These data are later correlated to the borehole log data.
3. Microbial and geochemical investigations, supported by Frank J. Wobber of the U.S. DOE, of 700-foot-deep sand/shale sequences of the Coastal Plain sediments in South Carolina indicated that the contacts between these sedimentary rock types, or facies, had yielded high concentrations of viable bacteria. See the *Geomicrobiology Journal* (1989), special volume. A recent paper by USGS microbiologists on the much

shallower (200 feet deep) Coastal Plain sediments came to the same conclusion: that the sand/shale interfaces were the places where subsurface microbial activity and thus biomass were the highest. McMahon et al. 1992.

4. Finding those interfaces was straightforward with the use of the spontaneous gamma log, which measured the inherent radioactivity of the formation. As the clay content increased, so did the gamma counts. In part this was related to increasing amounts of potassium in the clays, but it was also because in these Triassic basins, where the clay was deposited in an organic-rich, anoxic lake, the uranium content was also high, because uranium oxide had precipitated under these anoxic conditions. The sandstone, on the other hand, had little clay and a lot less uranium.

5. Gondwanaland, or Gondwana, was the supercontinent composed of South America, Africa, India, Antarctica, and Australia. Gondwana is a region of India named after the Gondi people who inhabit it, that is, the Land of the Gondis. Thus "land" in the name Gondwanaland is superfluous. Pangea was a supercontinent comprising Gondwana, North America, Europe, and Asia. Laurasia, composed of North America, Europe, and Asia, began to separate from Gondwana in the middle Jurassic.

6. The caliper logging tool does as the name implies: measures the shape and diameter of the borehole.

7. Perfluorocarbons (PFCs) are fluorinated hydrocarbons in which fluorine substitutes for hydrogen. They are extremely rare in nature and were developed as a fluid tracer in ocean circulation studies and atmospheric tracing studies. They are nontoxic, and their strong fluorine-carbon bond makes them resilient to reactions. Their detection limit on a gas chromatograph with an electron capture detector is one part per billion. Two of the problems with using them are their low solubility in water and high volatility. The PFCs used for microbial sampling in the past were perfluoromethylcyclohexane (PMCH) and perfluorodimethylcyclohexane (PDMCH). With some effort, the PFC concentration in the drilling mud can be as high as 2 ppm, providing a factor of 2,000 for tracking the penetration of drilling-mud filtrate into cores. McKinley and Colwell 1996.

8. There were two types of sidewall coring tools. The rotary sidewall corer contains one coring bit that rotates up to core perpendicular to the core tool, cores into the sidewall, then rotates back down again to the center where a rod pushes the core sample into a holding barrel. It repeats this process at each core point. The tool is then pulled to the surface and the cores removed from the holding barrel. The percussion

sidewall core "gun," has separate core "bullets" for each core point, each with its own explosive charge. When shot, the core barrel penetrates the sidewall. A chain connects the core barrel to the tool and is used to pull the core barrel with its core up from the surface.

9. Tommy Phelps was a microbiologist at ORNL, Rick Colwell was a microbiologist at INEL, and Todd Stevens was a microbiologist from PNL.

10. With percussion drilling, a technique that has been used for thousands of years, the core barrel is hammered into the underlying formation along with the surrounding casing to stabilize the hole as the core barrel is removed with sediment inside, but it is not as fast as rotary coring.

11. Appendix A provides a chronological tabulation of subsurface microbiological explorations beginning with these early studies.

12. An anoxic environment is one in which the O_2 concentrations of water are well below that found in water that is equilibrated with the atmosphere. The U.S. Geological Survey defines anoxic water as water where the O_2 concentrations are less than 0.5 milligrams O_2 per liter versus 8.7 milligrams O_2 per liter for water equilibrated with air at room temperature. Salty water and hot water equilibrated with air have lower dissolved O_2 concentrations.

13. A chronology of the discovery of subsurface life is given in appendix A. Bastin et al. 1926. Bastin was not the only one investigating bacteria in the brines associated with oil. Also in 1926, the Russian microbiologist T. L. Ginsburg-Karagitscheva identified sulfate-reducing bacteria in sulfide-rich water/oil samples collected from artesian springs and wells down to 330 meters depth in the oil region near Baku, Azerbaijan (then the Soviet Union). Unfortunately her work was published in an obscure Russian publication (T. L. Ginsburg-Karagitscheva 1926).

Neave and Buswell published a paper eight months after Bastin's note on the occurrence of denitrifiers in brines from Illinois oil wells. They focused on microbial calcite precipitation rather than on souring of oil. Gahl and Anderson (1928), however, also found evidence for sulfate-reducing bacteria producing the H_2S in oil wells, confirming Bastin's original discovery. Many publications by Russian microbiologists in Russian-language journals followed and focused not only on the occurrence of bacteria in oil-bearing formations but also on the proposition that the bacteria may be responsible for producing the oil. This Russian literature on subsurface microbiology is succinctly summarized in two books: *Introduction to Geological Microbiology* (Kuznetsov et al. 1962) and *Petroleum Microbiology* (Davis 1967). The proposal that bacteria could be involved in the generation of petroleum and that petroleum was generated from continental sources of organic matter put the Russian microbiologists in dispute with most non-Russian geologists, who at that time favored a strictly geological model of petroleum generation from marine sediments. The geologists disputed the claims of "indigenous" bacteria in oil-field waters for this very reason, even though H_2S and depleted sulfate existed in the water encountered upon drilling the reservoirs, and these compounds were not strictly the result of contamination afterward. The Russian microbiologists were the first to recognize that bacteria were also consuming the microbially generated H_2S. This discovery was first reported in 1935, when Volodin reported "pink water" from a depth of 1.8 km on the Absheron Peninsula (summarized in Gurevich 1962). The pink color was due to the red-pigmented sulfide-oxidizing *Chromatium* bacterium, which is an anoxygenic phototroph. In 1970, Rozanova and Khudyakova dismissed the pink water bacteria as an artifact created during oil-field flooding (Rozanova and Khudyakova 1970).

14. ZoBell 1945. This article led to news releases in *Time* magazine describing the oil-eating bacteria as "ferrets in the oil fields." ZoBell also reported finding SRBs up to 2.1 km depth in some oil-well brine samples but noted their absence in other oil wells and mines. Because ZoBell, like his Russian predecessors, was finding evidence of bacteria producing some alkanes, and because Cox firmly believed in a geological origin for petroleum, ZoBell's reported absence of SRBs from some petroleum deposits and not others was taken as evidence that the reported occurrences of bacteria in petroleum deposits were the result of contamination. The reader will find a good summary in Cox 1946 of the arguments against microbial petrogenesis. From the late '40s to the early '60s, ZoBell also studied the effects of high pressure on bacteria, but as far as the U.S. geological community was concerned, subsurface bacteria in oil fields were the result of contamination during drilling and development, and they had little to do with the origin and evolution of oil and gas.

15. This number is from the United States Energy Information Agency website and does not include wells drilled prior to 1949. "Petroleum and Other Liquids," http://www.eia.gov/dnav/pet/hist/LeafHandler.ashx?n=PET&s=E_ERTW0_XWC0_NUS_C&f=A.

16. Frank used this meeting as a basis for another round of proposals that would be built around a field site that would establish the age of subsurface microbial communities. Clear correlations between the viable communities and the physical and chemical properties of their subsurface environment, however, were wanting. The use of phospholipid fatty acids (PLFAs) was a valuable measure of the abundance of micro-

organisms, but the signature profiles depended upon both the composition of the microbial community and the environmental conditions. Disentangling one from the other had not been entirely successful yet. They had also started applying other tools such as MIDI, BioLog, and API-NFT. MIDI stands for Microbial Identification Inc., located in Newark, Delaware. It started in 1991 and had pioneered a rapid fatty-acid-based identification system. It represented a cheaper version of the type of analyses that D. C. White's group was performing. Biolog Inc. is a California company that started in the late '80s producing plates and arrays that perform rapid phenotype screens of cells from cultures and, in the case of SSP, environmental samples. API-NFT test strips (marketed by bioMérieux) expose cells from cultures to a variety of reagents. Based upon a colorimetric response, the identity of microorganisms could be established. These could identify bacterial species from an extensive databank based upon infectious disease-carrying bacteria, but they proved of limited use in identifying subsurface microbes. The past six years of analyses had thus raised more questions than answers.

In the meeting room, now filled with about fifty investigators, the geologists, hydrologists, and geochemists mostly sat on one side, while the microbiologists mostly sat on the other. It was left to Ellyn Murphy, of PNL, to summarize the outcome of a DOE Origins workshop held in October at Lewes, Delaware, and to lay out the theoretical basis for classifying field sites based upon their geological history, their hydrological properties, their paleoenvironmental history, and microbial evolutionary processes. Whether the sediments were first deposited in marine settings containing marine microbes or were paleosols containing soil bacteria, they would be buried to depth over geological time. Under increasing pressure and temperature, the sediments would undergo diagenesis, changing their permeability and porosity. For strata that were impermeable, the microbial communities would be trapped and have to adapt to the increasing pressure and temperature. They would also have to rely upon local sources of energy. If they survived they would represent living descendants of the original soil or seafloor community. On the other hand, for strata that were very permeable, such as the aquifers of Savannah River, many thousands of pore volumes of groundwater may have coursed through their sandy matrix. It depended upon the paleoclimate, how much precipitation occurred over the eons. As a result microbial immigrants may have replaced the original microbial community that was deposited with the sediments hundreds of millions of years ago. The nutrients for these microbial communities may have originated in the overlying photosphere and were transported down to the microbes with the groundwater flow. These two scenarios led to two end-member microbial communities. One end-member included a microbial community that had undergone in situ evolution over a long period of time. The opposing end-member was the microbial community that had experienced microbial transport and selection over a much longer distance, but a shorter period of time. The microorganisms found in the former were potentially more ancient in their origin than those found in the latter.

A great deal of discussion ensued after each speaker tried to place his or her ideas or proposed sampling sites within this context. During Rick's presentation to the group, "Deep Subsurface Microbiology Hypotheses," he started to rattle on about Tommy Gold and his recently reported discovery of archaebacteria and mantle oil at 6.1 km depth in the Siljan Ring meteorite impact structure. (The science reporter for the *New York Times*, Walter Sullivan, had reported on Gold's Siljan Ring project on March 22, 1987, "Natural Gas Well Is Believed Found." Geologists have estimated that the Siljan Ring's original crater was about 52 kilometers in diameter and formed 377 million years ago.) Tommy Gold's idea was that the impact was sufficiently large to fracture the continental crust down to the lower mantle. These fractures would have stabilized and acted as conduits for mantle hydrocarbons and helium to flow upward toward the surface. The drilling project did detect some methane, but the helium associated with it had an isotopic signature that indicated it was from radioactive decay of crustal uranium and not from the mantle. It made sense to Rick that this mantle oil could sustain deep subsurface communities indefinitely.

After he concluded his presentation the room became so quiet you could hear the proverbial pin drop, with some whispering occurring in the audience. When Rick returned to his seat he leaned over to Shirley Rawson, another PNL geologist working with Phil, and asked if it was something he had said. "Tommy Gold is what you said!" she whispered. Down to the last person, the geologists at the meeting discounted Gold's theories about abiogenic oil, and some seemed quite put out that Rick, a microbiologist, would even suggest such things. Caught practicing geology without a license, Rick didn't push the Goldian concepts after that meeting. This was the extent of Gold's input to the SSP.

But six months later, Tommy Gold would publish his paper "The Deep, Hot Biosphere," in the *Proceedings of the National Academy of Sciences*, outlining his evidence for a deep subsurface biosphere. The term "biosphere" refers to the region of the Earth where life is able to exist regardless of whether it is present or absent and was proposed by French naturalist Jean-Baptiste Lamarck (1744–1829). The term became more widely used after publication of the book *La Biosphere* in 1929 (the French edition of a 1926 Russian book) by Vladimir Vernadsky, a Russian geochemist. Gold cited an obscure Swedish technical report written by a K. Pedersen, an unknown at the time in the world of microbiology, but not citing any of the papers that had been published by DOE's SSP or by USGS. Apparently the SSP work had equally little influence on Tommy Gold's inductive thought processes as well. Gold even went so far as to suggest that life could have originated in the deep subsurface.

17. In the opinion of this author (and I could be wrong) Frank was in many respects like Charles "Swede" Momsen Jr., who was chief of Undersea Warfare at the Office of Naval Research and was described in William Broad's book, *The Universe Below*: "Momsen conducted a one-man war [to keep the Navy involved in deep-sea research] that bent rules, twisted arms, ignored directives, and spent money" (p. 56). Frank would employ some of the same tactics to keep the deep subsurface biosphere research alive in the U.S. DOE during the early days of the SSP. It was this same deep conviction on the part of both men in the right positions that led to pivotal moments in the history of the exploration of the deep biospheres of the oceans and the Earth's crust.

18. See chapter 2.

19. An interesting foreshadowing of the name that David Boone would give his unique *Bacillus* species that he would culture from the Thorn Hill samples.

20. As part of the agreement, Texaco and DOE signed off on liability clauses that removed legal responsibility for damage to each other's activities, and all of Frank's investigators going to site had signed confidentiality agreements that included exemptions for scientific publications.

21. Tommy had been part of a four-member team for the Environmental Management Office at the U.S. DOE that examined advancements in horizontal drilling technologies for use in environmental restorations. This office was at times competing with the DOE Office of Biological and Environmental Research, which supported the SSP, but in later years they collaborated.

22. The SRP was built in the early 1950s and managed by DuPont for a cost-plus-one-dollar operating contract. It ran five heavy water reactors spread over 300 square miles to produce enriched uranium and plutonium for nuclear weapons and heavy water for nuclear reactors. The

neutrino, an elementary particle that plays a role later in this book (see chapter 7), was discovered at SRP in 1956. In the mid-80s, SRP had started producing plutonium-238 for NASA's deep-space missions, for example, Cassini-Huygens. Well over 10,000 drums of trans-uranic waste produced during the Cold War now reside at the Waste Isolation Pilot Plant near Carlsbad, New Mexico (see chapter 2).

23. At the time, Jack Corey was deeply involved in the environmental cleanup program of the M Area, a settling basin in the SRP, under a recently launched DOE program. The settling basins were probably leaking contaminants into the underlying groundwater aquifers, and the scientists had been drilling a series of monitoring wells to determine the extent of that contamination plume and whether it would threaten the water supplies of communities down-gradient to the southeast, for example, Savannah, with its 100,000 residents. However, they were concerned about going deeper than 100 feet for fear of finding the deepest aquifers contaminated. What if trichloroethane (TCE) from the SRP was 200 meters beneath the surface in a drinking-water supply aquifer and slowly migrating toward Savannah? TCE is a degreasing agent that was commonly used after carbon tetrachloride was banned in 1970. TCE was linked to birth defects and subsequently banned in California in 1986 because of concerns regarding the drinking water. What could they or the DOE possibly do? Drive a 600-foot-deep weir wall surrounding SRP to keep the water from moving off the site? This would take at least $100 billion, four times the cost of the Apollo lunar landing program, and they were just one of many DOE labs, each with its own groundwater contaminant plumes. Jack had been recovering from a stroke and was pondering if his legacy to his kids and grandkids would be contaminated drinking water. Frank's proposition intrigued him. They were into their third phase of a baseline hydrogeological study that could involve drilling holes down to 1,200 feet. Frank was claiming that the bacteria down there could potentially be stimulated to clean up the toxic organic compounds. Even if they only removed 1% of the TCE, it could potentially save DOE $1 billion in clean-up costs. It was just possible that by cutting some corners on the well installations, he could spare a couple hundred thousand dollars to get Frank's cores.

24. The participants were given everything they needed to know about the geology and hydrology of the SRP from Horace Bledsoe, Van Price, and Carl Fliermans, all members of Savannah River Laboratory's environmental sciences group. Sitting in the audience were David Balkwill, Bill Ghiorse, and Joe Suflita, the microbiologists who had analyzed very

shallow aquifer sands with support from the U.S. Environmental Protection Agency (EPA). Also present were Eugene Madsen and Jean Marc Bollag, soil microbiologists from Penn State who had hosted a seminar the previous May that led to the formation of Frank's nascent Microbiology of Subsurface Environments group. The following morning the group drove around to the various sites at SRP and returned to the meeting room for a roundtable discussion of what should be done, how it would be done, where it should be done, when it could start to be done, and who would be doing it. Frank had outlined the previous night that he saw this as a two-phase exploration, with the first phase lasting only twelve to fifteen months and coinciding with the installation of what was referred to as the P-series wells. For the second phase he dangled the possibility of a long-term subsurface site with multiyear funding. The target zone for water sampling had been the Tuscaloosa aquifer, the regional drinking water supply, but the microbiologists wanted core samples from the surface down to and through this zone and from three different locations, making sure to collect the same formations at each, a classical microbial triplicate sampling approach. But for microbiologists who up to this point had seen samples from a depth of only eight meters, working on samples from 400 meters depth would be like working on samples from the moon! Approximately eighteen months earlier, Carl Fliermans and D. C. White had begun examining phospholipid biomarkers in SRP subsurface sediments. Under Carl's guidance, Tommy's postdoctoral research with D. C. White included retrieving subsurface core samples for the lipid analyses from several monitoring-well installations at SRP in late 1984. Those were samples of opportunity with no consideration as to contamination from drilling. Tommy had months of drilling experience at SRP under his belt, and he had figured out how to obtain the least contaiminated samples for this new campaign. At the meeting, a discussion followed his presentation about how to trace contamination from the drilling process. For this Phase I coring, they would measure the naturally occurring ionic species in drilling fluids and compare them to those of the groundwater and the fluids extracted from the cored materials. This exhaustive geochemistry effort was to be performed by the experienced field scientists from PNL. During the give-and-take, a laundry list of analyses was assembled that included radiolabeled aerobic and anaerobic activity measurements, cell counts, PLFA analyses, geochemistry, and carbon utilization.

25. D. C. White was in the process of moving his laboratory to ORNL and the University of Tennessee–Knoxville as one of the first of the jointly

appointed Distinguished Scientists for UT–Knoxville and ORNL. The current iteration of similar joint appointments is called the Governor's Chairs. Recent recipients include Terry Hazen and Frank Loeffler, who were SSP investigators.

26. Besides enriching plutonium, the SRP maintains vegetable farms within its borders on which are grown watermelons, corn, soybeans, and collard greens that are analyzed yearly for radionuclides.

27. This is not entirely true. In certain aspects, the SSP investigators, unbeknown to them, were following in the footsteps of Junta Sugiyama, a graduate student specializing in fungi and yeast from the Department of Botany at the University of Tokyo. In the mid-1960s he published a paper based upon part of his master's thesis research on the mycology of cores collected by rotary drilling of Late Tertiary sediments down to a depth of 3.7 km and temperatures of 80–90°C (Sugiyama 1965). The analyses of the drilling mud, the quick chilling of cores with ice bags, and processing in the laboratory by shaving the outer portions of the core under sterile conditions that he used were essentially the same procedures being used by SSP, minus the tracers and the anaerobic processing.

28. Collecting solid, intact cores was and still is an art. It requires just the right amount of pressure on the hundreds to thousands of feet of drill pipe, just the right rotation rate, and just the right flow rate for the drilling fluid. The drilling fluid lifts all the dirt, or "cuttings," from the hole and deposits them into a huge metal settling tank on the surface.

29. These tools—a Denison corer, Shelby tube samplers, a Pitcher barrel, and a phosphate barrel—took the approach of having the tube, or "boot," projecting down below the coring bit so that when the driller hit bottom of the hole, the tube would penetrate several centimeters into the formation before the driller started turning the bit above it. This approach helped to prevent drilling mud contamination of the core. The core would then slide up into a three-foot-long, steam-cleaned coring tube. This tube separated the drilling mud from the sediment as it flowed past down to the bit, picked up the cuttings and flowed upward outside of the core barrel and the drill pipe to the surface. The driller above kept track of the depth of penetration as he cored, to make sure he filled the tube, and then stopped the drilling and start pulling the core barrel up, hoping that the sediment remained inside the core tube. In sandy formations, a "core catcher" in the shoe of the Denison and Pitcher barrel coring tools held the core in place until it reached the surface. Phelps et al. 1989.

30. Susan Pfiffner had just finished her MS at the University of Oklahoma studying under the microbiologists Mike McInerney and Joe

Suflita and was well trained in the art of growing bacteria from field samples.

31. They used ethylene-oxide sterilized gloves, which they frequently changed, over the glove-bag gloves to reduce cross contamination.

32. Experiments were generally initiated in the respective laboratories within seventy-two hours of cores arriving at the surface. Subsamples were used by PNL investigators at the SRP lab for pore-fluid extraction, after which the geochemistry was measured and contrasted with the drilling fluids. These geochemical assays could detect 0.1% of drilling-fluid contamination within samples, but results were not available for several weeks, long after investigators had started their experiments. Although most samples proved to be of high quality (weeks later), adequate on-site quality assurance and quality control (QA/QC) were severe deficiencies in the rapidly implemented 1986 field campaign.

33. For those readers too young to have seen the TV show *MacGyver* (1985–1992), it was about a scientist/secret agent adept at saving the world with duct tape, a Swiss army knife, and an assortment of paper clips and bobby pins.

34. Even before the results from these samples were known to him, Frank spent the summer mapping out a research program on deep microbiology involving twenty-six categories of tasks that needed to be addressed and supported over the next five years. The spectacular abundance and diversity begged the question of whether contamination had occurred despite the best efforts of the DOE team. Joe Suflita reported the presence of enteric bacteria and bacteria that were resistant to penicillin in some samples, but not in all samples and not in the drilling mud. Nonetheless, their presence was surprising given that penicillin resistance was universally thought to be a recent attribute that developed after the widespread use of penicillin. It was left to Ray Wildung of PNL to play devil's advocate and criticize not only the results of the drilling campaign, but also Frank's proposed Deep Microbiology Program and its plan. Ray was an extraction chemist by training, had worked with transuranic elements, had developed a dosimeter, and had been at DOE and its predecessor, the AEC, longer than Frank had. New sampling technologies, certainly, and new sites, undoubtedly, would be required to confirm these stunning findings. Other scientists attending the meeting talked about their own sites in Nevada and Texas and about applying these tools to characterize the subsurface microbiology.

35. "Vadose zone" refers to the ground (rock and soil) that lies above the water table through which rainwater infiltrates to reach the water table and recharge the groundwater in aquifers.

By the mid-1970s, emerging groundwater quality issues stimulated scientists of the EPA to reevaluate the possibility that microorganisms inhabited water-yielding rock formations (aquifers). The scientists recognized the potential for microorganisms to modify the chemical properties of groundwater, but they also recognized that detection of truly subsurface microbial inhabitants was obfuscated by the contaminants introduced during the drilling of water wells. They realized that bacteria tend to be attached to the surfaces of mineral grains. How then could they be certain that the microbial composition of groundwater samples was representative of the in situ microbial communities? To make matters worse, the bacteria colonizing the surfaces of the submerged well casings may have been introduced during the drilling and casing of the well itself. Fed by nutrients from the groundwater, they could proliferate, much like the entire ecosystems that form around the pilings of oil platforms in the Gulf of Mexico. How could the scientists be certain that the bacteria recovered from groundwater samples bore any resemblance whatsoever to any bacteria that may be living in the aquifer sediments? McNabb and Dunlap (1975) pointed out the difficulty in obtaining uncontaminated subsurface samples. Their paper is a summary of a more detailed report submitted to the EPA on the potential of microbial activity in the subsurface and its possible role in remediating groundwater pollution (Dunlap and McNabb 1973). This report contains an extensive list of previously published studies of subsurface microbiology.

36. "Depth is another secondary ecological variable that affects the bacteria. In temperate zones, these organisms are almost all in the top meter, largely in the upper few centimeters" (Alexander 1977, p. 23).

37. Frank invited outside experts from different universities and DOE laboratories to critically evaluate the unpublished results of the P wells and the methods used to acquire them and to make recommendations on how to improve those methods in preparation for the "fourth hole." By the end of this planning workshop/review the recommendations were clear. They had to invest more time, money, and thought into the "tracers," the term now coined to describe the additives to the drilling system that would serve as an indicator of potential microbial contamination in the samples. They had performed some tests using rhodamine dye that suggested the latter could be added to the large volume of drilling mud to effectively trace its penetration into the cores. Gunnar Senum, a chemist from Brookhaven's Tracer Technology Center, offered that their PFC tracers, developed for atmospheric tracing, would be more sensitive than the rhodamine dye, but they had never been applied to soft sediments. Bill Ghiorse advocated strongly for physical

tracers like the fluorescent microbeads that the USGS was using on Cape Cod to track bacterial migration, or perhaps even specific bacterial strains that could be used in the drilling muds. John Rowley, an engineer from Los Alamos National Laboratory (LANL), recommended looking into wireline coring tools that would not require the drill pipes to be removed every time a core was collected and would shorten significantly the time between when the driller had stopped coring and when the core was being dissected in the MMEL. Finally, determining the location of the "fourth hole" would require a more comprehensive analysis of the hydrology and groundwater ages around the SRP. At the end of the meeting Frank broke the assembled participants into teams to provide a solution to each of these issues in time for drilling to be underway by December '87, "10 months from now." He also charged John Zachara of PNL to develop a scientific rationale for extending deep subsurface microbiology to other DOE sites, particularly those out in the arid west.

38. The Christensen engineers worked diligently with the drillers, SRP, and Tommy and crew to modify the C-P wireline coring shoe so that cores were actually recovered in front of the drill bit rather than behind it as typical of rock coring. Shoes at or behind the bit allowed faster coring and penetration rates but facilitated mixing of drilling fluids within the cored sediments. A series of extended shoes were designed and tested that cored one to three inches in front of the drill bit, dramatically reducing drilling fluid contamination of cores. Extended coring shoes remain as an optional coring accessory today.

39. "Gradient" in this case refers to the hydraulic gradient. The groundwater was flowing in the direction of the first borehole and away from the borehole where the microbial samples would be collected. The first core was taken to identify the target depths for microbiology since it was logistically impossible to perform microbial analyses on all the cores samples collected from 1,500 feet of core, that is, 300 cores.

40. Graves Environmental and Technical Services Company (Graves ETSC) had bought Profession Services, Inc. (PSI), which had been doing the environmental drilling for SRP, including the coring on the P-series wells.

41. In the SSP drilling projects subsequent to C10, the magnetic microspheres were replaced by yellow-green, micron-size fluorescent beads. Whenever these microspheres were observed in inner portions of a core (examined by fluorescence microscopy) it indicated that drilling fluid and any microorganism-sized particle from the drilling fluid had penetrated the core, a finding of contamination.

42. Coliform bacteria are rod-shaped, gram-negative, non-spore-forming bacteria that can ferment lactate at 35–37°C. They are easy to grow on standard media and, as such, are used to assess water purity.

43. The idea of using specific bacteria in drilling muds was neither new nor outlandish. Smirnova, a Russian microbiologist, during the 1950s seeded his drilling muds with *Serratia marcescens*, a red-pigmented bacterium that was easily visualized in the lab. He tested his microbe-spiked drilling muds on 75-millimeter-diameter cores of sand collected from 3 to 30 meters depth, of clay collected from 70 to 200 meters depth, and of dolomite collected from 100 to 400 meters depth. The more porous the rock the greater the depth of penetration of *Serratia marcescens*. The porous sands were completely penetrated by *Serratia marcescens*; whereas for the more endurated, that is, tightly cemented, low-porosity dolomites, sandstone, and claystones, *Serratia marcescens* penetrated only 1 to 7 millimeters. Smirnova 1957.

44. In 1950, Claude ZoBell with the help of Shell Oil strove to drill an oil well aseptically to two and a half miles in depth. The objective was to determine whether indigenous microbial bacteria were responsible for the corrosion of the oil-well casings that had been observed in oil fields. During the drilling campaign he added alkaline minerals to the drilling mud to raise the pH to 11.5 (very alkaline), to deactivate any soil microorganisms that might be present in the drilling mud. He also applied germicides and used blowtorches to sterilize the drill rods and casing as much as possible. Since all microbial assays at that time were culture based, as opposed to the culture-independent DNA analyses commonly practiced today, this was an effective strategy. They failed to detect any mesophilic, aerobic heterotrophs from cores collected at 4,265 and 14,435 feet, which you might expect to be surface contaminants, but were able to culture viable sulfate-reducing bacteria.

45. Rhodamine WT dye (WT for "water tracing") is a soluble organic salt that fluoresces in the ultraviolet (UV) region of the spectrum. It was cheap and easy to detect, though weakly mutagenic.

46. Dick Coy has just founded a small company in Michigan specializing in anaerobic glove bags and boxes. The C10 glove bag began a twenty-year relationship of specialty anaerobic chambers for subsurface biogeochemical investigations.

47. The geological stratigraphy of the coastal plain sediments of South Carolina near the SRP from the surface to depth is as follows: (1) Tobacco Road Formation, Eocene Epoch; (2) Dry Branch Formation, Eocene Epoch; (3) McBean Formation, Eocene Epoch; (4) Congaree Formation, Eocene Epoch; (5) Pee Dee Formation, Upper Cretaceous Period; (6) Black Creek Formation, Upper Cretaceous Period; (7) Mid-

dendorf Formation, Upper Cretaceous Period; and (7) Cape Fear Formation, Upper Cretaceous Period. Cape Fear sits unconformably on top of the Triassic Period sedimentary rock of the Dunbarton Basin and Paleozoic igneous and metamorphic rock. This is from Sargent and Fliermans 1989.

48. As articles began to appear in *Newsweek* and *Omni* magazines, Frank had begun to realize that the public was finally becoming intrigued with those weird DOE guys looking for life deep underground.

49. At an investigators meeting six months later in Tallahassee, Florida, the core analyses indicated that the extra effort on QA/QC at C10 had paid off and the insurance factor increased from 10^5 to 10^6. By this criterion, only some of the samples selected for detailed analyses were uncontaminated, and yet, these samples still contained a diverse community of aerobic heterotrophs, much like their stratigraphic equivalents in the P-series wells. The tens of millions of active heterotrophic bacteria found per gram of sediment were based on what could be grown in the lab, which meant that they represented a potent force for remediating polluted aquifers if stimulated with the right substrates. But this remarkable discovery raised a question. Was the massive and diverse community found in the subsurface sediments at SRP a fluke? Had the on-site well drilling and pumping activities at SRP somehow enriched the subsurface populations? Or had the scientists just been lucky in their first probe into the subsurface biosphere? The rich microflora present in the Cretaceous sediments may not be present at the other contaminated DOE sites out West, where the surface environments might be more austere. Everyone was in agreement with Frank on this point, which was a good thing because at the end of the meeting, he rolled out a preliminary plan for how they would transition the Deep Microbiology Program from SRP to other DOE field sites in the next year.

After the meeting was over, Tommy and Susan met John Zachara and Jim Fredrickson for dinner in a restaurant across the street from the biology department. Jim had been at PNL for less than a year before Ray Wildung and John Zachara sucked him into the SSP and the P-series research. Frank was always looking for young scientists who would be devoted to his program and upon whom he felt he could rely for technical productivity. Jim fulfilled both. He was a good communicator, both orally and in writing, and he could represent Frank's program at a technical level on microbiology that Frank could not. Frank appreciated Jim's candidness and the ability to banter with him. It seemed that both John and Jim could always joke openly with Frank and poke fun at aspects of his personality. Frank would reciprocate in spades. This gathering was Jim's first opportunity to meet Tommy and

Susan. Although Jim had worked on the P-series and C10 cores, he had never been in the field to help collect them and had written several papers without acknowledging Tommy at all. That had irked Tommy since Frank had promised their support with the field sampling. Tommy was in one of his less-than-congenial moods, complaining about the fact that not only were they overworked at C10 but that he was still out of pocket for the money he spent on the site. It had been six months and Frank still had not reimbursed him for the $10,000, and now he was talking about moving on to more drilling operations out West, probably at Hanford, John and Jim's home base. John was skeptically incredulous. "What do you mean out of pocket? Which pocket did you use to pull the $10,000 from?" he asked, thinking Tommy was referring to some other DOE grant. Tommy had had it. He pulled out his wallet. He slammed it on the table in front of them. He pointed at the wallet. "That pocket!"

50. The traveling block is the hanging motor that rotates the bushing that connects the motor driver above to the top of the drill pipe below.

51. Will Happer is the Cyrus Fogg Brackett Professor of Physics at Princeton University and who between 1991 and 1993 was director of DOE's Office of Science. He pioneered the development of adaptive optics in the 1970s and more recently has been a vocal critic of the claims made by the global-warming community.

52. Deuterated water is water enriched in deuterium, a heavy form of the hydrogen atom, so the formula for deuterated water would look like HDO rather than H_2O.

53. The sidewall coring tools are also much faster than conventional coring, which is a significant factor given the $50,000+ per day operational costs of a Murco 54. In addition to the PFC tracers, they had considered using tiny, bacteria-sized, fluorescent beads, but the Texaco geologists didn't want the fluorescent beads interfering with their analyses of the cuttings, the pea-sized chips of the rock formation that were carried up from the bit and filtered from the mud before it returned to the mud tank. Texaco's geologists scrutinized the cuttings underneath a field microscope so they could identify what formation they were drilling and to see if they would fluoresce under UV light. The aromatic hydrocarbons found in trace quantities associated with the oil, not the oil itself, fluoresce under a UV light.

54. Chapelle et al. 1987.

55. Methanogens are *Archaea* (not bacteria; no known bacteria produce CH_4, yet) and produce CH_4 using several different pathways and CO_2, carbon monoxide (CO), formate, acetate, methanol, trimethylamine,

and even dimethylsulfide. One particular metabolic reaction that figured prominently in the exploration of terrestrial subsurface life and in speculations about Martian subsurface ecosystems was the following reaction, utilized by autotrophic methanogens: $4H_2 + CO_2 \rightarrow CH_4 + 2H_2O$.

56. The committee is affiliated with *Bergey's Manual of Systematic Bacteriology*, which has been the main resource for defining prokaryote taxonomy. It was initially based upon physiological and morphological attributes but now incorporates genetic data. David Hendricks Bergey, a bacteriologist who led the hygiene laboratory at the University of Pennsylvania in the early 1930s, first published the manual in 1923.

57. Frank first began his SSP program with $40,000 in 1984. From this paltry sum he supported the microbiologists David Balkwill from FSU and Bill Ghiorse from Cornell University, who with prior support from the EPA had been publishing papers reporting on what they considered indigenous bacteria down to eight-meter depths in a shallow aquifer near Lula, Oklahoma, and other Plio-Pleistocene aquifers in the Midwest (Wilson et al. 1983; Balkwill and Ghiorse 1985). Frank had also supported Joe Suflita, a young microbiologist from the University of Oklahoma, who specialized in the biodegradation of noxious organic compounds by anaerobic bacteria.

58. Thousands of isolates had begun to fill David Balkwill's $-80°C$ freezers, officially recognized and supported by Frank as the SMCC . The results indicated that most of the isolates could be placed in well-established taxonomic groups (e.g., previously described genera), but many were new species within those groups. Moreover, some of these new species possessed characteristics not previously associated with the taxonomic group in which they were placed. The commercial sector was especially interested in the preserved cultures because they were a source of novel genetic information and of microbial forms that have not yet been screened for commercially useful properties. Of the small percentage of the organisms in the SMCC that had been examined in detail, a surprisingly high proportion had been found to possess potentially valuable metabolic capabilities. Examples of commercially useful metabolic traits included the ability to degrade toxic organic compounds (bioremediation of environmental contaminants) and to produce a variety of compounds: extracellular polymers such as polysaccharides (food additives, surfactants, and other applications), secondary metabolites (such as antibiotics of great interest to pharmaceutical and biotechnology companies), and novel metabolic products (e.g., a blue pigment-producing bacterium from the INEL borehole that could have applications as a

dye in the textile industry). Some showed the ability to bioaccumulate metals (bioremediation) and to function at high temperatures (producing heat-stable enzymes). One major pharmaceutical company had already screened 3,200 subsurface bacteria for the production of new antimicrobial products, and a major biotechnology company had screened 2,000 isolates for production of useful secondary metabolites. Other companies began requesting additional information about the strains in the collection before applying their own tests. Frank therefore decided to initiate an effort to characterize more fully a representative subset of the SMCC isolates while passing memos up the chain of command at DOE headquarters about technology transfers to the inquiring pharmaceutical firms.

59. Onstott et al. 1998.

60. Boone et al. 1995.

61. The bacterium premiered in a 2008 remake of the 1971 classic sci-fi film *Andromeda Strain*. In this version, *B. infernus* saves the planet (or at least Los Angeles) from the deadly Andromeda strain virus. *B. infernus* even figures in a hot and steamy romantic treadmill scene (thankfully the only one in this movie). Sadly, in the movie, the origin of *B. infernus* was attributed to a deep-sea vent, not to a buried Triassic rift basin. We cannot be too critical of the lack of fact-checking for this movie, given it was produced by Ridley Scott, one of my favorite producers. But we would like to encourage any producers to make more science fiction movies featuring subsurface life forms.

62. In microbiology the term "facultative" means "optional." Thus a facultative anaerobe is a microbe that can live as an aerobe, that is, use O_2 as its electron acceptor, or it can live without using O_2 and use an alternative electron acceptor. Until the discovery of *B. infernus*, that alternative electron acceptor for *Bacillus* was nitrate. The term "obligate" means "required." An obligate anaerobe is a microbe that cannot live in the presence of O_2. As far as we know, methanogens are obligate anaerobes.

63. The first speaker, Zachara from PNL, focused on organics and biodegradation, a topic about which I was totally ignorant. The only thing I could think of was maybe I could $^{40}Ar/^{39}Ar$ date some volcanic clasts from the "GeMHEx" core to which he kept referring. The $^{40}Ar/^{39}Ar$ dating method, which originated in the 1970s, is based upon the potassium-argon dating method, or K-Ar, which originated in the '50s. Potassium is mainly made up of one stable isotope, ^{39}K, but does have a radioactive isotope, ^{40}K, which has a half-life of 1.25 billion years. The daughter isotope formed by the decay of ^{40}K is ^{40}Ar. Argon-40 is stable and a noble gas, and it is the principal constituent of Ar in our atmosphere. To date a mineral or rock using the K-Ar technique, you split the min-

eral or rock in two, measure the potassium concentration of one piece and the ^{40}Ar concentration of the other piece, and calculate the age. In the ^{40}Ar/^{39}Ar dating method, you irradiate the mineral or rock with neutrons in a nuclear reactor. The neutrons collide with ^{39}K to create ^{39}Ar. You then take the irradiated mineral or rock piece and just measure the ^{40}Ar/^{39}Ar to determine the age. The big advantage of this approach is that it permits you to obtain an age by heating the mineral or rock with a laser beam, thereby reducing the atmospheric contamination. You can even map ^{40}Ar/^{39}Ar ages in thin sections.

64. The afternoon talks started with one on microbial transport and geophysical imaging by a geophysicist, Ernie Majer from Lawrence Berkeley Laboratory. He was followed by female speaker Ellyn Murphy, from PNL, whose talk on using isotopes to track groundwater flow and microbial activity fascinated me. She seemed so interdisciplinary. The last presenter, Ray Wildung, was a sage-sounding scientist, again from PNL, talking about the Environmental Science Research Center (ESRC).

65. Ellyn Murphy was a geochemist who had just started working at PNL when she started helping with the GeMHEx proposal. She had completed her Ph.D. with Stan Davis at the University of Arizona, where she used the Accelerator Mass Spectrometer (AMS) laboratory there to date groundwater using radiocarbon, ^{14}C, and was incredibly knowledgeable about how to constrain the subsurface residence times for groundwater with isotopic techniques. The AMS uses a small linear accelerator to send high-energy ions onto solid targets, in this case carbon, and deliver sputter ions, in this case of ^{14}C, to an ion-counting mass spectrometer. This method is far more sensitive than the standard approach for analyzing ^{14}C, which is to count the low-energy β particles slowly emitted by ^{14}C. The result is that it requires far less sample, which is important when you are dating artifacts like the Shroud of Turin. In 1989, the AMS facility at the University of Arizona had solved a centuries-old mystery by determining a medieval age for the Shroud of Turin, confirming that it was not an image of Jesus Christ but a fake.

66. George F. Pinder was the chairman of the Department of Civil Engineering at Princeton University before he moved to the University of Vermont. His role in Woburn, Massachusetts's groundwater contamination litigation is featured prominently in the book and movie, *A Civil Action*.

67. Among the types of isolates discovered in the P-series well samples on the SRP were aerobic methanotrophs. These types of bacteria oxidize CH_4, using O_2 as their electron acceptor, and produce CO_2. They are greedy when it comes to incorporating the CH_4 and CO_2 into their biomass, gobbling up to 50% of it, so they grow quite quickly. But their

value in remediation was recognized in the early 1980s for degrading TCE and chloroform. The enzyme these bacteria use to oxidize the CH_4 also reacts with TCE, forming a deactivated but stable protein structure. This is not good for the methanotroph, but as long as they continue to grow, the more the soluble TCE they mop up and turn into an insoluble sludge of methanotrophs. Soon after the P-well samples had been collected, Tony Palumbo, a research scientist at ORNL, and Dean Little, an FSU graduate student doing his research at ORNL, isolated a methanotroph in the lab and demonstrated that it degraded TCE. But how could they get them to grow in the subsurface aquifers of SRP? During their investigations of the P-well core samples, Susan, Tommy, and D. C. White discovered that the principal limitation to viable subsurface ecosystems was the low availability of phosphate. This deficiency limited the growth of biomass and thus the total activities in many subsurface aquifers within the eastern coastal plain. The reason was that phosphate was bound tightly to clay minerals. Although additions of phosphate alleviated nutrient limitations in test-tube experiments, this was not feasible for larger-scale subsurface implementation because the phosphate anion was too sticky and was difficult to disperse. Soon thereafter Terry Hazen led an SRP-integrated demonstration using DOE environmental management support (EMSP) money in which CH_4 was injected to stimulate biodegradation of chlorinated solvents. Phelps, Pfiffner, Hazen, Looney, and others complemented the CH_4 injections with the development of a gaseous triethylphosphate (TEP), which they called phoster, as a means of dispersing the phosphate with the CH_4. Once injection strategies were developed for CH_4 and TEP, it was an insignificant cost to additionally add ammonia to alleviate any nitrogen limitation. Using horizontal drilling technology, Terry's team was able to drill beneath the TCE plume at SRP. They then bubbled in CH_4/TEP/ammonia gas through the horizontal screened well, where the compounds migrated upward and were consumed by methanotrophs. A dramatic growth of methanotrophs occurred. The more they grew, the more the TCE declined, until it reached a level that it no longer left a threat down-gradient outside the SRP. The plan had worked. The CH_4/TEP/ammonia injections added about an 8% additional cost over conventional air stripping of the chlorinated solvents, while providing more than 30% more removal of the chlorinated solvents. Gaseous nutrient injections with Phoster have subsequently been used by numerous bioremediation practitioners at hundreds of sites. The Hazen-led SRP integrated demonstration was the first field-scale bioremediation project, and its roots derived from Frank Wobber's SSP program. Good to his word, Frank had delivered a solution to Jack Corey's concern and saved DOE a ton of money. Phoster resulted in a 1996 R&D 100 Award, one of several derived from SSP research.

68. These rod-shaped objects that looked like bacilli were found within fluid inclusions in quartz that had been trapped at temperatures above the current boiling point (Bargar et al. 1985). Fluid inclusions are microscopic cavitites inside minerals that trap tiny volumes of the fluid from which the mineral, in this case quartz, is precipitating. The diameter of fluid inclusions can range from less than a micron to tens of microns. The inclusion may form because of a defect in the crystal lattice or because of minerals, organic matter, or even perhaps bacteria adhering to the growth face, and then the crystal grows around them. In this case, hydrothermal water rising up toward the surface is saturated with dissolved SiO_2, and as the water cools, the silica becomes less soluble and the quartz precipitates rapidly. The fluid inclusions represent time capsules of the hydrothermal fluid and any objects it contained at the time the crystal grew around it. The rod-shaped objects, if they were bacteria, would have been coated with silica as the fluid inclusion cooled further and the silica precipitated out of solution.

69. Norton and Grant 1988.

70. Sidow et al. 1991.

71. Zvyagintsev et al. 1985.

72. In 1992, Raul Cano showed that he could recover intact bacterial DNA from insect inclusions in 15-million-year-old amber. The care with which he avoided contamination and the molecular evidence seemed to point to preservation of Miocene DNA. His evidence flew in the face of well-established chemical arguments as to the rate of DNA degradation in natural environments. His success spawned the Michael Crichton book and subsequent movie *Jurassic Park*. Later, in 1995, he would recover a viable bacterium from an amber inclusion (Cano and Borucki 1995), a stunning testimony to the endurance of bacteria completely isolated from exogenous nutrients. A comprehensive review of potential ancient, preserved microorganisms was published by Max Kennedy a couple of years later (Kennedy et al. 1994).

73. The gene for 16S rRNA (ribosomal RNA) is essentially the genealogy gene. Analysis of 16S rRNA gene sequences was a powerful new method for determining their taxonomic relatedness to previously described species. The idea that the DNA of living microorganisms holds a record of evolution, a genealogy gene, was first proposed by Emile Zuckerkandl and Linus Pauling (Zuckerkandl and Pauling 1965). But the question was, which genes evolved slowly enough and were present in all known forms of life? Then in 1972, two brothers, Steve and Mitch Sogin, working in Carl Woese's laboratory at the University of Illinois,

proposed that the 5S ribosomal RNA gene be used as a record of the evolution of prokaryotic life (Sogin et al. 1972). At that time, experimentally determining the nucleotide sequence of any strand of DNA was a tedious job involving two-dimensional gels and the use of radioactively labeled ^{32}P, but the 5S subunit of the ribosomal gene had only about 120 nucleotides and so was experimentally tractable. By 1977, the Woese lab had moved on to sequencing the 16S rRNA gene and had recognized that the 16S rRNA gene sequence of methane-producing bacteria was very different from that of all other bacterial isolates they had sequenced (Balch et al., 1977). In 1987, Carl Woese published a tree of life using the 16S rRNA genes and showed that the prokaryotes composed two separate kingdoms, which he called *Eubacteria* and *Archaea* (Woese 1987). In 1988 the National Institutes of Health, with funding from Congress, created the National Center for Biotechnology Information (NCBI), where the partial or complete 16S rRNA gene sequences were deposited. By the early '90s most microbiology laboratories had become proficient in sequencing the 16S rRNA gene of microbial isolates. The NCBI database began to act like a global fingerprint file that could be easily accessed by any police department on the planet for checking to see if a culprit had a prior arrest. Basically, a 16S rRNA sequence could be compared against the NCBI database to see if a microbe's species had been discovered elsewhere on the planet. Whether that sequence matched one in the database was determined by the Basic Local Alignment Search Tool (BLAST).

74. Restriction fragment length polymorphism is an inexpensive means of classifying the ribosomal RNA gene amplified from a microorganism's DNA, without the cost of sequencing it. rRNA probes are fluorescent dyes that contain a short fragment of rRNA that is complementary to that of rRNA of a ribosome. After permeabilizing a cell's wall, these probes penetrate the ribosomes to bind to a target region of the 16S rRNA of a ribosome. This provides a means of counting the number of cells that belong to a particular taxonomic grouping. Intergenic spacers (introns) are random bits of DNA that appear to have been inserted into protein-coding genes, which is what Sandra was talking about.

75. Girard and Onstott 1991.

76. Ray Wildung was a senior scientist at PNL's Environmental Science Research Center and worked closely with Frank, providing both sage advice and critical assistance in implementing the field research projects. He was our silver-haired Obi-Wan Kenobi.

77. The gas window is the temperature range over which sedimentary organic matter is thermocatalytically altered to CH_4. This temperature range is typically around 150–200°C.

78. The cell walls of bacteria and archaea are composed of a mixture of sugar and protein. Those with thick peptidoglycan layers are gram-positive, and those with very thin peptidoglycan layers are gram-negative bacteria. It was becoming apparent from studying the isolates from the DOE sites that gram-positive bacteria dominated the very deep subsurface collection. Hans Christian Gram, a Danish bacteriologist, developed the Gram staining method in 1884. Two stains are applied to a smear of bacteria on a glass microscope slide. The first stain is crystal violet, a dye that penetrates the cell wall and membrane of all the bacteria, turning them a very visible purple. Iodine, which reacts with the crystal violet to make it larger and less mobile, is added. The iodine is followed by acetone, which dissolves the outer membrane of gram-negative bacteria, releasing the crystal violet/iodide complexes but does not completely bleach the thick peptidoglycan layer of the gram-positive bacteria. This decolorization step is followed by a counterstain of safranin, a cyclic nitrogen and carbon chloride salt, which then binds to all the bacteria but colors the decolored gram-negative bacteria red. The protocol for Gram staining can be found readily on the Internet.

CHAPTER 2: THE TREASURE OF CERRO NEGRO

1. From a letter of March 29, 1951, from Claude ZoBell to Paul Tasch in response to a question about the viability of ancient bacteria. Quoted in Reiser and Tasch 1960.
2. *Desulfotomaculum putei* (Liu et al. 1999). This paper reports the isolation of a thermophilic sulfate-reducing bacterium from the same sidewall core samples as *B. infernus*.
3. As mentioned earlier in note 78 of chapter 1, as minerals grow, defects or adsorbed objects on the crystal surfaces can lead to voids that trap the fluid in which the mineral grows. If this growth occurs at very high temperatures, where the density of water is lower than it is at room temperature, then as the mineral cools the trapped water shrinks, leaving behind a vapor bubble and producing two phases in the fluid inclusion, liquid water and water vapor. Typically in a thin section the water-vapor bubble will bounce around due to the Brownian motion of the water molecules. In fact, Bargar's bugs bounced around inside the quartz fluid inclusions, although because of their mass they bounced more slowly than vapor bubbles do.
4. Hsin-Yi Tseng was a brilliant, ambitious graduate from Taiwan National University and a star student of my former, equally brilliant, graduate

student Chinghua Lo. Chinghua had recommended that Hsin-Yi attend Princeton for her Ph.D. to study $^{40}Ar/^{39}Ar$ geochronology. She had begun by studying radiation damage in feldspar during her first year. But after my meeting in Annapolis I sat down with her to convince her to change the direction of her thesis to work on the thermal history of Taylorsville Basin and the paleohydrology of sedimentary basins in general and how fluid inclusions could constrain the origin of subsurface microorganisms.

5. The participants included Keith Bargar of the USGS Menlo Park. At the meeting Keith showed us the remarkable bacteria-shaped forms (which we christened Bargar's bugs) bouncing around inside the quartz crystals found in drill cores from Yellowstone National Park. Keith pointed out that these inclusion bodies, as he called them, were only present in one out of twenty thin sections, and where they were present in a thin section, the inclusion bodies were found in only one out of twenty fluid inclusions. The frequency seemed about right compared with the concentrations of microbial cells in groundwater collected at the C10 borehole in South Carolina. But the temperatures of the fluid at the time of entrapment seemed to greatly exceed the maximum temperature at which life could survive.

6. Frank held an investigator meeting with David Galas, the director of OHER, and he wanted me there as the geologist reporting on the Thorn Hill no. 1 results. I had little to show other than the thin sections I had just obtained, so I thought, why not just show Galas the framboidal pyrites and fluid inclusions in the thin sections. Leaving early that morning with a Nikon petrographic scope in the back of my jeep, I made the three-hour journey to the Courtyard Marriott in Gaithersburg, Maryland, where I met the rest of the Taylorsville Basin group. We went over presentations on various aspects of the project, including a discussion by David Balkwill on the SSMC and how the Mars candy company had expressed interest in screening the several thousand isolates he was storing. His presentation pleased Frank immensely as further proof of the commercial potential of the culture collection he could exhibit to Galas. After driving over to the DOE headquarters, we, along with Frank, entered the main headquarters building, where getting the microscope through metal detectors and security turned out to be a nightmare. Once past security, we entered a wood-paneled auditorium, much smaller than the one I had been in about six months ago. Sitting in the middle of the room was Director Galas and an assistant who introduced himself as Ari Patrinos. Over the course of the next hour or so, the microbiologists presented some of the latest developments of the SSP, including the poster child of the Taylorsville Basin

project, *B. infernus*. I managed to squeak out my description of saline-tolerant, thermophilic, anaerobic, sulfate-reducing microorganisms as I showed Galas the tiny framboids in the microscope. Following that meeting and while we were waiting for the microscope to make its way back through security, we traversed down endless corridors painted gunship gray, passing tiny cubicle-sized offices with no windows until we finally arrived at Frank's office for a debriefing. By 2 p.m. I was back on the road heading home, but that would not be the last time I would make that commute.

7. At this time, the highest temperature known for microbial growth was 110°C for an archaeal species (Stetter et al. 1986).

8. Baross and Deming 1983. White also sent me a reprint of a paper he had just published with John Baross on the possible PLFA evidence for the existence of superthermophilic bacteria (Hedrick et al. 1992).

9. Carl Fliermans and Terry Hazen, another microbiologist working for the SRP at the time, had organized the conference with support from Westinghouse Savannah River Company, Graves ETSC, and Frank's program. The symposium was primarily a tribute to the efforts of those involved in Frank's program and an opportunity to showcase some of their results, including the new C10 data, in dozens of presentations. But the symposium attracted dozens of scientists from Russia, Finland, the United Kingdom, Germany, Denmark, Australia, France, and Canada. One of the attending Russian scientists was Mikhail Ivanov, who was by then a senior scientist in the new field of geomicrobiology. Mikhail, along with his colleagues Kuznetsov and Lyalikova, had led the Soviet studies into subsurface microbiology during the 1950s through to the '70s, when it had fallen out of vogue in the United States. Ivanov was accompanied by Sergey Beliaev, a noted Russian microbial ecologist experienced in petroleum microbiology. Scientists from the EPA and USGS, both organizations that had their own competing projects on subsurface microbiology, were also in attendance. Frank Chapelle, a hydrogeologist from the Columbia, South Carolina, office of USGS's Water Resources Program, had reported the presence of viable bacteria down to a depth of 130 meters in Maryland. His cores were extracted from sedimentary strata that were stratigraphically equivalent to those of the P-series wells, essentially scooping Wobber's SSP team and garnering him the USGS's Superior Service Award in 1988. At the time, Chapelle worked with a microbiologist from the University of Maryland, but now Chapelle had teamed up with an extremely talented USGS microbiologist from the Reston, Virginia, office by the name of Derek Lovley. While the Wobber SSP had been preparing and executing C10, Chapelle, with his much smaller team and budget, had cored

the same aquifers in the same state to the same depths and team members were quickly publishing their results in what could be described as a scientific race to take credit for discovering the deep subsurface microbiota. At this meeting Lovley and Chapelle presented their results on (1) using H_2 gas measurements in groundwater that correlated with the principal electron-accepting pathways, (2) the comparison of geochemical versus radiotracer measurements of microbial respiration of organic matter producing CO_2 in the groundwater, and (3) the production of iron-rich groundwater by ferric-iron-reducing bacteria. Ron Harvey, a hydrogeologist from USGS's Menlo Park, and his group reported on their shallow, contaminated aquifer near Falmouth, Massachusetts, on Cape Cod, where they performed microbial transport studies in a 450-well array of multilevel samplers spread across 3.5 kilometers. Despite the one-upmanship taking place at the meeting between the USGS and DOE investigators, Frank was pleased to see that the USGS results essentially validated the approaches and observations at SRP. The two completely independent groups of researchers had come up with essentially the same conclusions. They must be on the right track. The USGS scientists had also brought to the SRP's attention some methodologies they could incorporate in their upcoming drilling campaigns in the more challenging environments out West.

10. Frank was engaged with the first meeting of the newly formed International Continental Drilling Program, providing documentation on deep microbial life as an important scientific theme for that program.

11. This session was run by Ron Harvey of the USGS and Aaron Mills of the University of Virginia. Ron had been studying this topic for years at a field site in Cape Cod. Aaron had just started working at a field site on the DelMarVa Peninsula with support from one of Frank's subprograms. The question from the U.S. DOE's perspective was whether remediating microorganisms could be delivered to a contaminated site.

12. Stetter et al. 1993.

13. The Exxon Valdez oil spill occurred in Prince William Sound, Alaska, on March 24, 1989, and much of Hinton's work focused on biodegradation strategies of the floating oil, as well as that deposited onshore.

14. Those who believed that *B. infernus*'s ancestors could have survived or even grow at 200°C included: D. C. White and Jay Grimes, who said, "Keep an open mind;" Susan Pfiffner, Tommy Phelps, and Chuanlun Zhang, who said, "Maybe if they are dry;" and Rick Colwell, who commented jokingly, "Most likely, but glad I'm not in their shoes." Those who believed that *B. infernus*'s ancestors could not have survived 200°C and must have been transported down included, among others, Terry Beveridge, who adamantly stated, "No way!" The rest of the participants

who were not certain included: David Balkwill, who wryly commented, "Well maybe or maybe not;" Sandra Nierzwicki-Bauer and Jim Fredrickson, who suggested, "Maybe for a short time, but we will wait to see the data." Phil Long, however, was "very skeptical" that the bacteria could have been transported 2 km through low porosity/permeability rock that was a natural gas trap.

15. Liu et al. 1997.

16. During this meeting I also suggested that we should try to involve Lisa Pratt, a geochemist from Indiana University, in the project as she had recently published a paper on the organic composition of the lacustrine shale in the Triassic formations of New Jersey. We only knew the concentration of organic carbon in the sidewall cores. If we knew more about specific organic compounds that were present, then we could determine whether *B. infernus* could have been consuming those compounds over a period of 230 million years. Lisa could also determine the composition of neutral (or dead) lipids to see if they matched the composition of D. C.'s phospholipids. This could tell us whether the lipids represented the dead ancestors of microorganisms like *B. infernus*'s.

17. These arrived thanks to the help of Ric Vierbuchen, a former Ph.D. graduate of our department, I had met in Venezuela during my thesis research. He was very high up in the management at Exxon in Houston and pulled some strings to clear the release of the logs to me.

18. D. C. White used chemical extractants to separate the bacteria cell membrane constituents from the rest of the cell and from the rock matrix. He would then make it volatile enough that the chemical composition of the hydrocarbon backbone of the lipid could be determined using gas chromatograph mass spectrometry (GC-MS). He analyzed those lipids that still had phosphate groups attached to them, that is, the living bacteria, and those lipids that had lost their phosphate groups, the glycolipids, that is, recently dead or not so active bacteria. The composition of the lipids reflected both the physiological state of the organisms, and also its taxonomic identity. The abundance of lipids could also be converted into an approximate biomass concentration.

19. Project Gnome was the first of the Plowshare underground nuclear detonations. This one was detonated twenty-five miles southeast of Carlsbad, at a depth of 361 meters on December 10, 1961. Although the device had a small, 3.1-kiloton yield, there was a breach through the shaft to the atmosphere. The detonation left behind a 20-meter-wide by 50-meter-high cavity with a floor of melted salt. One of the goals of this project was to measure the radioisotope yields from neutron-induced reactions. See declassified U.S. Nuclear test film no. 34, https://www.youtube.com/watch?v=DFJ2MyWlXgs.

20. Carlsbad was only forty miles from Walker Air Force Base, which at this time was home to a Strategic Air Command wing of B-52 bombers that were on 24-hour alert and to a wing of Atlas intercontinental ballistic missiles. Walker AFB was also the location of the Roswell UFO incident in 1947, although at this time no one made much of it.

21. Tom had arrived in New Mexico as a graduate student working in the depths of the Carlsbad Caverns on his Ph.D. thesis. He had fallen in love with the state and never left, except when he did a two-year post-doc with Mary Firestone, an eminent soil microbiologist at the University of California, Berkeley. After that he returned and joined the faculty of the biology department at New Mexico Tech, in Socorro.

22. A headgear, also called headframe, gallows frame, or hoist frame, is a steel lattice work, sometimes enclosed, that sits above the vertical mine shaft and supports the wheel along which the steel cable that holds the cage is taken in or played out to raise or lower the cage.

23. Pump jacks, also called horse pumps or dinosaurs, mechanically pull oil up from a well by their seesaw action, and they litter the landscape of southwest Texas and southeastern New Mexico.

24. Dennis told us that we were not the first ones to collect samples for microbes here. A microbiologist friend of his, Russell Vreeland, of the University of Texas-El Paso, had also collected samples from WIPP, from which he was hoping to grow microorganisms.

25. Sergei Winogradsky was a nineteenth-century Ukrainian-Russian microbiologist who first discovered lithotrophic bacteria (meaning that they derive metabolic energy from inorganic compounds), specifically *Beggiatoa*, that harvested energy by oxidizing inorganic compounds, in this case H_2S, rather than oxidizing organic compounds.

26. These calculations were based upon two papers that Tommy had just written and were about to be published in *Microbial Ecology* (Phelps, Murphy et al. 1993; Phelps, Pfiffner, et al. 1993). The doubling time for a microbial cell (t_{cell}) can be calculated from the following equation,

$$t_{cell} = C_{cell}\, m_{cell}/(V\, Y_{growth})$$

where t_{cell} is in years, V is the sustaining metabolic rate estimated from geochemical constraints in moles of reactant per kilogram of sediment per year, Y_{growth} is the growth yield in grams dry weight of biomass per mole of reactant, C_{cell} is the concentration of microbial cells per kilogram of sediment, and m_{cell} is the average cell mass in grams dry weight per cell. If the cells that were included in the biomass concentration *were active and dividing*, then this would be their average age. Alternatively if the cells were not actively dividing but just repairing themselves in maintenance mode, then this calculation represents the amount of time required to replace all the cellular biomass. For example, Tommy found that the number of cultivatable bacteria from the Middendorf Aquifer in South Carolina was 4×10^8 colony forming units per kilogram of sediment. (For those readers unfamiliar with scientific notation here is an example. The Earth is 4.6×10^9 years old, which is the same as 4,600,000,000 years or 4.6 billion years.) Ellyn Murphy estimated that the rate of CO_2 production by microbial respiration was 1.1×10^{-8} moles of CO_2 per kilogram of sediment per year. Tommy assumed a growth yield of 1 gram of bacteria per mole of CO_2 respired and 10^{-13} grams of biomass per cell. Plugging the numbers into the equation above gives 3,640 years as a doubling-time estimate. Phelps, Murphy, et al. 1994.

27. Telomeres are a long string of repetitive TTAGGG base sequences that exist on the ends of chromosomes of Eukaryotes. Since during replication the chromosomes are not completely replicated to their ends, the telomeres act as noncoding junk DNA that is lost with each replication. Over time, however, these buffers get shorter and replicative senescence sets in. Without telomeres the chromosomes will fuse, leading to genetic damage. The shortening of telomeres, therefore, has been related to old age and old-age-related diseases.

28. These he inferred based upon UV and H_2O_2 experiments performed on isolates from the SRP samples (Arrage et al. 1993). Radiation resistance of microorganisms is usually reported as the total amount of radiation required to kill a certain percentage of a microbial culture. For example, one of the most radiation-resistant bacteria that we know is *Deinococcus radiodurans*. If *D. radiodurans* is exposed to 3000 Grays (300 krad) in the lab, then about 37% of the cells will still be alive and able to reproduce. For bacterial strains collected from the subsurface aquifers beneath Savannah River, Georgia, about 70 Grays dosage was required to kill 63% of the total population (Arrage et al. 1993). The equation for calculating the subsurface gamma-ray dosage from the Schlumberger logging tool is,

$$R = 1.27 \times 10^{-12} \text{ GAPI},$$

where R is the dosage rate in Grays per gram per second, and GAPI is the gamma-ray flux in API units (equivalent to 60 nanograms per ton of radium) as measured by the spontaneous gamma-ray logging tool. For the Thorn Hill no. 1 borehole, the GAPI was 200, which yields a dosage rate, R, of 2.5×10^{-10} Grays per gram per second, or 8×10^{-3} Grays per gram per year. If *B. infernus* or *D. putei* bore similar radiation resistance, then 63% of them would receive a lethal dose by about 9,000 years. The cell concentration in the Triassic rocks has to represent a balance of growth (see cell turnover time in note 25, and death by irradiation, or

$$dN/dt = (-N \times D) + (N \times G)$$

where N = cells per gram, D = death rate from irradiation = $8 \times 10^{-3} \times 0.45/70$, and G = growth rate.

Assuming that the cell concentrations in the Triassic sediments were at steady state, then $dN/dt = 0$ and $G = 5 \times 10^{-5}$ per year, or a cell turnover time of 20,000 years. This was longer than the cell turnover times Tommy estimated in the sedimentary formations beneath the SRP (see note 25), but the metabolic rates in the deeply buried Triassic sediments had to be much lower than in the relatively shallow sediments of the SRP. Cell death by irradiation was a significant selection factor for subsurface life, although as I would later learn, there were many other mechanisms leading to cell death in the subsurface, particularly at high temperatures.

29. The thermal aureole is the zone of metamorphic around an intrusive igneous rock. In thermal conduction models of the thermal aureole, the maximum temperature at the contact is half of the intrusive temperature and the country rock ambient temperature, and it decays with distance to the ambient temperature of the country rock. In the case of Cerro Negro, we needed to define on the ground the distance between the Cerro Negro contact and the Mancos Shale whose T_{max} did not exceed 80–120°C during the intrusion of Cerro Negro.

30. By common consensus, "Wobberization" was considered a process akin to the vulcanization of rubber: torturous to go through but at the end you were stronger and more resilient. It also had the effect of building an espirit de corps among the SSP investigators that led to long-lasting friendships. Perhaps Frank knew this, being a former captain in the U.S. Army.

31. Coleman et al. 1993.

32. "Contact" in geological parlance refers to the position on the ground surface and in the subsurface between two distinct geological units. The contact in the case of two sedimentary rock units, one overlying the other, could represent a gap in time during which no deposition took place. In other cases, the contact can represent a fault where the two geological units have been displaced from their original positions of formation and brought into juxtaposition sometime after their formation. In the case of intrusive igneous units, for example, dikes, the contact between the intrusive and the surrounding country rock is rep-

resented by the chilled margin of the igneous rock and the baked zone of the country rock.

33. A stratovolcano is a volcano that has the stereotypical conical shape that results from the high viscosity of the lava that is erupted. Typically the volcano is a composite of lava and pyroclastic ash deposits. In the case of Mt. Taylor, an extinct volcano, the upper portion of the cone was blown away by the last major eruption, and during the last Ice Age a glacier carved deep valleys into its interior so that it no longer retains its perfect conical shape.

34. In the parlance of paleomagnetism, this kind of a test would be called a contact test. The contact test used the same reasoning to determine whether the paleomagnetic directions recorded in a rock predated or postdated any adjacent igneous formation, either extrusive or intrusive, that thermally altered that rock. This was very much a geological approach to addressing a microbiological question.

35. The document was filed with the EPA and specified the environmental impact of the drilling activity and whether it threatened the drinking water supplies of the surrounding communities.

36. The oil trapped within the fluid inclusions of the diagenetic cements is a record of biological or hydrological processes that have occurred to the oil up to the time at which it was trapped. The composition of petroleum is altered by microbial degradation, with the lighter molecular-weight alkanes being preferentially degraded over the heavier molecular-weight alkanes. Detecting this pattern in the petroleum fluid inclusions would have meant that subsurface bacteria were active in the geological past just prior to the migration and entrapment of the petroleum in the fluid inclusions. "Water-washing" refers to the mixing of groundwater with petroleum. Because the lighter molecular-weight alkanes are more soluble than the heavier alkanes, repeated mixing of groundwater with petroleum will alter the composition of the petroleum by making it heavier. Detecting this pattern in the petroleum fluid inclusions would mean that groundwater had penetrated the formation in the geological past, and this groundwater could have transported subsurface microbes into the formation after T_{max}.

37. Brent also hired Bruce Hallett, who had just completed his Ph.D. at NMT and was the expert on these volcanic plugs that dotted the western flank of the Rio Grande rift. We also needed expertise on the sedimentology of these formations, and I thought of the ideal geologist for the job. John Lorenz, a sedimentologist who had been a graduate student in our department at the same time I was, worked at LANL, nearby. John had never encountered a rock in San Juan Basin that he didn't like, and he quickly agreed to join the project.

38. Those two months were filled with face-to-face meetings across the western United States. For airport/airplane reading, I was carrying with me Frank Chapelle's book *Ground-Water Microbiology and Geochemistry*, which had just been published. I studied this assiduously because it was at the right level of microbiology that I could understand. It also gave me a better appreciation of the importance of groundwater chemistry and hydrology and bioremediation.

39. This test was the Rulison-Mandrel nuclear test, which was part of the AEC Plowshare Program. It was followed by the Rio Blanco test in May 1973, in which three 33-kiloton devices were exploded simultaneously in the Mesaverde Formation approximately thirty-two miles northwest of the 1M-18 drill site. All three Plowshare nuclear explosions— Gasbuggy, which was in 1968, Rulison-Mandrel, and Rio Blanco— yielded 10–20% H_2, which from the DOE report was clearly created by irradiation of the formation water. The origin of this widely distributed observation was a report by H. A. Tewes in January 1979 (University of California Radiation Laboratory report UCRL-52656; http://www.osti. gov/scitech/servlets/purl/6276057), which was later released to the public in 2006.

40. John Lorenz had been heavily involved in the MWX experiment and could serve double duty assisting with the geological interpretation of the sedimentary and fracture structures at Cerro Negro.

41. This is true even today if you look at the Google Earth image. The drill site sits as high up along the steep southeast facing slope as is physically possible to get a big rig.

42. Barophiles (or peizophiles) are microorganisms that are active at pressures on the order of a few hundred to a thousand atmospheres (the pressures at the bottom of the deepest ocean trenches). Obligate barophiles have so adapted to their high-pressure existence that they cease functioning at low pressure. Barotolerant microorganisms can survive at high pressures but function optimally at low pressure.

43. Frank also brought in a dozen external reviewers. One of them was Barbara Sherwood Lollar from the University of Toronto, who had recently published a paper on the $\delta^{13}C$ and δ^2H of CH_4 and the δ^2H of H_2 from deep mine sites in Canada (Sherwood Lollar et al. 1993). I had read the paper and it was the first time I had seen someone perform isotopic analyses of H_2. I thought her isotopic analyses could be extremely valuable for telling us the source of the H_2 and perhaps whether bacteria were consuming it. This was her first report of abiogenic hydrocarbons in Precambrian Shield rocks. Intriguingly, some of her most abiogenic CH_4-rich gas samples were from the Kidd Creek mine

in Timmins, Ontario, where my dad had worked in the mid-60s, when we were living in Timmins.

44. Phil Long had been a postdoc at NASA's Johnson Space Center in Houston, working on the experimental petrology of granites and basalts, before moving to PNL, where he spent about ten years working for DOE on the Basalt Waste Isolation Project (BWIP). BWIP was DOE's plan to store high-level radioactive waste in the Columbia River Basalt. Congress precipitously changed things in December 1987, when they voted to focus all repository efforts on Yucca Mountain. He remembered it distinctly because everything he had been working on suddenly lost its significance and meaning. They had been doing some great hydrologic science but as soon as they heard about the congressional vote the context for the work evaporated. They had ninety days to shut down the project, which was funded at about $120 million per year! So he managed to get everyone in his group a job, and, in April 1988, he started working for Battelle, who then operated PNL, turning out the lights at his old job and walking across the street and up the block to start a new one. Late in the summer of 1988, while C10 was being cored, Ray Wildung and his boss at the time, Rick Skaggs, walked into Phil's office and asked if he was interested in working on the geoscience side of the subsurface microbiology studies then underway at Savannah River Site and whether he would like to lead the proposal for a new drill site to be located on the Hanford Site. He had been mainly working on cleanup activities at PNL and not finding it that exciting, so working for SSP was a huge opportunity, although he did not realize at the time to just what extent it would change his career. He first met Frank at the January 1989 meeting at FSU in Tallahassee.

45. The two proposed sites were adjacent to subsurface zones contaminated by DOE's Cold War activities, but they themselves were still pristine and readily accessible. They were different from the SRP sites in terms of the type and age of the rock. Located in the high desert country of eastern Washington and Idaho, the depth to the water table at the Hanford site and at the INEL was 200 meters, versus the 40 meters to the water table at C10. The sediment at Hanford was deposited in a 6- to 8-million-year-old lake, and the rock at INEL was Quaternary mafic volcanic rock, basalt. PNL and INEL were both well represented by the scientists and managers, but the PNLers outnumbered the INELers, the hometown favorites, and everyone thought that PNL had the inside track. The merits of the proposals were rated on an intricate point system with a complex weighting scheme crafted in meetings

earlier in the year. After each presentation the attendees rated the pros and cons of the site according to two dozen different required criteria. At the end of the day Brent Russel collected and tallied the numerical scores for PNL and INEL and the two proposals were rated equally at 249 points! All eyes turned to Frank sitting in the back row to find out what he would do. It had seemed that his carefully crafted rating scheme to select one site had backfired on him. Shaking his head and raising both his hands in the air, Frank reacted with an emphatic "We'll drill both!"

46. A paleosol is an ancient soil horizon that has been preserved by subsequent burial by sediments. It occurs when there is a gap in the deposition or a fall in sea level that exposes the sediment to subaerial weathering processes. Paleosols, therefore, provide a valuable record of ancient climates and have been used to determine ancient continental surface temperatures, climate processes, and atmospheric compositions of CO_2 and O_2.

47. Unfortunately Tommy was not present at this Lewes, Delaware, meeting, having suffered a severe head injury during a flight home to Knoxville. At the time we did not know just how severe that injury was. At the end of Tommy's flight, as the passengers were getting up and removing their overhead baggage, a man opened the overhead compartment above Tommy, and an all-metal briefcase fell down on the back of Tommy's skull, causing a severe concussion. Tommy was in the hospital for weeks, followed by months of recovery at ORNL while still struggling with cranial pressure. It was the result of this incident that airline attendants began announcing upon landing the warning to be cautious when opening the overhead compartments because baggage may have shifted during the flight.

48. Early in his career at DOE, Frank had dug into the reports on Project Gasbuggy. The AEC had collaborated with El Paso Natural Gas in 1968 to investigate ways to improve gas-well production from the low permeability Cretaceous sandstones. To do this, the AEC detonated a 29-kiloton nuclear warhead at 1,200 meters depth in a low-yield, natural gas well located about eighty kilometers from Farmington, New Mexico. In comparison, the atomic bomb dropped on Hiroshima during World War II produced a blast equivalent to 16 kilotons of TNT. The test was one of several subsurface nuclear tests as part of the Plowshare Program. In this instance, the desire was to fracture the natural-gas-bearing rock unit in order to increase the gas flow to the well. Essentially the U.S. government was doing atom-bomb fracking. Although the test proved that the technology could work, the gas yield over time

seemed lower than initially projected. Frank wondered whether micro-organisms had been at work consuming that same natural gas. He reasoned that if microorganisms were present at those depths oxidizing alkanes, then they might degrade the more toxic organic contaminants and/or, conversely, they might compromise the integrity of vaults containing radioactive waste.

49. Frank had set up a subsurface-microbiology meeting in San Diego for the Jasons to be held at GA Technologies in La Jolla and led by Gene McDonald. The JASON Society is a group of brilliant scientists, Nobel laureates, and members of the National Academy of Sciences who provide advice to the Department of Defense and the DOE. Will Happer is a member of the JASON Society. Gene McDonald was a chemist with interest in the origins of life. D. C. White, Jim, Ray, Ellyn, and an entourage of other scientists from PNL showed up. I was not quite sure why we were having this meeting and why I was there, but we had all received our marching orders from Frank about what to say and not to say and to be prepared for annoying questions from aloof gurus. It seemed as if Frank was getting quite concerned about the future funding of SSP and was trying to do what he could to backstop attempts to cut funding from the field program. D. C. and I talked about Taylorsville because Tommy was still recuperating from his accident. Jim presented some of the work that Todd Stevens was doing on basalt, something to do about H_2 generation, which I did not understand. Because the room was so crowded we were forced to leave after giving our presentations, and Jim and I took that time to finish designing a fluid inclusion stage we could use to extract DNA from Bargar's microbe inclusions. Jim also asked what I thought could be causing the H_2 production in the basalt. I confessed complete ignorance since I was still struggling to understand microbial redox reactions and had not realized that simple mineral reactions could dissociate water as well. I told him that the only abiotic H_2 generation process I understood was radiolysis. Soon after the meeting, we flew to the drill site because coring at the vertical control site, CNV, was just about to begin.

50. The borehole itself contains a cylindrical casing that sits against the rock and has slits in it to permit water to flow into the borehole. The pumping system used what we call a straddle packer, which contains two packers. A packer consists of an inflatable bladder that is wrapped around the pipe holding the pump. Upon inflation with either compressed air or water, the packers will inflate to form a hydraulically tight seal the borehole casing. In this case one packer sealed the top of the sandstone layer and the other packer sealed the bottom of the sand-

stone layer so that only water from the sandstone aquifer was being pumped out of the hole. The bladder pump is a steel cylinder with an inflated bladder on the inside and two ball check valves, one at the top and one at the bottom. Sitting in the borehole, the water will flow into the bladder forced by ambient water pressure through the lower check valve, thereby filling the bladder. Pumping compressed gas into the cylinder squeezes the bladder, which in turn seals the lower ball check valve and opens the upper ball check valve, allowing the water to be pumped to the surface. Pumping water is just a matter of repeating this cycle. Because the components of this system can be thoroughly cleaned, it provides an optimum methodology for microbial and dissolved gas samples. The MLS is a static equilibration method that does not use pumps but instead uses small-volume cylinders that contain distilled, degassed water sandwiched between two membrane filters. The pipe going down the hole contains a stack of these cylinders each separated from the other by a diaphragm that seals the MLS against the borehole casing. The MLS sits in the borehole, and over a period of several weeks the distilled water trapped within the membrane filters exchanges with the groundwater outside the membrane filters by diffusion. The MLS is then removed, the exchanged groundwater removed with a syringe from between the membrane filters, and the gaseous and aqueous species are then analyzed. In the case of this study, some of the MLS samplers were filled with crushed sandstone and shale and held between two membrane filters with pore sizes large enough to permit bacteria to colonize the rock chips with the MLS. This provided a means of capturing a gradient in the microbial composition of the sandstone/ shale interface. For a more detailed description of the MLS, the reader should consult Takai et al. 2003.

51. Phil Long's group sent Cerro Negro cores to a company that measured the size distribution of pores in a rock. That information was determined by a procedure known as mercury porosimetry measurement, a standard analysis done on cores collected by oil companies to evaluate the porosity and permeability of their sandstone reservoirs. Essentially a one-inch-diameter rock cylinder is placed inside a canister, and then mercury is progressively injected at increasing pressure up to about 2,000 pounds per square inch. The amount of mercury that can be added to the cylinder at low pressure is a measure of the volume of the larger pores. The amount of mercury added at very high pressures is a measure of the volume of the tiniest pores, some of which are much smaller than the size of a bacterium.

52. Brock and Madigan 1991.

53. Ibid., figure 6.19.

54. The article I had written for *EOS* on the Taylorsville Basin, with a great deal of help from Tommy, had just come out, but this was a trade journal for the American Geophysical Union and unlikely to have as big an impact at DOE headquarters as the *New York Times*, or so I naively believed.

55. The coring of the angled borehole was occasionally interrupted when the circulation of the N_2 was lost and N_2 began spurting from around the borehole, and we would have to reset the casing. In order to make sure the hole was advancing at the right angle, the rods also had to be pulled and the direction of the hole checked periodically with a gyroscopic system that measured the inclination.

56. See chapter 5.

57. The meteorite fell on May 14, 1864, near the town of Orgueil, France, and was a C1 carbonaceous chondrite, a type of meteorite that is very rich in fairly complex organic molecules, notably amino acids among others. Bartholomew Nagy and colleagues in 1962 using electron microscopy discovered structurally organized organic matter that he thought was indigenous to the meteorite. It turned out that these structures were either pollen or fungal spores, which was not surprising given the long time exposure of the meteorite to nonsterile conditions.

58. The cuttings were shipped to us by a Fluor Daniel geologist by the name of Lorraine Lafreniere.

59. Rick was there already with two of his research associates, Mark Delwiche and Mark Lehmann. Chuanlun Zhang, a postdoctoral researcher from Tommy's lab, had arrived to help process the cores. John Lorenz came up from LANL and Lorraine Lafreniere had come down from the Fluor Daniel office in Casper, Wyoming to join us. Jim McKinley arrived from PNL to set up the PFC tracers.

60. The production of petroleum from oil-rich shale in the Colorado region took off in the 1970s following the Arab oil boycott but then went bust in the early '80s when the price of oil dropped. Exxon had invested heavily in the Colony Shale Oil Project near Parachute, Colorado, but cancelled it on May 2, 1982, also known as Black Sunday, laying off more than two thousand workers. DOE continued encouraging the development of large-scale, in situ oil shale extraction through its Synthetic Liquid Fuels Program, but President Ronald Reagan abolished the program in 1985.

61. The gel coring system marketed by Baker Hughes uses a tube filled with gel, which they refer to as the CoreGel system.

62. For our gas chemistry, we relied on the gases captured from the drilling mud, which were injected into small evacuated tubes, like the ones

used to sample blood, and then shipped to Texas A&M University for $\delta^{13}C$ analyses of the CH_4. We also had gas composition data from Fluor Daniels for all the gas wells around us, but none reported H_2. H_2 gas measurements were never considered reliable because steel in contact with the water vapor in natural gas appeared to create H_2. Tim Meyers was going to track the logging at 1-M-18 to provide accurate temperature estimates as well as very detailed physical characterizations. Lorraine Lafreniere was able to obtain a formation fluid sample from a nearby well, 1-M-17, which came from 6,000 feet depth, and she shipped that to Jim McKinley to analyze geochemically and isotopically. We had far better samples, compared with the Texaco Thorn Hill no. 1 samples.

63. The U.S. Bureau of Mines was created in 1910 after a wave of mining disasters. It was the place to go for research on the environmental impacts of mining and, strangely enough, it was also in charge of the production, conservation, sale, and distribution of helium. It was shut down in 1995, and many of its functions were transferred to the U.S. DOE, USGS, and the Bureau of Land Management.

64. The RWMC was the radioactive waste dump where the radionuclides, organic solvents, acid, nitrate, and metal waste generated from the construction of fifty-one nuclear reactors were stored. INEL developed the reactors for the U.S. Navy's nuclear-powered submarines and also managed a nuclear reactor training facility for young submarine crews. It was for this reason that INEL was often considered the personal domain of Admiral Hyman Rickover, the commander of the submarine fleet. The organic solvents of the RWMC have leaked into the underlying Snake Rivers Plain basalt aquifer. To avoid further contamination, the transuranic waste has been exhumed, encapsulated, and shipped to WIPP in Carlsbad, New Mexico, for permanent storage.

65. Given that most of the microbiologists worked with less than 10 grams, this was an insurance factor of 5,000 against a false positive. Colwell et al. 1992.

66. Before the drilling began, Tommy had trained Jim McKinley in the art of PFC tracers. Because the drilling fluid was a gas, Jim had to design a PFC vaporizer that sat beside the drill rig; it looked like a distillation system from some science fiction movie. It was the same one that Jim had shown me at Cerro Negro a few months earlier.

67. In this case, sex is the one-way transfer of DNA between two prokaryotes by means of tiny protein tubules known as sex pili (even though it's not really sex in the animal sense).

68. Fimbriae are proteinaceous filaments that extend beyond the cell wall

of bacteria, both gram-negative and gram-positive bacteria. Fimbriae are shorter and more numerous than flagella and may serve the purpose of helping bacteria adhere to mineral surfaces; hence fimbriae are referred to as attachment pili.

69. "Tripping the drill rods (or pipes)" is drillers' lingo for pulling out or lowering in the drill rods.
70. Szewzyk et al. 1994.
71. Warren Miller directed, produced, and narrated some of the most hilarious extreme skiing and snowboarding films from the '70s through the '90s, back in a time when one could count on skiing on the East Coast.

CHAPTER 3: BIKERS, BOMBS, AND THE DEATH-O-METER

1. The conversation between Axel and his uncle, Professor Lidenbrock, in discussing the velocity of their descent to the center of the Earth.
2. Bob Griffith's enzyme-activity measurements and enrichments suggested that there was little microbial activity in the shale and that the viable microorganisms found there were not marine microorganisms.
3. Krumholz et al. 1997.
4. David Boone managed to isolate a mesophilic ferric-iron-reducing strain of *Shewanella*, CN32, which would become the poster child of the Cerro Negro project and the subject of future work at PNL.
5. Fredrickson et al. 1997.
6. The groundwater dating was performed by Fred Philips, a hydrologist from NMT, and his graduate student, Page Pegram, using ^{14}C dating of the DIC. Walvoord et al. 1999.
7. A geothermometer refers to any measurement that acts as a proxy for the paleotemperature, the temperature in the geological past. The basis of geothermometers usually is tied to the elemental or isotopic distributions within a mineral or mineral assemblage as these are governed by temperature-sensitive Arrhenius relationships. A thermochronometer refers to a subclass of radiometric tools that record the time when a rock passes below a particular temperature.
8. Thanks to Ray's help, Tommy had been able to get Lisa Pratt from Indiana University involved with characterizing the organic thermal maturity of the Mancos Shale samples. She had already been helping me with our Taylorsville samples out of charity, but now she could actually get some funding to perform the analyses to identify the thermal maxi-

mum, or T_{max}, of the vertical borehole and also that of the suspected sterilization zone near Cerro Negro.

9. Vitrinite reflectance is how it pretty much sounds. Vitrinite is a mineral name for the cellulose from trees and plants deposited in sediments and coal as detrital organic matter. Anthracite is very shiny because it reflects a lot of incident visible light. The vitrinite reflectance, expressed as percentage of reflected light, is very high for anthracite but much lower for bituminous coal. Vitrinite reflectance is, therefore, correlated with temperature as anthracite forms at higher temperatures from lower temperature bituminous coal and secondarily with time. Vitrinite reflectance (Ro) will typically range from 0.5 to 1.5% as the T_{max} varies from 60 to 125°C. This range covers the optimum temperature growth ranges of thermophilic and hyperthermophilic microorganisms and thus provides a convenient measure of sterilization, or "paleopasteurization." Hallett et al. 1999.

10. Clay minerals come in very many chemical compositions, but all contain sheets of aluminum, silicon, and oxygen. The elements and molecules, for example, H_2O, that occur between these sheets distinguish clay minerals from each other. In the case of illite, K^+ is the cation dominating its intersheet site. As explained above, ^{40}K decays to ^{40}Ar at a constant rate and is the basis of the K-Ar dating technique. ^{40}Ar is quantitatively retained in illite up to temperatures of 150°C, which means that it can be used to date the time of clay formation. Illite is formed progressively from smectitic clays that contain other elements between the sheets, at a rate that depends upon the temperature but is in the range of 50–150°C. The K-Ar age of illite/smectite mixtures therefore provides an integrated age and temperature constraint on the thermal history. This approach works well for clays in shale as long as the shale does not contain any reworked, older detrital illites, in which case the ages will be much older. One way to avoid this is to date the clays in altered volcanic tuffs, and fortunately the Cretaceous Period basins of the Rocky Mountains contain lots of volcanic tuffs interbedded with the shales.

11. Ray also brought in Crawford Elliott, from the University of Georgia (Athens), to K-Ar date on the K-bearing, illitic clay in shale to see where in the core they had been thermally impacted by the intrusion. Elliott et al. 1999; Hallett et al. 1999.

12. A more complete explanation of the opposing scientific philosophies of geology and microbiology can be found in Chapelle's excellent book, *Ground-Water Microbiology and Geochemistry*, first published in 1993.

13. From "The Crack-Up," *Esquire*, February 1936.

14. The data from these tools included (1) the smectite/illite composition; smectite clays become more illitic as they are exposed to higher temperatures, and their mean age could be determined by K-Ar dating; (2) vitrinite reflectance data; (3) fluid-inclusion analyses data; and (4) fission-track ages on apatite. Fission-track dating is based upon the spontaneous fission of ^{238}U, during which the fission fragments create a track of crystal damage. Depending upon the ambient temperature the crystal damage can be partially or even completely annealed. The tracks are imaged microscopically by etching the crystals in acid. The number of fission tracks can be used to calculate the age of annealing if the uranium concentration is determined for that specific crystal. The length of the tracks could be translated into a rate of annealing and thus a cooling rate. Different uranium-bearing minerals have different temperatures of annealing, but that for apatite coincides closely with the oil window of formation. These data helped to determine whether and when sediments had passed through the oil/gas window. Much of the development of this technique was supported by the oil companies.

15. The third sandstone body contained high concentrations of CH_4 and was overpressurized to 260 bars, compared with the hydrostatic gradient pressure of 210 bars. Tim Meyer completed the logging of the borehole just three months after we drilled it. His best estimates of the current temperature at the three depths were 43°C, 81°C, and 85°C. Lorraine was also able to obtain a water sample from 2 kilometer depth in the 1-M-17 well.

16. Coleman et al. 1993.

17. Lorenz et al. 1996.

18. Roh et al. 2002. The 16S rRNA gene of Tommy's isolates indicated that they belonged to the genus *Thermoanaerobacter*, the same genus found halfway around the world and three kilometers down in a drill hole in Sweden.

19. Wilkinson and Dampier 1990.

20. Kelley and Blackwell 1990.

21. This age was derived from the measurement of the abundance of chlorine-36 (^{36}Cl) in the water. Chlorine has two stable isotopes, ^{35}Cl and ^{37}Cl. In the atmosphere, the chloride in rainwater is irradiated by cosmic radiation. Low-energy neutrons from this radiation convert ^{35}Cl to ^{36}Cl while the water vapor is in the atmosphere. ^{36}Cl is a radioactive isotope that decays back to ^{35}Cl with a half-life of 250,000 years. Once the rain infiltrates beneath the surface, the ^{36}Cl concentration begins

to decline as the radioisotope decays. The measurement of the concentration of ^{36}Cl using an AMS can constrain the age of groundwater to up to 2.5 million years in age. The 1.6-million-year subsurface age for the groundwater was also supported by the δ^{18}O and δ^2H of multiple samples of Wasatch sandstone water that indicated mixing with paleometeoric water, not modern meteoric water. Apparently 1.6 million years ago the Colorado plateau was not as elevated as it is today.

22. Colwell et al. 1997, 2003.

23. To make the journey further down to the Mesa Verde Formation, cores would have required well over a million years, and those sandstone bodies still retained high concentrations of CH_4 gas and their water that was highly saline, indicating that little paleometeoric water had penetrated the sandstones. We wondered whether any of the bacteria in the thermal spring water in the overlying Green River formation on the Roan Plateau would include microorganisms similar to the iron-reducing bacteria found in the Wasatch core. With the last remaining money from our DOE grant, we embarked on a field sampling expedition in June 1995 to the thermal springs on the Roan Plateau north of the Parachute drill site where Yao's model indicated recharge should be occurring, but the bacteria found in our samples of the thermal springs were very distinct from Tommy's bacterium.

24. When the uranium present in minerals such as apatite or zircon undergoes spontaneous fission, the fragments of the nucleus recoil into the mineral matrix, creating a damaged crystal structure that can be imaged by the microscope after acid etching a polished surface of the mineral. By counting the number of tracks present per area and dividing by the uranium concentration, an age can be determined. Because the fission tracks will anneal upon heating the mineral, this age represents the time that the mineral has been below its annealing temperature. Fortunately for microbiologists, the annealing temperature of apatite is about 100–120°C, close to the maximum temperature limit of life. The length of the fission tracks can also be used to constrain the cooling rate through the annealing temperature zone. Fission-track apatite dates, therefore, provide the maximum age for microbial habitability.

25. Tseng et al. 1995.

26. Tseng et al. 1998a, 1998b.

27. The analyses were done by Harvey Cohen, a postdoctoral research associate of mine at the time, with the help of David Pevear at Exxon in Houston.

28. Fredrickson et al. 1995.
29. Stevens and Holbert 1995.
30. Further evidence that subsurface microorganisms were in a nutritionally challenged state was found by examining the structure of their membranes. Signature profiles of the PLFAs in some subsurface bacteria indicated that they possessed a higher degree of metabolic stress than was observed in typical soil microbial communities. The exceptions to this rule were the gram-positive Actinobacteria, which apparently were not stressed out at all. Kieft et al. 1994.
31. Sinclair and Ghiorse 1989.
32. An aquifer is a layer of rock or rock formation, for example, sandstone, with high porosity and permeability that allows water to flow readily. An aquitard, on the other hand, is a rock formation, for example, shale, where the low permeability restricts the flow of water. A rock formation that is completely impermeable, like the salt deposits at the WIPP site, completely occludes water and is called an aquiclude.
33. Saturation in this case means that all pores are filled with water. Undersaturated means that some of the porosity is filled with air or gas.
34. His data indicated that the concentration of cultivatable bacteria in a gram of sediment was many orders of magnitude less than the total counts per gram, and these sediments were only a few hundred thousand years old. Colwell 1989.
35. Acridine orange is a fluorescent DNA stain that was the most commonly used stain for microbial counts during the 1980s and early '90s.
36. Tobin et al. 2000. Propidium iodide binds to DNA and glows red under UV light, but blue background fluorescence from the minerals results in pink-colored cells.
37. As the samples from the INEL and GeMHEx sites were being processed in 1991, Frank used money to support a third field site, at NTS. Larry Hersman at LANL had brought this site to Frank's attention during the 1986 investigator meeting in Germantown. The DOE was investing heavily in developing storage sites for high-level nuclear waste at the NTS.
38. An Alpine miner is a 20–30-ton mining machine that consists of a tractor, usually on caterpillar treads, powered by electricity. It has a claw in front that contains a rotating wheel of cutters that chews into the rock and a mouth in front with shovels that move the rock debris onto a conveyor belt that dispenses the rock debris into a container mounted on the back of the Alpine miner. The Alpine miner also sprays water on the cutting heads to reduce sparks and dust. It is impossible to sterilize the cutting heads given their massive size, but the spray water in prin-

ciple could be amended with tracers and filter-sterilized. The larger chunks could be sampled for microbiology.

39. Penny Amy had obtained her doctorate with Richard Morita in marine microbiology at Oregon State University, focusing her research on the molecular and biochemical processes associated with bacterial starvation. Over a period of four years beginning in 1990, Penny Amy and Dana Haldeman, her Ph.D. student, developed techniques for sampling and processing microbial samples from this site.

40. Tritium is produced through the irradiation of H_2 by cosmic rays. It has a half-life of about 17 years. However, during the 1950s to 1960s, radiation from aboveground nuclear testing created high concentrations of tritiated water in our atmosphere that fell out as rain. This spike in the tritium concentration of rainwater can be detected in groundwater and has been used as a method of dating the time that has expired since the groundwater first entered the subsurface by infiltrating the soil.

41. Amy et al. 1992.

42. Matric potential is a measurement used to define the surface tension of water in soils. This surface tension will draw water up from the water table into the vadose zone, forming a transition zone called the capillary fringe zone. The amount of water and height of the capillary zone is equivalent to a negative pressure, as if the water was literally being sucked up by a straw. The units of matric potential are equivalent to negative pressure, usually –10 to –30 kilopascals, equivalent to 10–30% atmospheric pressure.

43. Amy et al. 1993.

44. Haldeman and Amy 1993.

45. The origin of this nitrate, however, was never explained. Russell et al. 1994.

46. Perhaps one of the most startling discoveries was that if the rock or core samples were simply stored in the refrigerator for days, weeks, or even months, the concentration of viable cells would increase, approaching that of the direct counts, but the total biomass concentration would not change and the biodiversity would decrease. This seemed to indicate that the extraction of the rock material from its environment altered it in such a way as to stimulate repair and growth of indigenous starved ultra-microorganisms and that growth occurred at the expense of other microorganisms. Brockman et al. 1992.

47. In geological parlance, "terrane" refers to an area where a particular type of rock formation or formations dominate or where a particular type of tectonic environment dominates.

48. Syntectonic fractures or faults are those fractures or faults that formed during a tectonic episode. Some fractures in rock formations can occur from structural unloading with erosion and decompression. This would enhance microbial transport from the surface downward, but if no topographic gradient is present, then deep penetration of the groundwater is unlikely. Topography generature during orogenic upwelling, folding, and faulting leads to deep fluid penetration into the crust along the faults that are active during deformation.

49. By the time we all met again for our final investigator meeting in Portland in mid-July, most of the details of the results that I summarized above had been documented. At the meeting Frank continucd to push for multidisciplinary research publications and encouraged the investigators not to give up on the project out of fear of the uncertain future for funding. As at all of the SSP meetings, the presentations reflected the wide range of tools brought to bear on the precious core samples. Some investigators were examining the sequences of specific genes, such as *recA* and *Trp*, from the isolates to see if they could detect a significant evolutionary difference between surface strains and subsurface strains. It would not be until the following year that complete genome sequencing of isolates would be undertaken in earnest. and reliable extraction of amplification of specific genes from environmental DNA from low-biomass samples was still in its infancy. No accessible database existed on the 16S rRNA gene. In low-biomass samples the best tools at this time were enrichments and isolates. At Cerro Negro, isolates were even used to capture bacterial phage from samples as a means of interrogating the viral population. These worked for the surface soil but did not yield phage from the Cerro Negro core samples. Sandra had focused on intergene spaces, or introns, in the genes adjacent to the very conservative 16S rRNA gene. Debates, discussions, and downright arguments broke out over whether any of the subsurface bacteria could really be called novel in an evolutionary sense, because we did not have enough sequence data on what existed on the surface of our planet. Joe Suflita pressed this argument further by claiming that dccp subsurface environments were not that different from surface environments and that was a good thing. Jim had started working with others to attempt direct extraction of DNA from some of the Cerro Negro samples for thc 16S rRNA gene rather than from isolates as this technique has just been reported for soils, but the extremely low DNA content was discouraging. He did report on successful extraction from our Laguna del Sol salt crystals, suggesting that fluid inclusions could potentially act as a repository of ancient DNA. But even in 250-million-year-old fluid in-

clusions of the WIPP salt, could you detect a significant difference in the 16S rRNA gene that would definitively say this was an ancient halophile?

50. The DOE and its predecessor, the AEC, unknowingly did engage in highly asymmetric nuclear warfare against the subsurface biosphere with the Gasbuggy, Gnome, Rulison-Mandrel, and Rio Blanco Plowshare atomic tests, along with the fifty-odd subsurface nuclear blasts performed on the Nevada Test Site. Certainly many billions of microbes were extinguished during these blasts, but the H_2 generated would have provided a valuable energy source to those potentially mutated bacteria that survived. Ironically it was the DOE that then led the discovery of the deep subsurface biosphere. It has even recently funded Duane Moser of the Desert Research Institute to study the subsurface microbiology of the Nevada nuclear blast sites.

51. A chemolithoautotrophic microorganism is one that obtains ("troph," meaning "nutritive" or "nourishment") its energy from inorganic chemicals ("chemo")—for example, sulfate—and rock ("litho")—for example H_2 generated from oxidation of Fe in the rock—and uses that energy to derive its biomass by fixing CO_2 itself ("auto").

52. The sequence was well exposed along the White Bluffs overlooking the Columbia River due to a recent irrigation-induced landslide near Savage Island. The withdrawal of the groundwater had dried the formation, causing the cliff face to give way. It provided a new exposure of the formation that had yet to be contaminated by surface vegetation, so it was quickly sampled for microbiology by SSP investigators. Brockman et al. 1992.

53. High porosity in this instance means that 40% of the volume of the rock is occupied by air or water. For land surfaces this is a typical starting value that diminishes with depth. The porosity of the Middendorf Aquifer at SRP was 25% at 200 meters depth.

54. Bjornstad et al. 1994.

55. By the end of 1992 they had finished drilling the GeMHEx and collected thirty closely spaced cores in the water-saturated zone below the water table down to 600 feet in depth. Kieft et al. 1995; Fredrickson et al. 1995.

56. Hanford and the neighboring towns had been booming agricultural communities in the late 1930s, and comprised 1,500 residents. Then on March 9, 1943, they were given thirty days to evacuate their town. The town was then razed to the ground with the exception of the high school, which was commandeered for the operations center of the new secret weapons plant. Unbeknown to the residents of nearby Richland, the U.S. government then installed three plutonium-producing reac-

tors for producing the first plutonium A-bomb, Fat Man, which was dropped on Nagasaki and ended WWII. During the cold war that followed, plutonium production increased, and the radioactive waste from processing the plutonium was stored in underground tanks. By the early '70s it was clear that the tanks were leaking into the underlying unconfined aquifer that discharges into the Columbia River. Wells were then drilled into the underlying Columbia River basalt aquifer to determine whether or not the source of the drinking and irrigation water in the region was under threat of contamination. Fortunately, it was not.

57. In collecting subsurface microbiology samples, we started classifying those bacteria found in the groundwater as being planktonic, a term widely employed in the description of marine microbiology. The term "sessile" was already employed to describe those microorganisms that were attached to the mineral surfaces of rocks, sediments, etc. By collecting core samples we could assay for both. In the collection of just borehole water samples, however, we had to always couch our descriptions as characterization of the planktonic community because of the likelihood that the planktonic community could be quite different from the sessile community.

58. Autotrophic acetogens are those bacteria (not *Archaea*) that metabolize H_2 and CO_2 using the following reaction: $4H_2 + 2CO_2 \rightarrow CH_3COO^-$ (acetate) + $H^+ + 2H_2O$. Intriguingly, the pathways utilized by autotrophic acetogens and methanogens bear a strong resemblance to each other, yet no bacteria have been found to date that utilize the methanogenic pathway and vice versa. A good review of the evolution of these two pathways that never crossed the divide between *Bacteria* and *Archaea* is provided by Martin and Russell 2006. Stevens et al. 1993.

59. Onstott et al. 1999.

60. In March 1995, Jim and I met again at the University of Tennessee Knoxville at another SSP meeting. With the imminent demise of the SSP, we thought we would make one last attempt at a high-profile publication. We had hoped that an article in the popular press could somehow rescue us. Jim and I cooked up the idea of approaching *Scientific American* about an article on the SSP, and the editor jumped at it. We discussed the idea with Tommy, Ellyn Murphy, Rick, and Jim McKinley. We agreed that Jim would focus on the history of the exploration of the deep biosphere by the SSP, that Ellyn would focus on microbial transport, that Tommy would focus on in situ origins of bacteria, that Rick and Jim McKinley would focus on the tracer technology, and that I might provide some global relationship to geology. Within a few months, between all the meetings, we managed to submit the text, and

Scientific American provided some mind-blowing graphics. Then, right before press time, the editor contacted me to tell me we could have only two authors on this article, not six. I tried to explain that the article was truly the joint effort of all us and I had the least to contribute to the actual science, but the editor was quite insistent. We couldn't even acknowledge their contributions. Jim and I discussed our options over the phone and agreed we should accept the editor's conditions, and I notified our coauthors. I felt like a traitor to them all and was determined to make it up to them.

61. Gold 1992.

CHAPTER 4: MICROBES IN METEORITES! WHERE DID THEY COME FROM? HOW DID THEY GET THERE? WHAT DID THEY WANT?

1. Axel, upon realizing they were traveling beneath the Atlantic along their granite lined passageway.

2. Boston et al. 1992. I found Chris's comment most heartening. My former undergraduate advisor, Gene Shoemaker, told me that he thought the only scientist at NASA Ames worth any beans was Chris, and high praise from Gene Shoemaker was a rarity. If Chris thought it was a significant finding, then that was good enough for me.

3. Gould 1996. Two review commentaries also appeared in *Science* in 1997.

4. Mafic rock refers to those rock types that are rich in magnesium and iron, or equivalently, those rock types rich in the minerals olivine, pyroxene, and hornblende, and for which quartz is generally absent as a mineral phase. Basalt (volcanic) and gabbro (plutonic) are examples of mafic rock. This rock type contrasts with felsic rock, which has less magnesium and iron and more silicon and usually contains quartz as a mineral phase. Rhyolites (volcanic) and granites (plutonic) are examples of felsic rock.

5. Cox (1946) reviewed studies showing the formation of oil from volatile hydrocarbons by radiolysis, but these reactions were dismissed as being irrelevant because the experiments released H_2 and high concentrations of H_2 were never observed in oil reservoirs. We know now that one reason high H_2 concentrations are not seen is because microorganisms are utilizing H_2 with a very high affinity. Zumberge et al. (1978) reported that the carbon seams of the Witwatersrand Au reefs contained ten to one hundred times the free radicals of coal and that these

free radicals likely were the result of irradiation from uranium trioxide (U_2O_3).

6. *Gold* was the title of a 1974 film starring Roger Moore and depicted a catastrophic flooding of the Western Driefontein mine in the late '60s.

7. von Humboldt 1793. Alexander von Humboldt was also one of the first biologists to characterize life in caves. In 1799, during a trip to Venezuela, he visited the Guácharo Cavern, where he described the oilbird, *Steatornis caripensis*, a nocturnal bird that nests in the cave and flies out to forage at night using echolocation much as bats do. His exploration of the Guácharo Cavern is mentioned in Jules Verne's *Journey to the Center of the Earth*.

8. First described from experiments performed by A. Bechamp (1868).

9. Lieske and Hoffman 1929; Lieske 1932. A good review of this early work is provided in Farrell and Turner 1932.

10. These papers were published soon after Bastin el al.'s *Science* paper of 1926. C. B. Lipman 1928.

11. Lipman 1931. The anthracite samples were collected for him from the Locust Summit Coal Mine, near Pottsville, Pennsylvania, at a depth of 600 meters.

12. Lieske and E. Hofmann 1928; Farrell and Turner 1932.

13. Lipman and Greenberg 1932.

14. Iizuka and Komagata 1965.

15. This was the former name of the Mponeng Gold Mine.

16. McKay et al. 1996.

17. In fact, in his paper on the deep hot biosphere (1992), Tommy Gold had suggested that signs of microbial life should be found in Martian meteorites.

18. At a pre-conference meeting held at the University of Reading, UK, Jim and I got the details of the study on ALH84001 from Chris Romanek, one of the study's coauthors. They had found their Martian fossils in carbonate concretions in the meteorite. At the time, that was enough to convince me that they had the real deal.

19. Stevens and McKinley had already shown during their microcosm experiments that granite produced little H_2, presumably because of the lack of sufficient Fe-bearing mineral of the right type. Radiolysis operates at a much slower rate.

20. See note 4 re. felsic rock.

21. You can find photographs of many of Karsten's microorganisms from Äspö and other locales in his book *Strolling Through the World of Microbes* (2004).

22. Dennis Powers, personal communication, June 7, 1993.

23. Dombrowski 1960, 1961; Dombrowski et al. 1963 and the papers cited therein; Dombrowski 1966. Dombrowski was not the only one investigating salt deposits in search of ancient life. Ralph Reiser and Paul Tasch (1960) had been performing similar experiments on Permian salt collected at 194 meters depth from the Carey Salt Mine in Hutchinson, Kansas, which was being investigated by the AEC for potential radioactive waste storage. They also analyzed salt samples ranging in age from Miocene salt domes in the gulf to the Silurian salt in New York State. They found few viable diplococcic forms that were gram-positive but did observe diplococcic forms in salt fluid inclusions. But the inability of another lab to replicate Dombrowski's findings did not help support his hypothesis regarding the indigenous nature of salt-bed bacteria.

24. De Ley et al. 1966. The authors used DNA/DNA hybridization techniques and physiological characterization to demonstrate that *Pseudomonas halocrenaea* was not novel by any means and was probably a contaminant. DNA/DNA hybridization was a technique developed before the age of genome sequencing to compare how similar two DNA genomes from microorganisms were to each other. It is based upon the assumption that genomes with similar base-pair sequences when denatured should hybridize to each other in a manner similar to when oligonucleotide sequences bind to target DNA in PCR reactions.

   It did not help that Dombrowski's paper had not been published in higher-profile journals like *Nature* or *Science* either. The fact that Dombrowski recovered a common bacterium and not a true archaeal halophile, which dominate evaporitic salterns, or salt pans, with high concentrations of 10^8 cells per milliliter, is also an argument against the autochthonous origin of *P. halocrenaea*. These high concentrations are sustainable thanks in part to the fact that halophiles are phototrophic, generating ATP directly from sunlight.

25. Larsen 1981.

26. Norton et al. 1993.

27. Norton et al. 1992.

28. One of the books in my stack of thirteen microbiology books that I recommend is an authoritative review of halophiles, *The Biology of Halophilic Bacteria*, by Russell Vreeland and Larry Hochstein (1992).

29. Bacteriorhodopsin is a protein found in the membranes of Halobacteria that acts as a cross-membrane proton pump when it absorbs a photon. This pumping action rejuvenates the proton gradient, which can then generate ATP.

30. De Ley et al. 1966.

31. Lowenstein et al. 1999.
32. In August 1994, I attended a NASA meeting on ancient DNA samples and salt deposits held at George Washington University in Washington, D.C. Despite being occupied with Cerro Negro and Parachute and Taylorsville basins, we had managed to examine the salt samples that Tom and I had collected in the fall. Using an optical microscope, we could clearly see that the fluid inclusions of the modern-day Laguna del Sol salt contained microbial cells, presumably halophiles, and Jim had detected DNA in these samples. He was struggling to detect anything in the old crystals from WIPP; however, dealing with the salt during the DNA extraction was proving challenging. It seemed like a good time to meet the experts. Michael Meyer, the program scientist for exobiology at that time, ran the NASA meeting. Michael had been a protégé of Imre Friedmann, who I learned from Michael had pioneered studies of Antarctic cryptoendoliths, bacterial and fungal communities that reside as a thin layer millimeters beneath the rock surface. I had never heard the term "exobiology" before and was curious as to why NASA was so keen on halophilic microbes. Jack Farmer, a NASA Ames scientist, answered that question by beginning the meeting with some Viking mission orbital imagery that revealed craters on Mars containing what appeared to be ancient salt deposits. These deposits would have been left behind by a long-since-evaporated crater lake. The most noteworthy of these is Pollack Crater's White Rock, located at 7.9° S and 334.8° W on Google Mars, which from Viking imagery has a very high albedo, or visible reflectivity, deposit at its center. The Mars Global Surveyor would later show that this feature was not salt, but just some type of silicate forming a mound at the crater's center. Salt deposits have, however, been discovered in various regions on Mars by the Odyssey orbiter using reflected infrared spectroscopy (Osterloo et al. 2008). Could those salt deposits still harbor Martian microbial life trapped in their crystalline matrix? Halophilic Archaea preserved in crystals from salt mines have been proposed and refuted for decades. But during the meeting, new results based upon the enrichment of living halophiles from salt were presented by Russell Vreeland for the WIPP salt. Raul Cano also presented his evidence for DNA extracted from amber-entombed, endosymbiotic bacteria from the Dominican Republic (the basis for *Jurassic Park*). Taken together, these findings provided a strong case supporting long-term survival of DNA and bacteria within the microscopic pockets of mineral deposits in which the activity of H_2O was low. If the salt deposits were anything like those at the WIPP site in terms of their preserva-

tion, then I could easily imagine that Martian halophiles might be frozen within them. Could they be resuscitated after 3 billion years in the deep freeze? Russell Vreeland's work seemed to suggest that it was a distinct possibility.

33. Many NASA scientists believed that Viking's failure to find life had removed a compelling reason for the public to support further NASA robotic exploration of Mars. An eminent scientist from the Johnson Space Flight Center told me this in the men's room at the Jet Propulsion Laboratory.

34. In planetary geology usage, the regolith is the uppermost surface layer and is composed of a heterogeneous mixture of rock with widely varying grain size that is produced by meteorite impacts over billions of years.

35. From the workshop we also learned that the first mission to start collecting rock samples for the MSR had just been awarded. It was the Athena mission, headed by Steve Squyres of Cornell University. Part of the proposed rover would core five centimeters into rock. Some form of planetary protection (PP) measures had to be designed and instrumented for this mission even though assembly was only a year away! I called Steve to ask if he needed any assistance with them on the Athena mission and he agreed. After that phone call I went through a rapid trial-by-fire introduction as to what takes place when trying to launch any space mission and to the internecine warfare that had erupted within NASA over the handling of any returned Martian samples, decades before they were due to arrive. My participation quickly and unceremoniously ended that spring in a Holiday Inn in Houston after I made a presentation to the Curation and Analysis Planning Team for Extraterrestrial Materials (CAPTEM) on how one could approach space missions with tracers to discriminate Athena contaminants from indigenous biomarkers. It was a critical step if the presence of potential Martian biological signals prevented the distribution of Martian return samples to planetary scientists. False positives had to be avoided. CAPTEM vetoed any use of tracers on MSRs even though I had not actually proposed any tracers, under Steve's advisement. But the choice was obvious given the highly variable sources of microbial contaminants, beginning with the assembly room and ending with Mars orbital entry. The IMT approach, like that used at Texaco Thorn Hill, was the obvious way. But as Todd Stevens, who was at the meeting, explained to me later, NASA has a "must be invented here" philosophy. Sure enough, the IMT did reappear as a PP protocol but now called "witness plates" in discussions of MSR missions.

36. Ken Nealson, a microbiologist from the Center for Great Lakes Studies in Milwaukee, Wisconsin, was hired by NASA's Jet Propulsion Laboratory to improve their PP and life detection programs in preparation for an MSR. NASA headquarters appointed a Planetary Protection officer, John Rummel, whose role was to supervise and approve plans to be emplaced to protect Earth from any possible Martian bug epidemic.

37. Mormile et al. 2003. Between 1998 and this book's publication, many other labs have reported successful isolation of archaeal halophiles from salt deposits. A very detailed summary of these studies was published by the Austrian microbiologist Helga Stan-Lotter in the *Encyclopedia of Geobiology* (2011). The many isolates should be prime targets for whole-genome sequencing to provide constraints on the rates of gene-specific evolution.

38. See Hoover 1997 for session papers.

39. Clearly I had misunderstood how he inferred there was a deep, hot biosphere but he good-naturedly corrected me. His evidence for the deep hot biosphere was not that the abiogenic hydrocarbons from the mantle were being eaten by the bacteria, although he believed they were. His evidence for the deep hot biosphere were the lipid biomarkers that petroleum geologists found in oil that Tommy believed were the dead remains of the deep subsurface bacteria that were eating mantle methane. It took me awhile to think about what he was saying. To comprehend that subsurface bacteria could change the organic biosignatures in a rock. Could it be that our carbon leader from South Africa was a layer of living and dead subsurface bacteria supported by radiolysis from the uraninite embedded in the organic-rich layer? I discussed this with Richard Hoover and he offered to look at some of our carbon leader samples with his SEM at Marshall Space Flight Center to see if he could see any signs of bacteria, living or dead.

CHAPTER 5: LIFE IN DEEPEST, DARKEST AFRICA

1. The Lidenbrock party makes its initial descent into Snaefells volcanic shaft.

2. See Edward Hooper's book, *The River: A Journey Back to the Source of HIV and AIDS*, in which she is featured. Hooper's book is based in part on her hypothesis that she published in the *Lancet* in 1992 that the AIDS virus and the polio vaccine developed in Africa were linked.

3. Even though SSP was gone, Frank still managed to develop projects that directly tackled bioremediation of contaminated DOE sites, essen-

tially developing the Phase II that he had conceived ten years prior. Frank worked through the Environmental Management Science Program (EMSP) of DOE to provide them with the insights garnered from the SSP effort. One of the projects on which SSP studies had a direct impact focused on a TCE plume that occurred in the Snake River basalt aquifer just below the facility at Test Area North of INEL. The hangar at Test Area North was the former home of America's atomic plane.

The atomic plane program began in the mid-'50s led by the U.S. Air Force. The concept was simple enough: take a turbojet engine and replace the chemical fuel source of heat with a nuclear reactor. Air enters the engine, is heated, and exhausted out the back. The program used a portable nuclear reactor and performed the test outside, because the jet exhaust was obviously radioactive. By the end of the '50s, an enormous hanger had been built for the 300-ton, 200-foot-long atomic plane prototype and a four-mile-long runway was planned, but the program was terminated by John Kennedy in 1961. A marvelous book, *Proving the Principle: A History of the Idaho National Engineering and Environmental Laboratory, 1949–1999*, by Susan Stacy, provides this and other titillating stories of the nuclear projects that took place behind the gates at INEL.

Rick Colwell was piggybacking on an operation to collect core during installation of wells at TAN-37, located in the heart of the TCE plume. During the first on-site meeting organized by Frank, Rick Colwell started with the site's groundwater hydrology, the TCE plume and its history. Rick mentioned the possibility that methanotrophs could be co-metabolizing the TCE. I, as usual, knew nothing about TCE, little about methanotrophs, and certainly had never heard of co-metabolism before. A voice in the audience asked Rick if anyone had measured the cis-to-trans ratio of the TCE. To the young woman who had asked the question, Rick confessed that he didn't know. She explained that it was a quick way of determining if the TCE was being biodegraded in situ because aerobic bacteria preferred cis over trans TCE and suggested that they should make these measurements. The cogent comment came from Mary DeFlaun, whose expertise was in bioaugmentation, the practice of injecting bacteria into the subsurface to remediate toxic organics like TCE. She was a senior microbiologist at Envirogen, a company that specialized in utilizing this approach at many contaminated sites. In fact, they had a patent on a specific bacterium that could degrade TCE. Frank introduced me to Mary during one of the few bathroom breaks. She was hilarious but no-nonsense when it came to work. Frank had brought her in as an external advisor during a review meeting for his microbial transport program and was impressed by the directness of her comments. She did not tolerate fools and did not think that DOE should, either. She quickly became my tutor when it came to environmental remediation and, given that Envirogen was just down Route 1 from Princeton, I took it upon myself to visit her and her laboratories frequently. Mary explained to me that if there were methanogens in the basalt, as Stevens and McKinley claimed, then methanotrophs were also likely to be in the basalt. They would then remove the TCE plume by epoxifying it, killing themselves in the process over time without requiring any pumping, extraction, and reinjection in the point source of the TCE. With the help of SSP investigators, methanotrophs had been stimulated in the subsurface at Savannah River to degrade the TCE plume there with great success. I commented that that seemed very noble of the methanotrophs to sacrifice themselves for environmental cleanup. This drew a cackle from Mary who shot back, "Do you think they had a choice?" Another way to determine whether the organics in the plume had been degraded since the time the TCE was first dumped into the well in the mid-1950s was to look for mineralogical signs, the role given to me. I quickly hired Ken Tobin, a geologist not just interested in, but positively obsessed with carbonate precipitation. Ken began looking for geological evidence of long-term microbial respiration of TCE in the Snake River basalts. He focused his microscope on the petrology and carbon isotopes of "dog-tooth spar" along the fractures in the basalt. Using a combination UV fluorescent/^{14}C microautoradiography, Ken was able to map the location of active heterotrophic bacteria in relationship to the calcite that was formed by microbial oxidation of the organic plume.

4. Hans Bjelke, the tall, stoic Icelandic guide for Axel and Professor Lidenbrock.

5. Fanagalo is a pidgin Zulu with a bit of Afrikaans and English thrown in that was widely used in the South African mining industry primarily because so many of the miners came from different tribal groups across southern Africa and spoke different languages. Its use has been slowly phased out in favor of English.

6. Duane et al. 1997.

7. The collar is the reinforced concrete deck that supports the headframe, or superstructure, of the shaft. Typically mine depths in South Africa are recorded as the depth below the elevation of the collar, or depth below a reference altitude. The barrel is the cased portion of the shaft right below the collar.

8. Shotcrete is low-slump concrete that is shot from a hose and sprayed onto walls. Mine tunnels will often be covered with a very coarse wire mesh that is held in place by rock bolts made of steel rebar. The shot-

crete is sprayed on top of the wire mesh. All of this is designed to keep in place the rock face, which is under an enormous stress or pressure gradient. The stress fractures the rocks on the tunnels walls and ceiling.

9. Biltong is the South African equivalent of what we would call beef jerky but is far more flavorful than jerky. It is dried cured beef from domesticated cows and from wild game.

CHAPTER 6: HUNTING FOR WATER AND CARBON

1. The Lidenbrock party's end to their desperate search for water after Hans intersects fissure water.

2. Proto teams are specially trained to fight mine fires and form rescue teams for the South African mines. Their name is derived from the name given to a self-contained breathing apparatus developed in 1902 for mines and used on the western front of World War I. The device would strip CO emitted by a mine fire and supplement the air with O_2 carried in a small cylinder.

3. At temperatures above 60°C the water-soaked mine socks would remove the skin from your legs when they were pulled off. In order to avoid this, they would coat their legs with Vaseline.

4. The West Driefontein flood was the result of ground that was close enough to the Transvaal dolomite aquifer giving way so that the aquifer poured into the mine. It was this flood that was the inspiration for the film *Gold*.

5. Microbial media recipes are regularly published in papers, and the most popular ones are found in encyclopedic volumes that are like microbiology cookbooks. You can grow your bacteria either on a plate of agar, a gelatinous material, or in a solution within a glass test tube sealed with gas-tight rubber stoppers. The media must contain electron donors—usually simple organic acids, methane, or glucose for heterotrophs—and electron acceptors. The electron acceptors can be something as simple as air (actually the O_2 in the air) for aerobes, or something more complex, like ferric iron hydroxides, for anaerobic iron-reducing bacteria. For chemotrophic bacteria you can use H_2 as an electron donor, which you can do with the liquid media in the test tubes. Usually you pick your favorite electron acceptor and then use multiple electron donors to increase your chances of hitting the right combination to grow your bacteria. If you are trying to grow chemolithoautotrophs, however, the only carbon substrate you can use is CO_2.

As the microbes eat their electron donors and acceptors, they produce H^+ or consume H^+, changing the pH. Bacteria often have a limited pH range at which they can grow, so the media recipe must also include a pH buffer to stabilize the pH range in which you think the optimum is located. The buffer is kind of like antacid and is usually a combination of bicarbonate and phosphate salts. The phosphate is always part of the buffer because microbes need phosphate for DNA and RNA, and for ATP and other metabolites. For proteins, which compose half the bacteria's biomass, you need amino acids. You can provide them by adding a healthy amount of yeast extract or the specific amino acids. If, on the other hand, you are trying to grow chemolithoautotrophs, you avoid adding amino acids to the media but add just ammonia instead. The gas in the headspace for anaerobes is mostly N_2, and if you are trying to isolate bacteria that are N_2 fixers, then this will be the only nitrogen compound in the media. The media will also contain some mineral salts that provide the potassium, magnesium, calcium, and sulfate that all life on Earth requires. Since microorganisms grow within a limited salt range, you can adjust the amount of salt to what you think should be the optimum salinity. For example, to grow archaeal halophiles you would add a lot of salt. All microorganisms require trace amounts of certain metals, like iron, manganese, tungsten, molybdenum, etc., and vitamins. You can purchase standard formulations of these, like Wolfe's mineral supplement, from the American Type Culture Collection, or you can try concocting your own combination. Once you have combined all the ingredients you pour the agar media onto plates or the liquid media into the test tubes, seal them, and autoclave them to kill any bacteria that hitched a ride with the ingredients. You usually add the vitamins after autoclaving just to be on the safe side. If you are making anaerobic media, you have to take extra care to degas the liquid media and fill the tubes with the appropriate gas combination. You usually add a little bit of dye, resazurin, which is sensitive to the O_2 levels in your anaerobic media and will be bright pink if the media is aerobic. To help remove any trace O_2, you can also add a little sodium sulfide (Na_2S), which the O_2 will quickly oxidize to sulfate. Finally, like a cook adding the tiniest amount of salt at the end, you adjust the pH with the tiniest amount of HCl or NaOH so the pH of the media it is just right. .

6. See appendix B.

7. The nomenclature commonly used to describe stable, cosmogenic and the initial ratios of radiogenic isotopes is the delta notation. This is defined by the following equation,

$$\delta^n X = \{[^nX/^mX]_{\text{unknown}}/[^nX/^mX]_{\text{standard}} - 1\} \times 1000,$$

where n is the minor and usually heavier isotope, and m is the more abundant and usually lighter isotope and X is the element letter symbol. $^nX/^mX$ is the ratio of the atomic abundance for the sample or unknown and for the standard. Because the isotopic ratios are typically similar to within a few percent of the international standard, one is subtracted from their measured ratios, hence the delta, and then multiplied by 1,000, hence the ‰ reported with the value. This last step is somewhat controversial as it seems superfluous. In this case,

$$\delta^{13}C = \{[^{13}C/^{12}C]_{\text{unknown}}/[^{13}C/^{12}C]_{\text{VPBD}} - 1\} \times 1000,$$

where VPDB stands for Vienna-Pee Dee Belmnite: Vienna is the home of IAEA and isotope standards, and Pee Dee Belmnite refers to the carbonate of fossil Belmnite from the Cretaceous Pee Dee Formation that was established as a standard in the early days of isotope mass spectrometry.

$$\delta^2H = \{[^2H/^1H]_{\text{unknown}}/[^2H/^1H]_{\text{VSMOW}} - 1\} \times 1000,$$

where VSMOW stands for Vienna Standard Mean Ocean Water, and ^2H is deuterium. VSMOW is used for both hydrogen and oxygen isotopes. The δ^2H and $\delta^{18}O$ of ocean water samples are typically close to zero, but during meteorological and hydrological processes, such as evaporation, freezing, and precipitation, the delta values will depart significantly from zero.

8. Uranium was a touchy subject with the South African mines because it used to be sold to the United States and was critical to the development of the first atom bomb in WWII. In 1987, however, when antiapartheid protests were on the rise in the States, Congress issued a ban on the purchase of South African uranium, much to the annoyance of that country's mining industry.

9. Equally important in the diplomacy battle, Duane and Joost consumed a lot of beer and braai (barbeque) with the various managers and geologists and lost a lot of pool games to them. Eventually they were allowed to go underground and follow the "samplers," whose job it was to collect samples of freshly blasted carbon leader for grade determination.

10. As the rock resettles after the daily blast, small earthquakes shake the region, slipping along ancient faults. On any given day, we felt several "bumps" of 1.5 on the Richter scale. These are just enough to rattle the windows or sometimes knock loose objects from shelves. In actuality, a slippage of only a few millimeters could cause magnitude 3–4 earth-

quakes, which happened once or twice a year and were very dangerous in the mines, causing significant subsurface damage. Recently, an earthquake in a nearby mine knocked out a main access shaft, trapping 1,000-plus men underground. The men were eventually rescued via a cross-connecting tunnel to another undamaged mine shaft. The only loss of life was one individual on the surface.

11. Duane Moser to various recipients, Oct. 30, 1998.

12. Any of you readers who have watched *The Simpsons* will know the meaning of this, or so Duane tells me. Shelbyville is always competing with its neighbor Springfield, with which it shares many similarities.

13. The Carletonville mining region is traversed by massive wall-like igneous intrusions (called dikes or dykes) that are hundreds of kilometers long, tens of meters wide, and kilometers deep and radiate from the Pilanesberg Mountains, which are formed by a massive intrusion, 200 kilometers to the north. Groundwater migrating to the southwest in the dolomites toward the Namaqua desert would hit these dykes and then bubble up to the surface, forming springs. The dykes act as underground dams dividing the dolomite aquifer into what were called compartments with each compartment given a name, usually after the name of the spring. As the mines attained greater and greater depths, their pumping would lower the water table around them, but just within its compartment.

14. During their two long recons in the deepest levels of West Driefontein, Duane and Joost identified no fewer than five uncemented, weeping boreholes. These were all dripping at rates at least equal to that of the hole that Nico had shown us at level 46 in the East Driefontein no. 5 shaft (more than 1 liter per hour). High-pH water (9 or 10) was flowing from all of them, and the temperatures ranged from 29°C to 37°C. All the boreholes were cased at the end and could be capped for the attachment of sand cartridges or water-sampling bottles.

15. Louise Gubb is a very well-known photojournalist in South Africa who chronicled Nelson Mandela's release and subsequent political life.

16. Kevin Krajick would later write the book *Barren Lands*, which documented the diamond mining industry in the Canadian Arctic.

17. Balch tubes are large, 30-milliliter, thick-walled test tubes commonly used for anaerobic culturing experiments. They can be sealed gas-tight using butyl-rubber stoppers and aluminum caps that are crimped close around the lip.

18. Over the next couple of weeks the team finished processing rock and water samples and began to return to their various countries for the Christmas holidays. We were able to get an extension on our house

lease into the New Year to allow time for Duane and Joost to come back and restore the house to its original condition and return the equipment to Jenn Alexander at Wits University.

19. Whitman et al. 1998.
20. Parkes et al. 1990. This work confirmed in many respects an estimate made in 1950 by Claude ZoBell based on seafloor sediment from the top ten meters. An earlier study of seafloor sediments was performed in 1986 by Whelan et al. . Their paper reported the results of ^{14}C- and ^{35}S-labeled anaerobic experiments but did not attempt to measure the abundance of microorganisms in the sediment down to a depth of 167 meters below the seafloor. These were collected from the submerged deltaic fan of the Mississippi River during leg 96. Prior to this, Ron Oremland, a USGS geomicrobiologist investigating methanogenesis in seafloor sediments, collected and analyzed samples from the Gulf of California on leg 64 in 1982. Ron was a student of Henry Ehrlich, the quintessential geomicrobiologist from Cornell University and author of many editions of the textbook, *Geomicrobiology*.
21. Parkes et al. 1994.
22. A much more complete review of marine subsurface biosphere discoveries can be found in *Microbial Life of the Deep Biosphere* (ed. Kallmeyer and Wagner 2014).
23. Although these alteration features had been reported in photomicrographs of volcanic glass by Morgenstein in 1969 in a master's thesis from Syracuse University, it was Thorseth et al. (1992) who first attributed them to a biotic origin.
24. It seems that subsurface life was quite infectious in the early '90s. The observations in 1992 led to a swarm of reports on these features in ocean floor basalts to great depths, in ophiolites of all ages, and even in Archean pillow basalts. The microorganisms have yet to be isolated after twenty years of investigation, but their biological origin seems to be in little doubt given the reported presence of DNA by Giovannoni et al. (1996).

CHAPTER 7: THE SUBTERRANAUTS

1. Axel Lidenbrock's discovery of oversized fungi in the deep subsurface.
2. Hoffman 1992.
3. One of the theories at the time as to why *Deinococcus radiodurans* was so radiation resistant was that it maintained two copies of its genome at all times, a phenomenon referred to as polyploidy.

4. We still do not know the answer to this question. Many studies have been done on the geographical distribution of thermophiles in hot springs and have found geographical proximity to be related to similarity in 16S rRNA of the same species. Wind is the presumed agent of transport of microorganisms across the surface of our planet, either as spore or on dust particles or ice/water droplets. Certainly deep ocean currents homogenize the pool of thermophiles that might inhabit a newly born deep-sea vent. But exactly how species of bacteria are geographically distributed beneath the terrestrial surface has not been resolved by any means. In the case of *P. abyssii*, it would had to have been transported by aerosol from the Southern Atlantic or Indian Oceans to land and survive long enough on the high veld of South Africa to infiltrate the groundwater that eventually would arrive at the fracture site intersected by the Kloof gold mine at three kilometers depth. Is this really more likely to have occurred than that a 2.5-billion-year-old descendant of *P. abyssii* was hanging around in the South African crust?

5. Takai et al. 2001.

6. The CBD was an international treaty adopted at the 1992 United Nations Conference on Environment and Development in Rio de Janeiro, Brazil. The three stated goals of the CBD were the conservation of biological diversity, the sustainable use of biological components, and the fair and equitable sharing of benefits arising from use of genetic resources. The main points of the treaty established the sovereign rights of states to their own natural resources and made access to the biological resources subject to national rules and legislations. The treaty imposed expectations of access by prior informed consent and of nations' fair and equitable sharing in the benefits of commercial exploitation of their biological resources. Of course, I had no prior informed consent, nor did I have any clue as to how to arrange fair and equitable sharing of the benefits because I had no interest in obtaining any commercial benefits. Each party nation was expected to establish legislation and policies regarding access and benefit sharing. I could find no information as to whether the South African government had done this yet, but the accusatory email implied that they had. Adequate protection of intellectual property rights (IPR) was granted, but the treaty also created expectations that developing nations would be granted access to technology arising from the use of their biological resources, and this could lead to challenges to the rights typically granted by patents. The issues of IPR raised by the CBD had increased the commercial risks involved in bioprospecting, which was widely practiced in the 1990s. The 1992 Rio Earth Summit resulted in more than 150 governments signing the

CBD, including the United States. More than 187 countries (not including the United States) ratified the agreement, providing global support and acceptance of the treaty. In 1994, the U.S. Senate Committee on Foreign Relations approved the treaty, but the full Senate failed to ratify it. The treaty was primarily intended for plants with traditional pharmaceutical applications, which were yielding critical ingredients driving a $100-billion-a-year global industry. The case for applying it to bacteria was straightforward. The use of thermally stable polymerase extracted from *Thermus aquaticus* found in Yellowstone National Park for use in PCR had become the standard method of DNA amplification in the biotech industry and academia. But none of the profits resulting from that discovery went to the National Park Service. That would soon change. The potential commercial applications of proteins derived from novel mine bacteria could be equally enormous, and the patents for the development of these applications would normally go to the discoverer.

7. In all fairness to my detractors, they were perfectly right. Many geologists from the United States had come to South Africa, collected rock samples from world-class geological sites, and returned to the States to produce papers in *Nature* and *Science* without bothering to communicate or collaborate with South African geologists. It was practicing geological colonialism.

8. The Pretoria Group is a sequence of shale and volcanic rocks that are 2.35 Ga (giga annum; 2.35×10^9 years) old and overlie the Transvaal dolomites. Near Kloof the sequence had been metamorphosed to temperatures of around 250 to 350°C, hence the term "metavolcanic."

9. Lihung Lin was one of my new Ph.D. graduate students, whose recent arrival from Taiwan's National Taiwan University coincided with Duane's departure to PNNL. Lihung had spent the summer at the Marine Biological Laboratory in Woods Hole, Massachusetts, taking their microbiology boot camp to become proficient in DNA extraction, PCR amplification, and clone sequencing. Soon he was training our lab in molecular techniques.

10. "Geezer" is the South African term for a hot-water heater.

11. In the South African system you spend three years completing your bachelor's degree and then a fourth year doing an honors program on a research topic selected by your advisor. The new department chairman was Mark's advisor. Mark was an extremely capable microbiologist and eager to learn anaerobic techniques. His father was a biologist on the faculty at the University of Wits so he came from solid academic stock. He was also very athletic with a mean golf handicap, and when the call was made to his cell to mobilize for a drill intersection he rock-

eted to Glenharvie in a Wits departmental van. The only problem with Mark was that in order to pay his university expenses, he also had to work at a video store renting videotapes. The fortunate aspect of twenty-first-century South Africa was that scholarships were more readily available to black South Africans. The unfortunate consequence was that scholarships were not as available to white South African students. So I decided to hire Mark full time to pay him to do his honors thesis so he wouldn't have to waste his time working at a video store. Mark also proved very reliable at keeping an eye on the lab in between team visits.

12. The cementation, or grouting process, in which cement or some other chemical sealant is injected into fractures to seal the fracture, is fairly common practice in hard rock mines. Cementation Products was one of the companies we used that provided the required implements to the mines. Premier Valve was another. The packers, sometimes called Margot type packers, come in a variety of lengths, from a convenient 1.5-foot length up to 10 feet. The packers come in diameters ranging from 1 inch for diamond drill holes, to 96 millimeters (3.75 inches) for HQ-size boreholes. The packers are normally just temporary because at these great depths the boreholes themselves would deform and partially seal themselves over a period of years.

13. Takai et al. 2001.

14. Sean McCuddy, a graduate student of Tom Kieft's, and Amy Welte, a graduate student of Gordon Southam's.

15. We saw this contraption only at Kloof Gold Mine. It was a small railroad car fitted with a seat and bicycle chain and pedals for the driver, two seats in the front for passengers, and a small platform in the back for gear. It was like an underground version of the rickshaw.

16. Early one morning while awaiting an email response from the lab, I searched for the contact information for John Bahcall, a noted astrophysicist at the Institute of Advanced Studies in Princeton. John Bahcall was a good friend of Storey Musgraves, a former astronaut whom I wanted to invite to Princeton to talk to students taking my class "Terraforming Mars." My pupils dilated coming across his website, where it mentioned in bold capital letters, "National Underground Laboratory." I couldn't believe what I was reading. I knew nothing about neutrinos or dark matter, but the details revealed that a group of physicists from across the United States was pushing NSF to submit a monumental proposal to support the development of a two-mile-deep underground laboratory in the United States. The proposed lab would be located at Homestake Mine in Lead, South Dakota, The same mine that Duane had mentioned three years earlier as our backup option. Later that day

I called John, who fortunately was in his office. I explained what I had done and what we were doing in South Africa, and he quickly became quite excited. He told me that in December they were meeting in Lead, South Dakota, where the Homestake Mine was located. He suggested that I gather together a group of like-minded microbiologists to attend the meeting and get to know the physicists. The timing was perfect, because we needed a face-to-face meeting of all investigators to go over the results from our first year in South Africa and to plan for the next year's fieldwork.

17. Omar et al. 2003. The dating was part of Joost Hoek's undergraduate thesis at the University of Pennsylvania.

18. The warm little pond is in reference to a letter from Charles Darwin to Joseph Hooker on Feb. 1, 1871. "It is often said that all the conditions for the first production of a living organism are now present, which could ever have been present. But if (& oh what a big if) we could conceive in some warm little pond with all sorts of ammonia & phosphoric salts, light, heat, electricity & carbon present, that a protein compound was chemically formed, ready to undergo still more complex changes, at the present day such matter would be instantly devoured, or absorbed, which would not have been the case before living creatures were formed."

19. The Fischer-Tropsch process generates hydrocarbons from catalytic conversion of H_2 and CO at high temperatures and pressures and was developed in 1925 in Germany by Frans Fischer and Hans Tropsch. The chemical reaction is described by the following equation for producing alkanes: $(2n+1)H_2 + nCO \rightarrow C_nH_{2n+2} + nH_2O$. It was Barbara Sherwood Lollar who first discovered that a similar process was producing hydrocarbons naturally in the Earth.

20. Günther Wächerhäuser was a former patent lawyer who in 1988 proposed a hypothesis for the first energy source for the origin of life that involved the following reaction, $FeS + H_2S \rightarrow pyrite + H_2$. The H_2 could provide additional energy for the formation of prebiotic compound that eventually led to an autotrophic microorganism. His hypothesis was greatly influenced by Karl Stetter's discovery of pyrite coating the hyperthermophilic archaean, *Pyrodictium occultum*.

21. Sagan 1995, p. 200.

22. In this instance of PCR, short segments of DNA that are complementary to some portion of the target 16S rRNA gene DNA (oligonucleotide primers) are annealed to the denatured DNA of the sample. The polymerase then extends the DNA to copy the targeted 16S rRNA gene, and this process is repeated by thermal cycling, leading to millions of

copies of the target 16S rRNA gene sequence in a sea of other DNA sequences. But when the target DNA concentration is low, nonspecific binding may occur, which leads to copies of the nontargeted DNA mixed in with the copies of the 16S rRNA gene. Taking the amplified DNA product and then subjecting it to another round of PCR using different primers that target a smaller segment of the 16S rRNA gene that is more concentrated in the DNA mixture should, and often does, yield a purer 16S rRNA amplicon library that is better at detecting minor members of the microbial community.

23. Positive circulation of the drilling fluid is illustrated in figure 2.6. In reverse circulation the fluid moves down the outside of the coring barrel and up into the interior of the core barrel, carrying with it any broken core. Reverse circulation can work in hard rocks where the drilling fluid is not lost and where the water pressure can carry the rock chips or core to the top of the core barrel, which in the case of underground drilling is fairly easy, particularly if the coring is subhorizontal.

24. Another problem for us was that with several groups performing 16S rRNA gene analyses on either filters or enrichments, we really needed someone to act as a central collection and curation site for all the 16S rRNA gene sequences in the project and to make sure that everyone was using a common set of primers. In this way we could more readily correlate the findings from one group's borehole with another group's enrichments. Tom Gihring, being a young aspiring microbial ecologist with a penchant for molecular work, volunteered to act as the 16S rDNA guru for the project. A final problem was that the enrichments were proving problematic, as usual, so Tom and Tommy came up with the idea of using nutrient implants in the boreholes to see if the yield from these in situ enrichments would be more successful. But we had to find an existing low-pressure borehole that was not cemented and would be left available to us for these experiments.

25. By this time Russell Vreeland had moved from the University of Texas at El Paso to West Chester University and had just published his first efforts of extracting living bacteria from the Permian age WIPP salt. He had not attempted direct analysis of DNA extracted from the salt crystal as Jim had done. Instead he enriched for halophilic microorganisms using organic-lean mineral media. The strategy seemed to work because he was recovering only about one cell per kilogram of salt, if that. As we had, Dennis Powers pointed out the primary Hopper crystals to him. Russell improved upon the sterilization methods that Dombrowski had used in the 1960s and was able to obtain QA/QC assurance against contamination of 1 in 10^9, three orders of magnitude better

than the SSP had achieved at C10. Nonetheless, his publication received the same criticism that had been directed at Dombrowski's claim, only more so. Vreeland et al. 2000.

26. When we were working on the Parachute Basin coring campaign, I had come across a paper by the geochemist Hal Helgeson the previous year (Helgeson et al. 1993) describing the thermodynamic equilibrium between hydrocarbons and their organic acids. Organic acids are food for bacteria, and we thought we might be able to use his model to determine whether the bacteria were eating them. The model was based upon the second law of thermodynamics, which calculates the Gibbs free energy of any chemical reaction using the following formula,

$$\Delta G = \Delta G^\circ + RT \ln[Q],$$

where ΔG° is the Gibbs free energy of the reaction at standard conditions of temperature and pressure and at a pH of zero, R is the real gas constant, T is the temperature, and Q is the ionic product ratio. If you have a reaction represented by the following

$$aA + bB = cC + dD,$$

where A and B represent the reactants, C and D represent the products, and a, b, c and d are the stoichiometric coefficients of the reaction. Q is given by

$$Q = \{[C]^c[D]^d\}/\{[A]^a[B]^b\},$$

where [A] represents the concentration of that species. Well, not really the concentration, but what chemists call the "activity" of the species. If the organic acids were in thermodynamic equilibrium with the alkanes in the Parachute borehole, then the $\Delta G = 0$ for the relevant reactions. If they were not, then it was possible that their concentrations were being perturbed by microbial utilization of the organic acids. I really did not understand most of that paper at that time, because I hated thermodynamics. It seemed the most boring topic in my college physics classes so I never really understood it. But in the appendix of Thomas Brock's textbook that Tom had given me was an easy-to-understand application of Gibbs's relationship to microbial redox reactions. At this time Everitt Shock, a protégé of Helgeson's, began producing papers on the stabilities of organic compounds as a function of temperature, and this could be important for thermophilic microorganisms. Helgeson's and Shock's students began publishing a thermodynamic database for organic and inorganic molecules that were the food sources and waste products of microbial metabolisms. How this chemical energy was utilized was elegantly described in a paper by Rudolf Thauer for anaerobic bacteria

(Thauer et al. 1977). By performing experiments on anaerobic isolates, Thauer was able to demonstrate that bacteria conserved energy using the chemiosmotic mechanism to maintain a proton gradient and produced ATP as a result of the protons crossing that gradient from the exterior to the interior of the cell. ATP is produced from ADP and phosphate by the ATP synthase protein, and it requires two to four protons to synthesize one ATP molecule. The reaction can be written as

$$ADP + P = ATP,$$

and it too has associated with it a ΔG. Intriguingly, Thauer claimed that the free energy released by the biologically catalyzed chemical reaction was conserved to produce ATP. This meant that if the ΔG of the chemical reaction was less than the ΔG required to produce ATP, then the bacterial metabolism would stop. The Thauer paper; the thermodynamic database produced by Helgeson, Shock, and others; and Gibbs's simple formula provided us with a straightforward tool to evaluate the potential bioenergetics of any metabolic reaction in our deep-fracture environments as long as we diligently determined the concentrations of all the relevant reactants and products. We could now use our geochemical data to determine which metabolic reactions were favorable and which were not. As a result, thermodynamics had become very interesting to me.

27. To see the You Tube version of this 1952 move clip, check out https://www.youtube.com/watch?v=rqy4xcXgtxM.

28. In South Africa, cage inspections occurred on an almost weekly basis. We would often see men carrying ropes and chains and dressed like they were going mountain climbing, hitching a ride in the cage.

29. We would learn in a few weeks that Johanna's analyses revealed that the water contained a large amount of atmospheric noble gases, consistent with a very young age for the fracture water.

30. Tommy was more prophetic than he probably would have wished to be. Within two years time, Barrick Mining, a Canadian mining firm, would shut down the Homestake mine. With the pumps turned off, Homestake began to fill with water. This happened just after several of us had produced a white paper for the Geosciences Division at NSF describing a new earth lab that would include microbiology experiments. With the situation at Homestake unresolved, NSF developed a process whereby the physics and geoscience communities would propose several possible sites for a Deep Underground Science and Engineering Laboratory (DUSEL). But in the end, Homestake was selected again for the location of DUSEL in part because T. Denny Sanford had

donated multimillions of dollars to support the development of a lab in his name there. By the time DUSEL could turn the pumps back on, the water had risen to the 4,500-foot level, 3,500 feet above the 8,000-foot level we had visited in 2000. However, with the collapse of the U.S. economy in 2009, NSF and the executive branch backed out of supporting any research at DUSEL. The lab then changed its name to the Sanford Underground Research Lab (SURF). DOE has supported the installation and operation of several astrophysical experiments including searches for dark matter and double beta decay on the 4,850-foot level. NASA has supported microbial investigations of these upper levels. The levels below 4,850 feet, however, remain flooded to this day.

CHAPTER 8: A LOT OF BREAKS AND ONE LUCKY STRIKE

1. Axel's ruminations on the origin of the Lidenbrock Sea, deep beneath the surface of the Earth.
2. SASOL operates five coal mines immediately surrounding the town of Secunda. Each mine delivers multimillion tons of coal per annum to the SASOL plant. The plant burns the coal and the CH_4 gas from coal to produce steam to generate electricity. Most of this electricity is then used to gasify the coal at 1,300°C. This syngas is then used in reactors to produce synthetic gasoline and ammonia.
3. As we approached the boom gate, the guard came out with a clipboard and a white box with a red button on it. Last year, every time we drove into and out of the property we were obliged to write our names and the registration number of the vehicle. Eric had done this a lot and was convinced that there was absolutely no real security value in this exercise at all. As if to demonstrate this point to me, when the guard handed him the clipboard he signed "Eleanor Roosevelt" and nonchalantly handed it back to the security guard. The guard looked carefully at the signature and then looked at our faces and then tucked the clipboard under his arm. We had never seen the box before. Kloof clearly had instituted new security procedures to cut down on theft in the mines. The guard held out the box to Eric, who was the driver. He didn't tell us to do anything, but his intention seemed quite clear to Eric that he needed to press that red button. Eric reached out and pressed the red button. This pleased the guard immensely until he looked at the back end of the box and a wave of disappointment swept his face. He then returned to the boom gate and opened it for us to drive through. Lisa and I

started cracking up as Eric drove onward ranting about the absurdity of the security regulations in South Africa. You have to understand that Eric reminded me of a young, tan Rodney Dangerfield, only funnier. He was exactly what we needed at the Glenharvie field house because not only did he look out for everyone to make sure they were safe, but he could defuse any tense moment with a joke.

4. The Witwatersrand Supergroup is almost entirely dominated by fluvial sandstones and conglomerates deposited in the channels of streams eroding mountains to the north 2.9 billion years ago. The sequence is several kilometers thick but contains only one thin, approximately 100-meter-thick shale layer that probably represents marine clay deposited during a minor transgression of the Archean ocean.

5. The next day we, along with Peter Roberts, visited the Level 19 borehole that Colin had "donated" to Bianca for her experiments. Two weeks earlier she had installed her in situ cultivators deeply into the borehole and had tried to seal the borehole with a compression fitting that we had designed in the lab. Unfortunately, old, deep boreholes deform as they get older and loose their circularity. This one was no exception. When we arrived at the cubby where the horizontal borehole resided, water was pouring out past our carefully machined compression seals. The individual cartridges Bianca had inserted some three meters into the hole were still above water. We took measurements and returned to the surface after collecting some biofilms deep inside a ceiling borehole for T. J.'s senior thesis work. Peter suggested we use a wooden plug that would swell as it became saturated with water. We returned to Glenharvie to manufacture the plug, drilling holes into it for the tubes coming from each in situ cultivator and one larger hole for the support rod. The next day we returned to level 19. We slid the cultivator partway out of the hole, slid on the wooden plug, threaded the tubing through the plug, and pushed it back into the hole. Peter then hammered the plug into place until the water stopped leaking. The water level started rising in the borehole and after about a half hour, water started leaking through all the tubes coming from all the cartridges. We let them flow out of the tubes for another half hour before closing their tiny valves. As long as the plug held and the pressure relief valve did what it was supposed to do, we had solved the last problem from the Lead workshop, that of in situ cultivation.

6. In pursuit of funding for this student training program, Susan arranged a meeting with several program managers at NSF headquarters in February 2001. The cherry trees had just begun to blossom as Wes, Esta, who had just officially joined the faculty by this time, Derek, Susan, and

I met in Washington, D.C., and pitched Susan's idea. Wes had prepared a Nike-style "Just do it" presentation outlining the salient points of the proposal with his charisma gene switched full on. With Rich Lane's help, NSF's International Program endorsed the idea. We would start with some seed money from the REU's program to be followed by a full proposal with Susan as the lead PI.

7. Although we did not know it at the time, the isotopic analyses of the water from these boreholes would reveal that the water actually originated not from the Sand River, but from the 8,500-foot-high mountains of Lesotho, 140 miles to our southeast. The noble gas isotope data would tell us that the water had taken more than a million years to make the journey.

8. This event is referred to as the Merriespruit tailings dam disaster. It occurred in 1994 and led to new regulations regarding the zoning around tailings dams across South Africa and to the implementation of hazard management strategies at each tailings dam.

9. Kieft et al. 2005.

10. Lippmann et al. 2003.

CHAPTER 9: LIFE BENEATH THE ICE

1. The Latin (with English translation) ancient coded message from the Icelandic alchemist, Arne Saknussemm, discovered by Professor Lidenbrock, describing how to enter the subsurface beneath the frozen volcanic landscape of Iceland.

2. Rossbacher and Judson 1981.

3. Clathrate (or gas hydrate) is a water-ice lattice that contains a molecule that above the melting point of the clathrate is a gas. CO_2 forms its own ice at low enough temperatures, but in the presence of water and high pressure, a CO_2 clathrate will form that will stabilize the CO_2 in a solid phase to a temperature well above the sublimation temperature of dry CO_2. Similarly, beneath the ocean floor at low temperatures and high pressures, CH_4 gas will combine with water to form a clathrate.

4. Both the European Space Agency Mars Express and the Mars Global Satellite use GPR to probe the Martian subsurface.

5. During the previous six weeks, I had managed to assemble the requisite sixty pages of text, patching together text and figures emailed from Terry Hazen at LBNL, Fred Brockman at PNNL, Tommy, Lisa, Steve, and Susan. Unfortunately, we were only two weeks out from the deadline, and I had completely failed in acquiring the institutional support

from Princeton University that was required to submit the proposal. Fortunately for all of us, Lisa, a former assistant dean at Indiana University, was able to twist arms at the university to contribute institutional matches. I handed the NAI proposal over to her with ten days to spare, and she submitted it with herself as the lead PI and Indiana University as the lead institution. To cover our bases, I had managed to include email letters from the mine managers of Lupin, Con, Ekati, and Nunasivik expressing their willingness to host our sampling teams, but Lupin was our primary hope. I had not worked with Terry Hazen before, but he had just been funded by the U.S. DOE to support gene sequencing and DNA microarray analyses of environmental samples. These were new molecular tools we had not tested on our subsurface samples. Fred wanted to try a new microgel, single-cell isolation and enrichment technique that a biotech firm Diversa had pioneered.

6. Issatchenko 1940.
7. Vernadsky 1926.
8. Rozanova and Khudyakova 1970.
9. Bjergbakke et al. 1989.
10. Vovk 1987.
11. Madsen et al. 1996. By this time Derek Lovley had moved from the USGS to the University of Massachusetts, Amherst.
12. Fry et al. 1997.
13. Anderson et al. 1998.
14. Stevens and McKinley 2000.
15. Chapelle et al. 2002.
16. Kita et al. 1982.
17. Johanna Lippmann-Pipke, however, would find evidence in 2012 that H_2 was released during man-made mine quakes in Tau Tona Gold Mine, South Africa (Lippmann-Pipke et al. 2012), so perhaps the Japanese theory had some legs to it after all.
18. It was most closely related to *Desulfotomaculum kuznetsovii*, which was isolated from 3,000-meter-deep thermal water from an oil reservoir in western Siberia by Nazina et al. (1988).
19. Gihring et al. 2006.
20. Chapelle et al. 2002.
21. Mass-independent isotopic fractionation worked as follows. In the Archean rocks, the amount of O_2 in the atmosphere and thus the amount of O_2 in the surface water was very limited. This meant that most of the sulfate that was present in Archean water did not originate from oxidation of sulfide minerals on the continental surface or from oxidation of dissolved sulfide emitted by deep-sea vents. Most of the sulfate in the

water originated from the SO_2 emitted by volcanoes into the atmosphere. Due to the absence of significant ozone, UV photolysis of SO_2 molecules meant that the H_2SO_4 molecules that formed in the atmosphere and precipitated into the ocean as acid rain had an isotopic distribution of ^{32}S, ^{33}S, ^{34}S, and ^{36}S that did not follow the mass-dependent pattern associated with chemical reactions. When pyrite or barite minerals formed from this sulfate, that mineral inherited the oddball isotopic signature.

22. Lefticariu et al. 2006.
23. Lin et al. 2006.
24. The bacteria can also penetrate pore sizes smaller in diameter than their rods by protoplasmic extrusion during growth. This was noted by ZoBell (1951).
25. Fisk and Giovannoni 1999.
26. Gaidos et al. 1999.
27. Onstott et al. 2006.
28. Chyba and Hand 2001.
29. Just to give the reader a sense of how transient the precious-metal mining industry is, I will provide a quick update on the history of ownership of this extreme Arctic mine. In 2006 Wolfden Resources acquired Lupin from Kinross primarily to support its zinc mines in the region, including nearby Gondor (yes, that is right, you can never escape the Lord of the Rings in the mining world). I thought it rather appropriate that a company named Wolfden would own a mine named Lupin, but alas, Wolfden was acquired by Zinifex in 2007. In 2008 Zinifex merged with Oxiana and formed OZ Minerals. In 2009 most of OZ Minerals' assets, including Lupin, were acquired by China's Minmetals, now known as MMG Resources. Then in 2011, Elgin Mining purchased Lupin from MMG. Elgin Mining then suffered heavy losses and sold Lupin to Mandalay Resources in 2014. All these changes were governed by fluctuations in the price of gold and exchange rates with the U.S. dollar, and the owners couldn't decide whether to reopen Lupin or not.
30. An excellent historical review is provided by Gilichinsky and Wagener (1994). See McKay 2012 for a tribute to David Gilichinsky and a summary of his career.
31. Rivkina et al. 1998.
32. Priscu et al. 1999; Karl et al. 1999. But prior to these publications, a Russian paper (Abyzov et al. 1998) published the detection of viable bacteria in the Vostok core between 1,500 and 2,700 meters depth.
33. Skidmore et al. 2000.
34. Christner 2002.

35. Jakosky et al. 2003.
36. This is about as close as I can get to the Finnish pronunciation of "sauna."
37. The Tibbitt to Contwoyto Winter Road was the location of the first season of *Ice Road Truckers*.
38. They would start in a tunnel parallel to the vertical gold ore zone. They would then make a short tunnel, a draw point, sideways until they intersected the ore zone. Then at the end of the short tunnel they would start drilling upward, creating a raise with the Alimak rising up into the raise. The rock they removed would simply fall down to the bottom of the raise, from which it was removed. They would then break through to the tunnel overhead, where they would set an anchor bolt in its ceiling. They then connected the cable of the Alimak to the ceiling of the top tunnel and disconnected the cable at the bottom. The Alimak was now suspended from the upper tunnel. The miner then simply pushed "down" on the Alimak and it lowered him safely down to the bottom of the raise. He could then drill and blast into the ore zone on opposite sides of the raise, starting at the bottom and working his way up, pulling up his cable with him. The ore would fall to the bottom of the raise where it would be scooped out. When he finally reached the top tunnel, he would have completely removed the ore in a fifteen- to twenty-meter vertical slice. The loaders then brought in waste rock and poured it down into the hole from the upper tunnel. This backfilled the vertical slice.
39. Clifford 1993.
40. Ibid.
41. Taliks come in both through and closed forms. They arise because the heat released by water during freezing is released into the residual underlying water layer and keeps the latter from freezing.
42. As our sun has been evolving along the main sequence, its intensity has been slowly increasing. Four billion years ago its intensity was only 75% of the current intensity, which means that our planet and Mars were receiving less solar radiation and would have been colder. If not for a CO_2-rich atmosphere, Earth would have been covered with ice. Mars would have been so cold that even with a CO_2-rich atmosphere, its surface would have still been frozen. Models of surface geology that are interpreted as indicating long-standing water have to be reconciled with this dim-sun fact.
43. I could not imagine that after having put so much money into building this tunnel and installing this infrastructure, Kinross would close down Lupin. It would make a great Arctic amusement park. They could hold

mini-NASCAR-like races up and down the tunnel or perhaps use it for paintball war games. The profits, of course, would support environmental and scientific research. Just a suggestion to any entrepreneurs reading this book.

44. Boart/Longyear is a mineral exploration company that offers drilling rigs for hire, including ones especially designed for underground hard rock mines.

45. Chivian et al. 2008.

46. Horizontal gene transfer, also called lateral gene transfer, is a process in which phylogenetically distinct lineages, be they different species or even from different domains, exchange genes that become integrated into the chromosomal, or core, genome that is replicated during reproduction. In the case of prokaryotes, a variety of processes exist that can move foreign DNA from a donor to a recipient cell. In bacterial transformation, the recipient acquires fragments of intact DNA from the environment, from lysed cells of the donor. In bacterial transduction, the DNA is transferred from the donor to the recipient via a virus. In bacterial conjugation, the donor transfers small loops of DNA, called plasmids, to a recipient through a proteinaceous tube (the equivalent of bacterial sex). This is the process generally deemed responsible for the spread of antibiotic resistance among pathogens in hospitals. The DNA of the plasmid would then have to be incorporated into the recipient's genome.

47. In the twenty years that had passed since Carl Woese's group first proposed using the rRNA genes, especially the 16S rRNA gene, to categorize prokaryote life forms, entire genomes had been sequenced. From this sequencing effort it was discovered that almost all the genes related to forming the proteins that were also part of the ribosome were fairly conservative and not exchanged between species, yet they were universally distributed. See, for example, Ciccarelli et al. 2006. Concatenating these sequences into one long sequence of 8,090 base pairs provided much higher resolution of the evolutionary differences between species. In our case, "*Candidatus* Desulforudis audaxviator" was 88% identical to the closest *Desulfotomaculum kuznetsovii* and thus a different genus. Appropriately, *D. kuznetsovii* was named after the Russian microbiologist who had been an early advocate of deep life. This new tree also placed "*Candidatus* Desulforudis audaxviator" into the earliest, deepest branch in the bacterial domain. From an evolutionary point of view, the ancestor of "*Candidatus* Desulforudis audaxviator" was one of the earliest bacteria to form, but its genome had probably evolved greatly since then, given its foreign elements.

48. It would take more than two years of battling with editors and referees before it would finally appear in print in *Science*.

49. The term "thermokarst" is the name given to irregular topography formed on slopes by landslides caused when the ice that cements together the permafrost soil matrix thaws. Without that icy framework, the soil loses its cohesion and collapses. Thermokarst is a feature common in any polar region where the soils are underlain by permanently frozen ground, aka permafrost.

50. Gossan is the reddish oxidized soil that develops over sulfide ore deposits that are exposed at the surface. The red color is due to ferric iron, but yellowish and orange colors are also present and reflect other ferric iron mineral phases. Because the primary mineral is sulfide, iron and sulfide-oxidizing aerobic bacteria will convert the sulfide to ferric iron minerals and sulfate, which forms sulfuric acid. The acidity of the water in the gossan dissolves the other silicate phases, leaving behind quartz. The remarkable thing about gossan in permafrost regions is that it penetrates only down to the active layer, the seasonally thawed layer, but not into the permafrost itself, where water never forms. The name "gossan" originated as Cornish miner's slang and comes from the Cornish *gōs*, meaning "blood."

51. Both sulfide- and ferrous-iron-oxidizing acidophilic bacteria and Archaea are involved in this complex process.

52. The mess hall was the largest building on site and also contained the toilets and showers and water supply for the site.

53. Ice-wedge polygons are ubiquitous surface features in the Arctic and are formed by the seasonal freeze-thaw of ice. The polygons are typically square or trapezoidal shaped, and the winter ice wedge penetrates down about one foot, to the permafrost. During the summer the ice wedge melts, leaving behind a trough that separates the hummock. Depending upon how far north you are, the polygons are more or less vegetated. Similar features appear on the surface of Mars without any vegetation, of course.

54. Cryopegs are saline pockets of water that are formed during the fractional freezing of water. Upon freezing of saline or brackish groundwater, the salts become concentrated in a residual, unfrozen brine pocket. In the Siberian permafrost deposits, bacteria were isolated from brine pockets that were believed to have formed more than a million years ago. This is the icy version of the salt-crystal fluid inclusions that serve as repositories for ancient life forms.

55. The position and length of each investigator's core was noted. Whirl-Pak bags for DNA and lipid samples were taken out of the glove bag and placed on dry ice in a cooler. Whirl-Paks for activity or biological

samples were placed in sterile mason jars, taken out of the glove bag, and placed in a cooler with Blue Ice. Core chips and internal nuggets were collected in centrifuge tubes for microsphere analysis and in serum bottles for PFC analysis. Cores collected to determine the chemical, mineralogical, and physical properties of the rock core were placed in separate plastic bags and required no special storage. These included mineralogical and physical properties of the rock core, fracture mineralogy, fluid inclusion, porosity, geophysical properties, $\delta^{13}C$ and $\delta^{18}O$ isotopes on calcite and total organic carbon, $\delta^{34}S$ on sulfides, $\delta^{34}S$ and $\delta^{18}O$ isotopes on sulfate, crush-and-leach chemistry for organics, inorganics, and $\delta^{81}Br$ and $\delta^{37}Cl$ isotopes, and chemical diffusion experiments. Cores for voltametric analyses were placed in plastic bags and stored by two methods, one refrigerated in anaerobic jars and the other frozen with dry ice. Microbial diffusion experiment cores were stored in anaerobic jars and refrigerated.

56. Stotler et al. 2009; Onstott et al. 2009.
57. These results were obtained by Bruno Soffentino, at the time a postdoc associated with Steve D'Hondt at the University of Rhode Island. Bruno employed the same highly sensitive methods developed by the oceanographic community to count bacterial cells in seawater.
58. Abyzov et al. 1998.
59. Rivkina et al. 1998.
60. Miteva et al. 2004.
61. Pfiffner et al. 2008.
62. Stotler et al. 2010.
63. The U.S. Army's Cold Regions Research and Engineering Laboratory is in charge of developing the technology for the Army and other branches of the Department of Defense to support operations in winter or cold (i.e., alpine or polar) regions that range from military operational needs to environmental issues associated with those operations and also to engineering for construction of facilities.
64. Classic line by Slim Pickens (as Major T. J. "King" Kong) from the movie *Dr. Strangelove*.
65. This was in reference to the "Naysaw" (NASA) guys who own space in the much-discussed Canadian comedy show *Trailer Park Boys*, which none of us Americans had seen. I've only seen YouTube clips, so my apologies to my Canadian colleagues.
66. Our team comprised Adam Johnson; Randy Stotler, a videographer whom Lisa Pratt had sent along to document the expedition; a graduate student of Sean Frape's; and Sean Frape's field technician.
67. The drillers monitored the salinity closely and constantly added salt to

keep the water at a concentration at which the melting temperature was –20°C. The coldest spot in the borehole was only –6.5°C.
68. Freifeld et al. 2008.

CHAPTER 10: THE WORM FROM HELL

1. The end of a battle between aquatic dinosaurs encountered by the Lidenbrock party aboard a raft on the Lidenbrock Sea.
2. Even Chris McKay was reconciled to the fact that NASA could not transport a drill that could penetrate more than one to two yards without great cost. A mission called THOR was even proposed to send a half-metric-ton sphere of copper on a ballistic trajectory into the Martian surface that would produce a four-yard-deep impact crater. If the copper sphere collided near one of the rovers, but not too near it, NASA could send it over to look at the impact debris or even enter the crater and look for signs of organics.
3. Malin et al. 2006.
4. It has been proposed in popular blogs that H-bombs detonated in the polar regions of Mars could also be an relatively inexpensive means of terraforming Mars.
5. National Research Council 2006. Chris Chyba, who had been a physicist at the SETI Institute and Stanford University and who had just moved to Princeton University, headed the committee that drafted the report. The current NASA sterilization procedures are less stringent that those used in the days of the Viking missions. The contamination assays—cultivating bacteria using a single medium—also drew scorn from the committee, which believed this approach highly underestimated the bioload being carried by NASA Rovers. At that time, the committee was also concerned about the discoveries made of microbial life in extreme environments, particularly those microorganisms found in the deep biosphere, ice sheets, and salt deposits. NASA had not been paying serious attention to the possibilities of extant extremophilic life on the Martian surface.
6. The MEPAG Special Regions-Science Analysis Group 2006, p. 677.
7. A very eloquent and logical summary of the Committee on Space Research (COSPAR) PP policy concerning "harmful contamination" of potential Martian habitats by Earth spacecraft has been provided by Chris McKay (2009). The remarkable images and data produced by NASA and European Space Agency spacecraft in the previous ten years had unraveled the history of water on Mars to the point that we could

identify certain regions on the surface that might contain extant life, but budgetary restrictions prevented us from ever going there to explore for life as NASA had in the 1970s. Even if the budgets were not an issue, there is a fear that any mission sent to a special region on Mars to search for life might fail to find it, as the Viking missions had failed to definitively find life. In my opinion, in the case of the Mars Exploration Program, mission failure was never an option.

8. This was the first meeting of what would become the ISSM series.

9. Boston et al. 1992.

10. At this time *Mariner 9* and Viking orbital imagery indicated features on the oldest surface areas of Mars that appeared to be related to erosion by surface water (the subject is well reviewed by Michael Carr 1996). The Martian stratigraphic system has been subdivided into three time periods. The oldest is the Noachian, which dated from the formation of Mars at 4.66 Ga, or giga annum, until about 4.0 Ga. This is followed by the Hesperian, which spans the time from 4.0 Ga until about 2.7 Ga. The youngest time period is the Amazonian, which runs from about 2.7 Ga until today. This system was first proposed by Ken Tanaka of the USGS (Tanaka 1986).

11. This was just two months after Tom Kieft, Dennis Powers, and I made our first visit to the WIPP tunnels southeast of Carlsbad. Because Lechuguilla had been sealed, it should have been free of bats and surface life forms that hopelessly contaminate most caves. This, plus its sulfuric acid history, made it more interesting for microbiology and motivated the 1994 expedition. This was originally suggested to Chris McKay by Larry Lemke (an engineer at Ames) after he saw a TV show on the cave. Chris immediately called Penny, and the three of them were the guests of Kim Cunningham. Although Penny became an avid cave scientist, Chris decided it was not for him because he wanted to focus on year-round autonomous environmental monitoring (hard to do in the 1980's) of extreme environments where variability in the environmental conditions strongly influenced the microbial ecosystems. Not much annual variability occurred in Lechuguilla.

12. Moonmilk is an often white, almost alabaster, coating on cave walls that has a knobby, botryoidal texture. It is composed of extremely small crystals of calcium carbonate.

13. Back-reef facies refers to the type of sediments deposited in the shallow ocean water separating the reef from the shoreline.

14. "Evaporitic" means formed during evaporation of seawater. During the progressive evaporation of seawater, carbonates precipitate first, followed by gypsum and other sulfates, which are then followed by halide salts.

15. "Speleothem" is derived from the Greek *spelaion* for "cave" and *thema* for "deposit." "Speleothem" usually refers to the stalagmites and stalactites that form from dripping water seeps and that are layered because of the seasonality of the water seepage. As a result, speleothems have been used as continental paleoclimate records.

The $\delta^{34}S$ of the cave sulfur and gypsum was as light as –26‰, which matched the isotopic composition of the H_2S gas from the Delaware Basin oil and gas fields and left no doubt that it represented oxidized, isotopically ^{34}S-depleted H_2S. This H_2S was produced by subsurface SRBs that had reduced isotopically heavy Permian Period gypsum that had a $\delta^{34}S$ of about +15‰. (Hill 1990). The difference between the $\delta^{34}S$ of the H_2S and the sulfate of the gypsum is consistent with the isotopic fractionation of SRBs. The reaction proposed by Hill for the production of the isotopically depleted H_2S, was $Ca^{2+} + 2CH_4 + 2SO_4^{2-} + 2H^+ \rightarrow 2H_2S + CaCO_3 + CO_2 + 3H_2O$, and not due to hydrocarbon oxidizing SRBs. This is a good indication of just how long the opinion was held that oil field SRBs were contaminants. These days we would look at Hill's reaction and consider anaerobic CH_4 oxidizing Archaea as the agents that are enzymatically catalyzing this reaction.

16. Morehouse 1968; Egemeier 1981.

17. An extensive review of the controversial topic of biological deposition and erosion of carbonate is provided by Viles 1984.

18. Troglophilic means liking caves or adapted to cave-like environments. A springtail is a small flea-like organism that belongs to the subphylum Hexapoda. Members of this subphylum hold the world's record for being the deepest-dwelling animal. It was discovered in 2010 existing two kilometers beneath the surface in Krubera Cave, located in Abkhazia, Georgia.

19. Speleosol is the soil that forms on cave floors. In this case the Fe/Mn deposits form on the walls by oxidation of the reduced forms of Fe and Mn that are present as trace elements in the calcium carbonate. As the calcium carbonate is dissolved, it leaves behind a fluff formed by Fe and Mn oxides. This fluff then falls to the floor.

20. Cunningham et al. 1995.

21. Sarbu et al. 1996. These authors found that the $\delta^{13}C$ and $\delta^{15}N$ of the biomass of the carnivores and grazing meiofauna was very similar to that of the sulfide-oxidizing bacteria, demonstrating that the chemoautolithotrophs were supporting the food web.

22. Riess et al. 1999. Nematodes, or roundworms, are distinguished by having a tubular digestive track with openings at both ends. They occur both as free-living species and as parasites. They vary in length from half a millimeter up to a meter in length for parasitic nematodes.

23. An excellent review of the biodiversity of sulfidic caves is provided by Annette Summers (2007). At that time the ability of multicellular animals to live in water with less than 18 micromolar O_2 had become well established. In this paper she also summarizes multiple findings that indicate that the rates of autotrophy exceed the rates of heterotrophy in sulfidic cave systems.

24. Thermophilic chemolithoautotrophs have been found more recently in the fracture water at depth beneath the crystal cave by Ragon et al. (2013).

25. Hose et al. 2000.

26. This segment was first broadcast in March 2006 by the BBC.

27. These were crystals of selenite: $CaSO_4 \cdot 2H_2O$. The calcium originates from the limestone. The snotites are composed of filamentous *Acidothiobacillus* bacteria.

28. Borgonie et al. 2010.

29. Nematodes have a life cycle that includes an ability to enter an alternative larval stage during which it can survive much harsher environmental conditions in a stasis state than can the actively reproducing adult and actively growing juvenile.

30. The cage rides at Tau Tona had considerably mellowed since my experiences five years earlier. I attributed this to the presence of a considerable number of female underground workers. Their presence in the cage seemed to have put an end to the hazing and horsing around that had been common practice in the late '90s. The integration of both women in general and black South African men and women, more specifically, into all levels of management as well as the workforce at Tau Tona was a signature that even the mining community, usually a bastion of male authority in any country, had been transformed in South Africa. It was quite an amazing sight.

31. The details of the remarkable space journey by *C. elegans* and its relationship to the tragic events surrounding the Columbia shuttle disaster were published by Szewczyk and McLamb (2005) and by Szewczyk et al. (2005).

32. Armand Bester, Marianas Erasmus, and Errol Cason and her postdoc, Antonio García-Moyano.

33. You'll find a deliciously written account of Marc's visit to the shipwreck tunnel in Northam platinum mine with Gaetan and Esta in Marc's book *First Contact* (Kaufman 2011).

34. Dr. Seuss (1960) *One Fish, Two Fish, Red Fish, Blue Fish*. Random House, New York.

35. It would, in fact, take two years before we could convince reviewers at *Nature* that *H. mephisto* was not a contaminant. We first had to collect

40,000 liters of mining water to determine whether it carried *H. mephisto*, which it did not. In fact the DNA in the mining water was highly degraded. A key piece of the evidence was an experiment performed by Esta's graduate student Marianas Erasmus, who tried to get *H. mephisto* to grow in the mining water, but it died every time. The mines in South Africa heavily treat the water with chlorine, bromine, and hydrogen peroxide in order to remove any potential infectious agent, including roundworms. For the miners, the mining water is a source of drinking water. If it had carried roundworms, the consequences would be disastrous for all.

EPILOGUE

1. Olson et al. 1981.
2. Wilson et al. 1983.
3. Ekzertsev 1951.
4. Meshkov 1958.
5. Kuznetsov et al. 1962.
6. Parkes et al. 1994.
7. Parkes et al. 2014.
8. Rinke et al. 2013.
9. See the review by Edwards et al. (2012).
10. Orsi et al. 2013a.
11. Orsi et al. 2013b.
12. Labonté et al. 2015.
13. Onstott et al. 2014.
14. For just one example, the reader should look at Rothman et al. 2014.

APPENDIX A: CHRONOLOGY OF THE EXPLORATION OF SUBSURFACE LIFE

1. Many other geological, geochemical, hydrological, and isotopic studies have contributed to the study of Earth's deep biosphere. This chronology focuses just on biological exploration primarily of the continental crust.

REFERENCES

Abyzov, S. S., et al. (1998) *Advances in Space Research* 22:363–368.

Alexander, M. (1977) *Introduction to Soil Microbiology*. R. E. Krieger, Malabar, FL.

Al'tovskii, M. E., et al. (1961) *Origin of Oil and Oil Deposits*. Consultants Bureau, New York.

Amy, P. S., et al. (1992) *Applied and Environmental Microbiology* 58:3367–3373.

———. (1993) *Current Microbiology* 26:345–352.

Anderson, R. T., et al. (1998) *Science* 281:976–977.

Arrage, A. A., et al. (1993) *Applied and Environmental Microbiology* 59: 3545–3550.

Back, W. (1989) *Groundwater* 27:618–622.

Balch, W. E., et al. (1977) *Journal of Molecular Evolution* 9:305–311.

Balkwill, D. L., and Ghiorse, W. C. (1985) *Applied Environmental Microbiology* 50:580–588.

Bargar, K. E., et al. (1985) *Geology* 13:483–486.

Baross, J. A., and Deming, J. W. (1983) *Nature* 303:423–426.

Bastin, E. S., and F. E. Greer (1930) *Bulletin of the American Association of Petroleum Geologists* 14:153–159.

Bastin, E. S., et al. (1926) *Science* 63:21–24.

Bechamp, A. (1868) *Annales de chimie et de physique* 13:103.

Belyaev, S. S., et al. (1983) *Applied Environmental Microbiology* 45:691–697.

Bjergbakke, E., et al. (1989) *Radiochimica Acta* 48:65–77.

Bjornstad, B. N., et al. (1994) *Ground Water Monitoring and Remediation* 14:140–147.

Bogdanova, V. M. (1965) *Microbiologiya* 34:300–303.

Boone, D. R., et al. (1995) *International Journal of Systematic Bacteriology* 45:441–447.

Borgonie, G., et al. (2010) *Biological Bulletin* 219:268–276.

———. (2011) *Nature* 474:79–82.

Boston, P. J., et al. (1992) *Icarus* 95:300–308.

Brock, T. D. (1986) *Thermophiles: General, Molecular, and Applied Microbiology*. John Wiley & Sons, New York.

Brock, T. D., and Madigan, M. T. (1991) *Biology of Microorganisms*. 6th ed. Prentice Hall, Englewood Cliffs, N.J.

Brockman, F. J., et al. (1992) *Microbial Ecology* 23:279–301.

Cano, R. J., and Borucki, M. K. (1995) *Science* 268 (5213):1060–1064.

Carr, M. H. (1996) *Water on Mars*. Oxford University Press, New York.

Chapelle, F. H. (1993) *Ground-Water Microbiology and Geochemistry*. John Wiley & Sons, New York.

Chapelle, F. H., et al. (1987) *Water Resources Research* 23:1625–1632.

———. (2002) *Nature* 415:312–315.

Chivian, D., et al. (2008) *Science* 322:275–278.

Christner, B. C. (2002) *Applied and Environmental Microbiology* 68:6435–6438.

Chyba, C. F., and Hand, K. P. (2001) *Science* 292:2026–2027.

Ciccarelli, F. D., et al. (2006) *Science* 311:1283–1287.

Clifford, S. M. (1993) *Geophysical Research Letters* 18:2055–2058.

Clifford, S. M., and Parker, T. J. (2001) *Icarus* 154:40–79.

Cockell, C. S., et al. (2012) *Astrobiology* 12:231–246.

Coleman, M. L., et al. (1993) *Nature* 361:436–438.

Colwell, F. S. (1989) *Applied Environmental Microbiology* 55:2420–2423.

Colwell, F. S., et al. (1992) *Journal of Microbiological Methods* 15: 297–392.

———. (1997) *FEMS Microbiology Reviews* 20:425–435.

———. (2003) *Encyclopedia of Environmental Microbiology*, 2047–2057. John Wiley & Sons, New York.

Cox, B. B. (1946) *AAPG Bulletin* 30:645–659.

Cunningham, K. I., et al. (1995) *Environmental Geology* 25:2–8.

Cushing, G. E. (2012) *Journal of Caves and Karst Studies* 74:33–47.

Davis, J. B. (1967) *Petroleum Microbiology*. Elsevier, New York.

De Ley, J., et al. (1966) *Antonie van Leeuwenhoek Journal of Microbiology* 32:315–331.

Dockins, W. S., et al. (1980) *Geomicrobiology Journal* 2:83–98.

Dombrowski, H. (1960) *Zentralblatt für Bakteriologie, Parasitenkunde, Infektionskrankheiten und Hygiene* 178:83–90.

———. (1961) *Zentralblatt für Bakteriologie, Parasitenkunde, Infektionskrankheiten und Hygiene* 183:173–179.

———. (1966) *Second Symposium on Salt*. Ed. J. L. Rau, 1:215–220. Northern Ohio Geological Society, Cleveland.

Dombrowski, H., et al. (1963) *Annals of the New York Academy of Science* 108:453–460.

Duane, M. J., et al. (1997) *African Earth Sciences* 24:102–123.

Dunlap, J. F., and McNabb, W. J. (1973) *Subsurface Biological Activity in Relation to Ground-Water Pollution*. EPA Report EPA-660/2-73-014.

Edwards, K. J., et al. (2012) *Annual Reviews of Earth and Planetary Sciences* 40:551–568.

Egemeier, S. (1981) *National Speleological Society* 43:31–51.

Ehrlich, H. *Geomicrobiology*. CRC Press, Boca Raton, FL.

Ekzertsev, B. A. (1951) *Mikrobiologiya* 20:324–329.

Elliott, W. C., et al. (1999) *Clays and Clay Minerals* 47286–96.

Farrell, M. A., and Turner, H. G. (1932) *Journal of Bacteriology* 23:155–162.

Fisk, M. R., and Giovannoni, S. J. (1999) *Journal of Geophysical Research* 104:11805–11815.

Fredrickson, J. K., et al. (1995) *Molecular Ecology* 4:619–626.

———. (1997) *Geomicrobiology Journal* 14:183–202.

Fredrickson, J. K., and Phelps, T. J. (1997) Subsurface drilling and sampling. In: *Manual of Environmental Microbiology*, ed. C. J. Hurst et al., 526–540. ASM Press, Washington, D.C.

Freifeld, B. M., et al. (2008) *Geophysical Research Letters* 35:L14309.

Frimmel, H. E. (2005) *Earth-Science Reviews* 70:1–46.

Fry, N. K., et al. (1997) *Applied and Environmental Microbiology* 63:1498–1504.

Gahl, R., and Anderson, B. (1928) *Zentralblatt für Bakteriologie, Parasitenkunde, und Infektionskrankheiten*, Abt. 2, 73:331–338.

Gaidos, E., et al. (1999) *Science* 284:1631–1633.

Galle, E. (1910) *Zentralblatt für Bakteriologie, Parasitenkunde, und Infektionskrankheiten*, Abt. 2, 28:461.

Gihring, T. M., et al. (2006) *Geomicrobiology Journal* 23:415–430.

Gilichinsky, D. A., and Wagener, S. (1994) Historical review. In: *Viable Microorganisms in Permafrost*, ed. D. A. Gilichinsky, 7–20. Russian Academy of Sciences, Pushchino.

Ginsburg-Karagitscheva, T. L. (1926) *Azerbajdzanskoe Neftjanoe Khozjajstvo*, nos. 6–7. Baku

———. (1933) *Bulletin of the American Association of Petroleum Geologists* 17: 52–65.

Giovannoni, S. J., et al. (1996) *Proceedings of the Ocean Drilling Program* 148: 207–214.

Girard, J.-P., and Onstott, T. C. (1991) *Geochimica Cosmochimica Acta* 55: 3777–3793.

Gold, T. (1992) *Proceedings of the National Academy of Sciences USA* 89: 6045–6049.

Gold, T. (1999) *The Deep Hot Biosphere*. Springer, New York.

Gould, S. J. (1996) *Natural History*, March 1.

Gurevich, M. S. (1962) The role of microorganisms in producing the chemical composition of ground water. In: *Geological Activity of Microorganisms*, ed. S. I. Kuznetsov, 65–75. Transactions of the Institute of Microbiology, No. 9, Consultants Bureau, New York.

Haldeman, D. L., and Amy, P. S. (1993) *Microbial Ecology* 25:183–194.

Hallett, R. B., et al. (1999) Geology and thermal history of the Pliocene Cerro Negro volcanic neck and adjacent Cretaceous sedimentary rocks, west-central New Mexico. In: *Albuquerque Geology*. New Mexico Geological Society Guidebook, 50th Field Conference, ed. F. J. Pazzaglia and S. G. Lucas, 235–246. New Mexico Geological Society, Albuquerque, NM.

Hedrick, D. B., et al. (1992) *FEMS Microbiology Ecology* 10:1–10.

Helgeson, H. C., et al. (1993) *Geochimica Cosmochimica Acta* 57:3295–3339.

Hill, C. A. (1990) *American Association of Petroleum Geologists Bulletin* 74: 1685–1694.

Hoffman, B. A. (1992) Isolated reduction phenomena in red-beds: a result of porewater radiolysis? In: *Proceedings of the 7th International Symposium on Water-Rock Interaction*, ed. Y. K. Kharaka and A. S. Maest, 502–506. Balkema, Rotterdam.

Hooper, E. J. (1999) *The River: A Journey Back to the Source of HIV and AIDS*. Allen Lane, London.

Hoover, R. B., ed. (1997) *Instruments, Methods, and Missions for the Investigation of Extraterrestrial Microorganisms*. Proceedings of SPIE, vol. 3111. SPIE, Bellingham, WA.

Hose, L. D., et al. (2000) *Chemical Geology* 169:399–423.

Iizuka, H., and Komagata, K. (1965) *Journal of General Applied Microbiology* 11:15–23.

Issatchenko, V. (1940) *Journal of Bacteriology* 40:379–381.

Jakosky, B. M., et al. (2003) *Astrobiology* 3:343–350.

James, N., and Sutherland, M. L. (1942) *Canadian Journal of Research Sect C Botanical Science* 20:228–235.

Kallmeyer, J., and Wagner, D., eds. (2014) *Microbial Life of the Deep Biosphere*. De Gruyter, Berlin.

Kaufman, M. (2011) *First Contact*. Simon & Schuster, New York.

Karl, D. M., et al. (1999) *Science* 286:2144–2146.

Kelley, S. A., and D. D. Blackwell (1990) *Nuclear Tracks Radiation Measurements* 17:331–353.

Kennedy, M. J., ct al. (1994) *Microbiology* 140:2513–2529.

Kieft, T. L., et al. (1994) *Applied Environmental Microbiology* 60:3292–3299.

———. (1995) *Applied Environmental Microbiology* 61:749–757.

———. (2005) *Geomicrobiology Journal* 22:325–355.

Kita, I., et al. (1982) *Journal of Geophysical Research* 87:10789–10795.

Kolbel-Boelke, J., et al. (1988) *Microbial Ecology* 16:31–48.

Krumholz, L. R., et al. (1997) *Nature* 386:64–66.

Kuznetsov, S. I. (1950) *Mikrobiologiya* 19:3.

Kuznetsov, S. I., et al. (1962) *Introduction to Geological Microbiology*. McGraw-Hill, New York.

Kuznetsova, Z. I. (1962) Distribution and ecology of microorganisms in the deep ground waters of some regions of the USSR. In: *Geologic Activity of Microorganisms*, ed. S. I. Kuznetsov, 94–99. Trans. Inst. Microbiol. no. 9, Consultants Bureau, New York.

Labonté, J. M., et al. (2015) *Frontiers in Microbiology* 1–14.

Larsen, H. (1981) The family Halobacteriaceae. In: *The Prokaryotes. A Handbook on Habitat, Isolation and Identification of Bacteria*, ed. M. P. Starr et al., 1:985–994. Springer, Berlin.

Lefticariu, L., et al. (2006) *Geochimica Cosmochimica Acta* 70:4889–4905.

Lieske, R. von. (1932) *Biochemischen Zeitschrift* 250:1–6.

Lieske, R. von, and Hoffman, E. (1928) *Brennstoff-Chemie* 9:74.

———. (1929) *Zentralblatt für Bakteriologie, Parasitenkunde, und Infektionskrankheiten*, Abt. 2, 77:305–309.

Lin, L.-H., et al. (2006) *Science* 314:479–482.

Lindblom, G. P., and Lupton, M. D. (1961) In *Developments in Industrial Microbiology*, 9–22. Plenum Press, New York.

Lipman, C. B. (1928) *Science* 68:272–273.

———. (1931) *Journal of Bacteriology* 22:183–198.

Lipman, C. B., and Greenberg, L. (1932) *Nature* 129:204–205.

Lippmann, J., et al. (2003) *Geochimica Cosmochimica Acta* 67:4597–4619.

Lippmann-Pipke, J., et al. (2012) *Applied Geochemistry* 26:2134–2146.

Liu, S. V., et al. (1997) *Science* 277:1106–1109.

Liu, Y., et al. (1999) *International Journal of Systematic Bacteriology* 47:615–621.

Lorenz, J., et al. (1996) Sandia Report SAND96-1135, UC-132. Sandia National Laboratories, Albuquerque, NM.

Lowenstein, H., et al. (1999) *Geology* 27:3–6.

Madigan, M. T., et al. (2009) *Brock Biology of Microorganisms*. 12th ed. Pearson Education, New York.

Madsen, E. L., et al. (1996) *Science* 272:896–897.

Malin, M. C., et al. (2006) *Science* 314:1573–1576.

Martin, W., and Russell, M. J. (2006) *Philosophical Transactions of the Royal Society B* 362:1887–1926.

McKay, D. S., et al. (1996) *Science* 273:924–930.

McKay, C. P. (2009) *Science* 323:718.

———. (2012) *Astrobiology* 12:169.

McKinley, J. P., and Colwell, F. S. (1996) *Journal of Microbiological Methods* 26:1–9.

McMahon, P. B., et al. (1992) *Journal of Sedimentary Petrology* 62:1–10.

McNabb, W. J., and Dunlap, J. F. (1975) *Groundwater* 13:33–44.

Mekhtieva V. L. (1962) *Geokhimiya* 8:707–719.

MEPAG Special Regions Science Analysis Group (2006) *Astrobiology* 6:677–732.

Meshkov, A. N. (1958) *Mikrobiologiya* 27:3.

Miller, L. P. (1949) *Contributions from the Boyce Thompson Institute* 15: 437–465.

Miteva, V., et al. (2004) *Applied and Environmental Microbiology* 70:202–213.

Morehouse, D. F. (1968) *National Speleological Society* 30:1–10.

Morita, Y., and ZoBell, C. E. (1955) *Deep-Sea Research* 3:6673.

Mormile, M. R., et al. (2003) *Environmental Microbiology* 5:1094–1102.

National Research Council (2006) *Preventing the Forward Contamination of Mars*. National Academies Press, Washington, D.C.

Naumov, V. B., et al. (2013) *Geochemistry International* 51:417–420.

Naumova, R. P. (1960) *Microbiologiya* 29:415–418.

Nazina, T. N., et al. (1988) *Microbiologiya* 57:832–827.

Norton, C. F., and Grant, W. D. (1988) *Journal of General Microbiology* 134:1365–1737.

Norton, C. F., et al. (1992) *ASM News* 58:363–367.

———. (1993) *Journal of General Microbiology* 139:1077–1081.

Ollivier, B., et al. (1994) *Microbiological Reviews* 58:27–38.

Olson, G. J., et al. (1981) *Geomicrobiology Journal* 2:327–340.

Omar, G. I., et al. (2003) *Geofluids* 3:69–80.

Omelyansky, V. L. (1911) *Arkhiv Biologicheskikh Nauk* 16:335–340 (in Russian).

Onstott, T. C., et al. (1998) *Geomicrobiology Journal* 15:353–385.

———. (1999) A global perspective on the microbial abundance and activity in the deep subsurface. In: *Enigmatic Microorganisms and Life in Extreme Environments*, ed. J. Seckbach, 487–499. Kluwer, Dordrecht.

———. (2006) *Astrobiology* 6:377–395.

———. (2009) *Microbial Ecology* 58:786–707.

———. (2014) *Geobiology* 12:1–19.

Oremland, R. S., et al. (1982) *Initial Reports of the Deep Sea Drilling Program* 64:759–762.

Orphan, V. J., et al. (2002) *Proceedings of the National Academy of Science USA* 99:7663–7668.

Orsi, W., et al. (2013a) *Nature* 499:205–208.

———. (2013b) PLoS ONE 8:e53665.

Osterloo, M. M., et al. (2008) *Science* 319:1651–1654.

Oxley, J. (1989) *Down Where No Lion Walked. The Story of Western Deep Levels*. Southern Book Publishers, Johannesburg.

Parkes, R. J., et al. (1990) *Philosophical Transactions of the Royal Society A* 331:139–153.

———. (1994) *Nature* 37:410–413.

————. (2014) *Marine Geology*. doi: 10.1016/j.margeo.2014.02.009.

Pedersen, K. (1989) *Deep Ground Water Microbiology in Swedish Granitic Rock*. Tech. Rep. 89-23. Swed. Nuclear Fuel Waste Manage. Co., Stockholm.

————. (2004) *Strolling through the World of Microbes*. Palmedblads Tryckeri AB, Göteborg, Sweden.

Pedersen, K., and Ekendahl, S. (1990) *Microbial Ecology* 20:37–52.

————. (1992) *Microbial Ecology* 23:1–14.

Pfiffner, S. M., et al. (2008) *Astrobiology Journal* 8:623–638.

Phelps, T. J., et al. (1989) *Journal of Microbiological Methods* 9:267–279.

————. (1994a) *Microbial Ecology* 28:335–349.

————. (1994b) *Microbial Ecology* 28:351–364.

Plumb, J. J., et al. (2007) *International Journal of Systematic Evolutionary Microbiology* 57:1418–1423.

Priscu, J. C., et al. (1999) *Science* 286:2141–2143.

Ragon, M., et al. (2013) *Frontiers in Microbiology* 4: 37.

Reiser, R., and Tasch, P. (1960) *Transactions of the Kansas Academy of Science* 63:31–34.

Riess, W., et al. (1999) *Aquatic Microbial Ecology* 18:157–164.

Rinke, C., et al. (2013) *Nature* 499:431–437.

Rivkina, E., et al. (1998) *Geomicrobiology Journal* 15:187–193.

Roh, Y., et al. (2002) *Applied and Environmental Microbiology* 68:6013–6020.

Rossbacher, L. A., and Judson, S. (1981) *Icarus* 45:39–59.

Rosso, L., et al. (1995) *Applied and Environmental Microbiology* 61:610–616.

Rothman, D. H., et al. (2014) *Proceedings of the National Academy of Sciences USA* 111:5462–5467.

Rozanova, E. P., and Khudyakova, A. I. (1970) *Mikrobiologiya* 39:321–326.

Russell, C. E., et al. (1994) *Geomicrobiology Journal* 12:37–51.

Sagan, C. 1995. *The Demon-Haunted World: Science as a Candle in the Dark*. Random House, New York.

Sargent, K. A., and Fliermans, C. B. (1989) *Geomicrobiology Journal* 7:3–13.

Sarbu, S. M., et al. (1996) *Science* 272:1953–1955.

Schröder, H. (1914) *Zentralblatt für Bakteriologie, Parasitenkunde, und Infektionskrankheiten*, Abt. 2, 41:460.

Sherwood Lollar, B., et al. (1993) *Geochimica Cosmochimica Acta* 57:5087–5097.

Sidow, A., et al. (1991) *Philosophical Transactions of the Royal Society B* 333: 420–433.

Sinclair, J. L., and Ghiorse, W. C. (1989) *Geomicrobiology Journal* 7:15–31.

Skidmore, M. L., et al. (2000) *Applied and Environmental Microbiology* 66:3214–3220.

Smirnova, Z. S. (1957) *Mikrobiologiya* 26:717–721.

Sogin, S. J., et al. (1972) *Journal of Molecular Evolution* 1:173–184

Stan-Lotter, H. (2011) *Encyclopedia of Geobiology*, 313–317. Springer Science +Business Media, New York.

Stetter, K. O., et al. (1986) Diversity of extremely thermophilic *Archaebacteria*. In: *Thermophiles: General, Molecular and Applied Microbiology*, ed. T. D. Brock, 39–74. John Wiley & Sons, New York

Stetter, K. O., et al. (1993) *Nature* 365:743–745.

Stevens, T. O., et al. (1993) *Microbial Ecology* 25:35–50.

Stevens, T. O., and Holbert, B. S. (1995) *Journal of Microbiological Methods* 21:283–292.

Stevens, T. O., and McKinley, J. P. (1995) *Science* 270:450–454.

———. (2000) *Environmental Science and Technology* 34:826–831.

Stotler, R. L., et al. (2009) *Journal of Hydrology* 373:80–95.

———. (2010) *Groundwater* 49:348–364.

Sugiyama, J. (1965) *Journal of General and Applied Microbiology* 11:15–23.

———. (1966) *Journal of the Faculty of Science, University of Tokyo*, ser. 3, 9:287–311.

Summers, A. (2007) *Journal of Cave and Karst Studies* 69:187–206.

Szewczyk, N. J., and McLamb, W. (2005) *Wilderness and Environmental Medicine* 16:27–32.

Szewczyk, N. J., et al. (2005) *Astrobiology* 5:663–689.

Szewzyk, U., et al. (1994) *Proceedings of the National Academy of Sciences USA* 91:1810–1813.

Takai, K., et al. (2001) *International Journal of Systematic and Evolutionary Microbiology* 51:1245–1256.

———. (2003) *Environmental Microbiology* 5: 309–320.

———. (2008) *Proceedings of the National Academy of Sciences USA* 105: 10949–10954.

Tanaka, K. L. (1986) *Journal of Geophysical Research* 91:E139–E158.

Telegina, Z. P. (1961) Distribution and specific content of bacteria that oxidize gaseous hydrocarbons in the ground water at gas deposits in the Azov-Kuban basin. In: *Geologic Activity of Microorganisms*, ed. S. I. Kuznetsov, 99–101. Trans. Inst. Microbiol. No. 9. Consultants Bureau, New York.

Thauer, R. K., et al. (1977) *Bacteriological Reviews* 41:100–180.

Thorseth, I. H., et al. (1992) *Geochimica et Cosmochimica Acta* 56:845–850.

Tobin, K. J., et al. (2000) *Chemical Geology* 169:449–460.

Tseng, H.-Y., et al. (1995) *Chemical Geology* 127:297–311.

———. (1998a) *Water Resources Research* 34:937–948.

———. (1998b) *Geological Society of America Bulletin* 111:275-290.

Vernadsky, V. E. (1926) *The Biosphere*. Moscow-Leningrad.

Viles, H. A. (1984) *Progress in Physical Geography* 8:523–542.

von Humboldt, A. (1793) *Florae Fribergensis, Specimen, Plantas Cryptogamicas Praesertim Subterraneas Exhibens . . . accedunt Aphorismi ex Doctrina, Physiologiae Chemicae Plantarum.* Berolini Apud Henr. Augustum Rottmann.

von Lieske, R. (1932) *Biochemischen Zeitschrift* 250:1–6.

von Lieske, R., and Hofmann, E. (1928) *Brennstoff-Chemie* 174–176.

———. (1929) *Zentralblatt für Bakteriologie*, ser. 2, 77:305–309.

von Wolzogen Kühr, C.A.H. (1922) *Proceedings of the Koninklijke Akademie van Welenschappen te Amsterdam* 25:189–198.

Vovk, I. F. (1987) Radiolytic salt enrichment and brines in the crystalline basement of the East European Platform. In: *Saline Water and Gases in Crystalline Rocks*, ed. P. Fritz and S. K. Frape, S. K., 197–210. Geological Association of Canada, Ottawa.

Vreeland, R. H., and Hochstein, L. I. (1992) *The Biology of Halophilic Bacteria.* CRC Press, Boca Raton, FL.

Vreeland, R. H., et al. (2000) *Nature* 407:899–900.

Walvoord, M. A., et al. (1999) *Water Resources Research* 35:1409–1424.

Whelan, J. K., et al. (1986) Evidence for sulfate-reducing and methane-producing microorganisms in sediments from sites 618, 619, and 622. In: *Initial Reports of the Deep Sea Drilling Program*, ed. A. H. Bouma et al., 767–775. U.S. Govt. Printing Office, Washington D.C.

White, D. C., and Ringelberg, D. B. (1997) Utility of the lipid signature biomarker analysis in determining the in situ viable biomass, community structure, and nutritional/physiological status of deep subsurface microbiota. In: *The Microbiology of the Terrestrial Subsurface*, ed. P. S. Amy and D. L. Haldeman, 119–136. CRC Press, Boca Raton, FL.

Whitelaw, K., and Rees, J. F. (1980) *Geomicrobiology Journal* 2:179–187.

Whitman, W. B., et al. (1998) *Proceedings of the National Academy of Sciences USA* 95:6578–6583.

Wilkinson, M., and Dampier, M. D. (1990) *Geochimica et Cosmochimica Acta* 54:3391–3399.

Wilson, J. T., et al. (1983) *Groundwater* 21:134–142.

Woese, C. R. (1987) *Microbiological Reviews* 51:221–271.

Woese, C. R., et al. (1990) *Proceedings of the National Academy of Sciences USA* 87:4576–4579.

Zhang, G., et al. (2005) *Applied and Environmental Microbiology* 71:3213–3227.

Zheng, M., and Kellogg, S. T. (1994) *Canadian Journal of Microbiology* 40:944–954.

Zinger, A. S. (1966) *Microbiologiya* 35:357–364.

ZoBell, C. E. (1943) *Petroleum World* 40:30–43.

———. (1945) *Science* 102:364–369.

———. (1951) The rôle of microörganisms in petroleum formation. In: *Fundamental Research on Occurrence and Recovery of Petroleum*, American Petroleum Institute Research Project 43a, pp. 98–100.

Zuckerkandl, E., and Pauling, L. (1965) Molecules as documents of evolutionary history. In: *Evolving Genes and Proteins*, ed. V. Bryson and H. Vogel, 97–166. Academic Press, London.

Zumberge, J. E., et al. (1978) *Minerals Science Engineering* 10:223–246.

Zvyagintsev, D. G., et al. (1985) *Mikrobiologiya* 54:155–161.

INDEX